A SHORT HISTORY

OF

GREEK MATHEMATICS

BY

JAMES GOW, M.A.

FELLOW OF TRINITY COLLEGE, CAMBRIDGE.

Παρ' Εὐκλείδῃ τις ἀρξάμενος γεωμετρεῖν, ὡς τὸ πρῶτον θεώρημα ἔμαθεν,
ἤρετο τὸν Εὐκλείδην· τί δέ μοι πλέον ἔσται ταῦτα μανθάνοντι; καὶ ὁ
Εὐκλείδης τὸν παῖδα καλέσας· δός, ἔφη, αὐτῷ τριώβολον, ἐπειδὴ δεῖ
αὐτῷ, ἐξ ὧν μανθάνει, κερδαίνειν.
STOBAEUS, Flor. IV. p. 205.

CHELSEA PUBLISHING COMPANY
NEW YORK, N. Y.

THE PRESENT WORK IS A REPRINT, TEXTUALLY UNALTERED
EXCEPT FOR THE INCORPORATION INTO THE TEXT OF CROSS-
REFERENCES TO THE ADDENDA AND THE CORRECTION OF
ERRATA, OF A WORK ORIGINALLY PUBLISHED IN CAMBRIDGE

ORIGINALLY PUBLISHED, CAMBRIDGE, 1884
REPRINTED (IN AUSTRIA), 1923
REVISED REPRINT, NEW YORK, 1968

LIBRARY OF CONGRESS CATALOG CARD NO. 68-21639
STANDARD BOOK NUMBER 8284-0218-3

PRINTED IN THE UNITED STATES OF AMERICA

PREFACE.

THE history of Greek mathematics is, for the most part, only the history of such mathematics as are learnt daily in all our public schools. And very singular it is that, though England is the only European country which still retains Euclid as its teacher of elementary geometry, and though Cambridge, at least, has, for more than a century, required from all candidates for any degree as much Greek and mathematics together as should make this book intelligible and interesting, yet no Englishman has been at the pains of writing, or even of translating, such a treatise. If it was not wanted, as it ought to have been, by our classical professors and our mathematicians, it would have served at any rate to quicken, with some human interest, the melancholy labours of our schoolboys.

The work, as usual, has been left to Germany and even to France, and it has been done there with more than usual excellence. It demanded a combination of learning, scholarship and common sense which we used, absurdly enough, to regard as peculiarly English. If anyone still cherishes this patriotic delusion, I would advise him to look at the works of Nesselmann, Bretschneider, Hankel, Hultsch, Heiberg and Cantor, or, again, of Montucla, Delambre and Chasles, which are so frequently cited in the following pages. To match them we can show only an ill-arranged treatise of Dean Peacock, many brilliant but scattered articles of Prof. De Morgan, and three essays by Dr Allman. I have treated all these writers with freedom and have myself added matter which is not to be found in any of them, but they strike me still with humiliation such as a classical scholar feels when he edits a text which Bentley has edited before him.

My own book represents part of a collection of notes which I have for many years been making with a view to a general history of the great city of Alexandria. The fact that the history of Alexandrian mathematics begins with the Elements of Euclid and closes with the Algebra of Diophantus, both of which are founded on the discoveries of several preceding centuries, made it necessary that I should extend my inquiries over the whole field of Greek mathematics. In this way, the materials for an account of the Alexandrian Mathematical School grew to exceed the reasonable limits of a chapter, and I have thought it desirable to publish them as a separate essay. I shall treat the Literary School with the same fulness.

As a history of Alexandria ought to be interesting to most people, I took especial pains that my treatment of the Mathematical School, which was the oldest, the most conspicuous and the longest-lived of them all, should not be excessively technical. I have tried to put my account of it generally in such a form as should be useful and attractive to readers of various tastes. As a matter of fact, mathematicians will here find some account of every extant Greek mathematical book and a great number of pretty proofs translated from the *ipsissima verba* of the ancients. Greek scholars will find nomenclature and all manner of arithmetical symbols more fully treated than in any other work. A student of history, who cares little for Greek or mathematics in particular, but who likes to watch how things grow, will be able to extract from these pages a notion of the whole history of mathematical science down to Newton's time, and will find some very curious questions raised which it is his especial duty to answer. It was impossible to satisfy the requirements of all readers, but each will perhaps be willing to concede something to the claims of the others, and wherever a subject is introduced but inadequately treated, I have at least given references to sources of fuller information, if any such exist to my knowledge.

As the whole book is an endeavour to compromise between conflicting claims, I have allowed myself, with the same intent, some inconsistency in two details. In the first place, I have not drawn a strict line between pure and mixed mathematics, but have given an account of the *Phaenomena* and *Optics* of Euclid and the mechanical books of Aristotle and Archimedes, while I have omitted any summary of Ptolemy's astronomical theories. The former are books which are little known, which are short, which came in my way and which are almost purely deductive. A complete history of Greek astronomy is tolerably common, is long, is founded for the most part on non-mathematical writers and would consist largely of a history of astronomical observations. In the second place, I have tried to write proper names (following indeed Smith's *Dictionary of Greek and Roman Biography*) in a way which should generally indicate the Greek form and pronunciation without offending the ordinary eye. I have always written *c* for *κ* and final *-us* for *-ος*. I have generally not Latinized names ending in *-ων*, because it is sometimes inconvenient and the Latin usage was inconsistent with itself; for instance, it retained *Conon* but altered *Platon*. I have left in their English form, the names of well-known writers. Thus the reader will find Plato, Aristotle and Euclid side by side with Heron, Nicoteles and Neocleides. If I must offend somebody, I would soonest offend a pedant.

The complete MS. of this book left my hands last January and the whole edition has been printed off, sheet by sheet, at various intervals, since that time. I have therefore been unable to correct any errors or omissions which I observed too late or to incorporate new matter which appeared after the sheet to which it was relevant had gone to press. The chief notices which I wish to insert are given in the *Addenda* which immediately follow this preface.

My work, dreary as it has often been, has been enlivened by one constant pleasure, the interest and unselfish assistance of many friends. Two of them, in particular, deserve recognition very near the title-page. The first is Mr F. T. Swanwick, late scholar of Trinity College, Cambridge, and now Mathematical Lecturer in the Owens College. He has, with incredible care and patience, read through the whole book from page 66, has made a hundred valuable suggestions and has saved me a hundred times from myself. The other is Mr Joseph Jacobs, late scholar of St John's College, Cambridge, whose wide intellectual interests and unsurpassed knowledge of bibliography have made the book far more useful and entertaining than it otherwise would have been. I would say more of their kindness to me but that I would not have them held responsible for any slips which they may have overlooked in my work but would not have made in their own.

JAMES GOW.

LINCOLN'S INN.
October, 1884.

TABLE OF CONTENTS.

CHAPTER IV.

GREEK THEORY OF NUMBERS (*ARITHMETICA*)   66—122

PART III.   GREEK GEOMETRY   . 123—313

CHAPTER V.   PREHISTORIC AND EGYPTIAN GEOMETRY 123—133

CHAPTER VI.   GREEK GEOMETRY TO EUCLID   134—191

(a) *Preliminary.*

PART I. PROLEGOMENA TO ARITHMETIC.

CHAPTER I.

THE DECIMAL SCALE.

1. IN the book of *Problemata*, attributed to Aristotle, the following question is asked (xv. 3): "Why do all men, both barbarians and Hellenes, count up to 10 and not to some other number?" It is suggested, among several answers of great absurdity, that the true reason may be that all men have ten fingers[1]: "using these, then, as symbols of their proper number (viz. 10), they count everything else by this scale." The writer then adds "Alone among men, a certain tribe of Thracians count up to 4, because, like children, they cannot remember a long sum neither have they any need for a great quantity of anything."

It is natural to regret that an author who at so early a date was capable of writing this passage, was not induced to ask himself more questions and to collect more facts on the same and similar subjects. Had he done so, he might have anticipated, by some two thousand years, the modern method of research into prehistoric times and might have attempted, with every chance of success, a hundred problems which cannot now be satisfactorily treated[2]. In the fourth century B.C. and for long after, half the Aryan peoples were still barbarous and there must still have survived, even among Greeks and Italians, countless relics of primitive manners, forming a sure tradi-

---

[1] Cf. Ovid, *Fasti* III. v. 121 sqq.

[2] It seems probable that Aristotle himself was inclined to reconstruct primitive history on much the same lines as would be taken by a modern evolutionist. See Sir H. Maine, *Early Law and Custom*, pp. 196, 197.

tion of the past. Nearly all these materials, so abundant in Aristotle's day, are irretrievably lost to us and the primeval history of Aryan culture depends now chiefly on the evidence supplied by comparative philology. It is so with the art of calculation. We may assume evolution and, by careful comparison of the habits of the existing lower races, we may form *ab extra* a theoretical history of arithmetic among our forefathers; but almost the whole (so to say) *internal* evidence is concealed in a few numeral words. To the etymology of these a few pages may here be fitly devoted, not only because it is habitual with our generation to commence every inquiry from the beginning of things, but also because Greek arithmetic offers no other prehistoric inquiry but this, because, in fact, ordinary Greek calculation remained to the last so clumsy and primitive, that if any progress in the art is to be ascribed to the Greeks, it can be exhibited only by going back to the beginning.

2. The words for 1000, and every higher power of 10, are different in all the great branches of the Aryan family of languages, and the cardinal numerals up to that limit are manifestly derived, by mere addition or composition, from the first ten. These last, therefore, are of by far the greatest interest and importance and the present inquiry may be confined to them. Before examining the individual words, however, it will be well to consider the whole group. The first three are adjectives, agreeing with only casual and partial exceptions (*e. g.* δύο) in gender and case with the substantives which they qualify. The same might be said of the fourth, but that in Latin *quattuor* is wholly indeclinable. The rest, from five to ten, are generally uninflected and have or had originally the form of a neuter singular[1]. In Sanskrit, indeed, these six numerals are declined as adjectives but they do not take the gender signs and in older writers are often employed without any inflexions at all[2]. In old Slavonic they are extended by a suffix (as in Semitic tongues) into abstract nouns of the

---

[1] The final sibilant of 'six,' 'sex,' etc. is part of the root, and the apparently dual ending of *octo* is not primeval. Schleicher, *Vergl. Gram.* § 237. 6 and 8.

[2] Whitney, *Sanskrit Gram.* § 486 c.

feminine singular and are so declined[1] (cf. Gr. πεμπάς etc.), but this usage is also obviously late and may be ignored in a general discussion of the origin of the words. For the present purpose, it may be stated broadly that the first three are adjectives, the fourth is generally an adjective but sometimes an uninflected noun, the remaining six are uninflected nouns only. All of them, in all Aryan tongues, are constructed of the same materials, which, moreover, seem familiar enough in different connexions. The difficulty is how to adapt the apparent meanings of the roots to a numerical signification. Some metaphor probably underlies each word, but though metaphor, as we shall see, is competent to make numerals, it is not able to extend their application. Things are not eight or ten by a metaphor. They are so as a pure matter of fact, and we are thus debarred from inferring the original meanings of the numerals from any subsequent usage of them by transference. The propriety of each numeral to its signification must be explained *a priori* or not at all. And this, apart from any linguistic difficulties, constitutes the chief objection to the etymologies hitherto proposed by Bopp, Lepsius, Pott and others[2]. Sometimes they do not explain the choice of the particular name, sometimes they involve patent anachronisms. When for instance they say that *pankan* and *saptan*, 'five' and ' seven,' mean 'following,' because they follow 'four' and 'six'

---

[1] Bopp, *Comp. Gram.* § 313. Stade, *Lehrb. der Hebr. Spr.* p. 216.

[2] Bopp, § 308 sqq. Pott, *Die quinäre etc. Zählmethode*, pp. 130 sqq. On p. 142 Pott, discussing Lepsius' derivations, points out that he ascribes to 1 (besides its original *êka*) the forms *k*, *tsh*, *p* and *ç* in the composition of the other numerals. The common derivations, taken chiefly from Bopp, are set out in Morris, *Hist. Outlines of Eng. Accidence*, p. 110 n. The following only need be cited :

Three = 'what goes beyond ' (root *tri*, *tar*, to go beyond).

Four (quattuor) = 'and three,' i.e. 1

and 3 : or else *ka*, *qua* = 1.

Five = 'that which comes after' (four). Sk. *pashchât* = after.

Six, Sk. *shash* is probably a compound of *two* and *four*.

Seven = 'that which follows ' (six).

Eight, Sk. *ashtân* = 1 + and + 3.

Nine = *new* = that which comes after 8 and begins a new quartette.

Ten = two and eight.

Pott, in *Etym. Forschungen*, 2nd ed., 1859, I. p. 61 n., declares his opinion that the numerals are derived from names of concrete objects, but suggests no particular etymologies.

respectively, they suggest no reason why any other numeral above 1 should not have been called by either or both of these names: so when they say that *navan,* 'nine,' means 'new' (*νέος* etc.) because it begins a new quartette, they assume a primeval quaternary notation and do not explain why 'five' was not called *navan:* so again when they say that *navan* means 'last' (*νέατος* etc.) because it is the last of the units, they evidently speak from the point of view of an arithmetician who has learnt to use written symbols. What one really wants, in this as in so many other problems of philology, is to get at the point of view of the primitive language-maker and to see from what sources he was likely to get his numerals. And this can only be done by a careful examination of the habits and languages of modern savages.

3. It is probably familiar enough to most readers that many savage tribes are really unable to count, or at least have no numerals, above 2 or 3 or 4, and express all higher numbers by a word meaning 'heap' or 'plenty,' and that every nation, which can count further than this, uses a quinary or decimal or vigesimal notation or a combination of these[1], which is generally founded on, and expressly referred to, the number of the fingers and toes. These facts, which are beyond dispute[2], suggest two initial questions, first, what is the real difficulty which a savage finds in separating the units which go to make a multitude? and

[1] No nation has a *purely* quinary or vigesimal notation at all. The Mayas of Yucatan, however, and the Aztecs have special words and signs for 20, 400 and 8000. Pott, *Zählmethode* supr. cit. pp. 93, 97, 98. Wilson, *Prehistoric Man,* ii. p. 61.

[2] See, for instance, Pott, *Die quin. Zählmethode,* etc., with an appendix on Finger-names, *supra cit.* : Pott, *Die Sprachverschiedenheit in Europa* etc., on the same subject, being a *Festgabe zur* xxv. *Philologenversammlung :* and an article by the same writer in *Zeitschr. für Völkerpsychologie,* Vol. xii. These will be cited hereafter as Pott, *Zählm., Festgabe* and *Zeitschr.* respectively. Also Tylor, *Primitive Culture,* i. ch. 7 : Lubbock, *Orig. of Civilisation,* ch. 8, and *Prehist. Times* (4th ed.), p. 588. Another collection of similar facts will be found in Dean Peacock's article 'Arithmetic' in the *Ency. Metropolitana.* It may be added here that the Maoris are said to use an *undenary* scale (Pott, *Zählm.* pp. 75 and 76) in which rests are taken at 11, 121, 1331, etc., but this is doubted. The Bolans or Buramans of W. Africa are also said to use a *senary* scale (Pott, *Festgabe,* p. 30). These cases, if correctly reported, seem to be the only complete exceptions to the rule stated in the text.

secondly, why are the fingers—which, one would say *a priori*, are as hard to count as any other collection of five or ten things—always adopted as the means and basis of calculation ? The answer seems to be, at least in part, as follows. A savage knows large and familiar things by special distinguishing marks and these special peculiarities prevent him from forming, in the case of such things, the generalizations which are essential to arithmetic. A black cow and a dun cow, a tall child and a short one, a wood-chopper and a battle-axe, his own hut and his neighbour's, are not, to him, essentially similar, but essentially different from one another and from everything else, to be spoken of by proper and not by generic names, not forming part of a class and therefore not requiring to be counted. In respect of these things, he does not count, he *enumerates :* as if a man, when asked how many children he has, should say, not that he has 3 or any other number, but that he has Tom and Susan and Harry and so on, naming each individual[1]. With small or unfamiliar things, on the other hand, with beans or fruits, for instance, or strangers from another tribe, the savage, though he is compelled to generalize, is not necessarily compelled to count, for there are many ways of roughly indicating a quantity, without knowing its component parts. Everybody has tried the difficulty of counting quickly a number of spots irregularly disposed, and what we are unable to do quickly, a savage may well be unable to do at all. In order to count a heap correctly, it is essential that the same thing be not counted twice, and

[1] Compare Lubbock, *Orig. of Civil.* pp. 292—294, and the quotations from Galton and Lichtenstein, as to the special knowledge of individual animals by which a savage, unable to count a high number, keeps his herds together and conducts his barter. Tylor (*Prim. Cult.* I. p. 303 and *passim*) gives abundant examples of proper names applied by savages to inanimate objects. The same writer (I. p. 254) relates that some Australians, as well as other tribes, have a series of nine proper names which they give to their children in order of seniority and which might well serve for numerals, yet they cannot actually count above 2. So also, according to Dr Rae (cit. Lubbock, *Prehist. Times*, p. 525), many Eskimos, who are said by Parry (also *loc. cit.*) to have numerals up to 10 at least, cannot count their children correctly even when they have only four or five. But *quaere*, whether a man, who could not count his own children, would find the same difficulty in counting a number of strangers.

this, to the unpractised calculator, can be secured only by arranging the things in such a form that the counting may follow a definite direction from a fixed beginning to a fixed end. Given such an arrangement, it is further necessary that the calculator should have words or other symbols to serve as a *memoria technica* of each successive total, otherwise he will be as ignorant at the end of the counting as he was at the beginning. But a savage, who *ex hypothesi* is making his first essay in counting, can hardly be expected both to arrange his units and to invent his symbols immediately. Time and practice and some hard thinking are obviously necessary before he can master both operations.

4. These difficulties, however, are soonest surmounted with very small numbers, of which any arrangement is bound to be more or less symmetrical and of which so definite an image may be retained in the memory that names or symbols are unnecessary for the mere operation of counting. But some means of communicating a total, and, with a higher number, some *memoria technica* of the arrangement adopted are still wanting. For both these purposes, the fingers and toes are especially well adapted. They are a moderate number of similar things, easily generalized, symmetrically disposed and arranged in four groups of small contents. They can be so moved, shown, concealed or divided, that they will exhibit any number under 21 : they are so familiar that the eye is constantly practised in counting them, and they are so universally supplied to human beings that they can be used to communicate arithmetical results.

But men did not arrive at this use of the fingers till they had already made some little progress in calculation without them. That this is the true history of the art of counting is evident if we consider the following facts in order. First, there is hardly any language in the world in which the first three or four numerals bear, on the face of them, any reference to the fingers. Secondly, there are many savage languages in which these numerals are obviously taken (not from the fingers but) from small symmetrical groups of common objects. Thus, ' two ' is, among the Chinese, *ny* and *ceul*, which also mean

'ears:' in Thibet *paksha* 'wing:' in Hottentot *t'Koam* 'hand:'
and so also among the Javanese, Samoyeds, Sioux and other
peoples. So again with the Abipones, 'four' is *geyenkñaté*,
'ostrich-toes:' 'five' is *neenhalèk*, 'a hide spotted with five
colours:' with the Marquesans 'four' is *pona*, 'a bunch of four
fruits,' etc.[1] Thirdly, there are also many savages who, having
only a very few low numerals, count to much higher numbers
dumbly by means of the fingers[2].

5. But just as, in the examples quoted above, the name
of the pattern group (e.g. *ears* or *hands*) becomes the name of
the number which that group contains, so with finger-counting
the savage, advancing in intelligence, begins to name the
gesture with or without performing it, and this name becomes
the symbol of the number which the gesture is meant to
indicate. Hence all the world over, in nearly every language
under the sun where names for the higher units exist and show
a clear etymology, the word for 'five' means 'hand,' and the
other numbers, up to 10 or 20, as the case may be, are merely
descriptive of finger-and-toe-counting. In Greenland, on the
Orinoco, and in Australia alike, 'six' is 'one on the other hand,'
'ten' is 'two hands,' 'eleven' is 'two hands and a toe' and
'twenty' is 'one man[3].' In some cases, we find even greater
definiteness. Among the Eskimos of Hudson's Bay the names
of the numerals 'eight,' 'nine' and 'ten' mean respectively the
'middle,' 'fourth' and 'little finger,' and the same use of actual
finger-names is observed also among the Algonquin Indians of
North America, the Abipones and Guarani of the South,
the Zulus of Africa and the Malays of the Asiatic islands[4].

6. Enough has now been said, or at least references
enough have been given, to show that wherever a quinary,
decimal or vigesimal notation is adopted in counting, there

<hr/>

[1] For other examples, and especially
for a curious set of Indian poetic
numerals, in which e.g. 'moon' stands
for 1 and 'teeth' for 32, see Tylor,
Prim. Cult. i. pp. 252, 253, 256, 259,
and reff. Civilised peoples (*ib.* p. 257)
sometimes employ a similar nomenclature, though not often with very
low numbers (e.g. *couple*). See also
Farrar, *Chaps. on Lang.* pp. 198—201.

[2] Tylor, *Prim. Cult.* i. p. 244.

[3] Tylor, *Prim. Cult.* i. pp. 247—
251.

[4] Pott, *Zählm.* pp. 190 and 301,
Zeitschr. pp. 182, 183, *Festg.* pp. 47, 48,
83.

is the strongest possible presumption that the notation is founded on the number of the fingers and toes : and secondly, that wherever these scales are used and the etymology of the numerals is obscure, the most likely explanation will connect the higher units with the gestures used in finger-and-toe-counting.

If we turn, then, to the languages of the Aryan peoples, we shall find many signs that they acquired the art of calculation slowly and by precisely the same modes as we see in practice among modern savages. There is no word for 'counting' common to all the Aryan tongues, but the special words generally mean 'to arrange' or 'to group' ($\dot{a}\rho\iota\theta\mu\epsilon\hat{\iota}\nu$[1], numerare, rechnen) and a similar notion must underlie the double uses of tell, putare, $\lambda\acute{\epsilon}\gamma\epsilon\iota\nu$. Again, three numbers only are distinguished in the inflexions of nouns and verbs, viz. the singular, dual and plural. This, like the three strokes which mark plurality in Egyptian hieroglyphics, seems to point to a time when 3 was the limit of possible counting. It is noticeable also, in this regard, that 'three' always retained a notion of great multitude [2]: that Sanskrit employs, for this numeral, two distinct roots, tri- and tisar-[3]; and that, after 'three,' the first divergence appears in the grammar of the Aryan numerals. The common use of a duodecimal notation in measurements of length and capacity and the sudden variation in the grammatical position of 'four' may be taken as evidence that 'four' was a separate addition to the numerals and that 3 and 4 were for some time used together as limits of the groups used in counting [4]. The use of $\chi\epsilon\acute{\iota}\rho$ and manus to signify 'a number,'

---

[1] Cf. Odyssey x. 204. $\delta\acute{\iota}\chi a$ $\dot{a}\rho\iota\theta\mu\epsilon\hat{\iota}\nu$ must originally have meant 'to arrange in two groups.'

[2] Cf. $\tau\rho\iota\sigma\acute{a}\theta\lambda\iota\sigma$, ter felix, $\delta\iota\grave{a}$ $\tau\rho\iota\hat{\omega}\nu$ in Eur. Or. 434. The suggestion was W. von Humboldt's. That about the dual and plural was Dr Wilson's. See Tylor, Prim. Cult. I. p. 265.

[3] Cf. Irish tri, masc. : teoir, teoira, fem. : Welsh tri, masc. : teir, fem.

[4] Hence also Pindar's $\tau\rho\grave{\iota}s$ $\tau\epsilon\tau\rho\acute{a}\kappa\iota$ $\tau\epsilon$, Horace's terque quaterque beati, etc.

(cf.n.2). In hieroglyphic numeral-signs, though the system is denary, units, tens, etc. are grouped by threes and fours and not by fives, (e.g. 7 is written, ||| invariably, and similarly for 70, |||| etc.). Observe also that Egyptian, like Aryan, had a dual. The occasional use of more than one group-limit in counting is common enough. Thus, beside the examples given a little later in the text, the Bas-Bretons use trio-

the common phrases ἐπὶ δακτύλων συμβάλλεσθαι and *digitis computare* and many references in ancient authorities[1] sufficiently attest the practice of finger-counting in the earliest historical times, and there are some signs that the practice was not yet settled before the separation of the Aryan races. Thus the Homeric πεμπάζειν (lit. 'to five'), meaning 'to count,' and the form of the Latin numeral signs imply the occasional use of a quinary notation, while the Kelts and Danes use, to some extent, a vigesimal (e.g. *quatre-vingts*), from which we derive our habit of counting by scores.

7. A further question still remains, whether any connexion can be traced, in the Aryan languages, as there certainly can in most savage tongues, between the first ten numeral words and the gestures used in counting with the fingers. It has been already pointed out that in Aryan languages there is a difference in kind between the first three or four numerals and the last seven or six. The former are adjectives and are so inflected: the latter are nouns neuter in form and uninflected[2]; interjections, as it were, thrust into the sentence in brackets, like the dates in a history-book. This difference in kind seems to point to a difference in etymology and also in antiquity. The higher numerals, being nouns, are names of things and, being uninflected, are names of things which are not really connected with, and subject to the same relations as, the other things mentioned in the same sentence. Secondly, the general abruptness of the transition from low inflected numerals to higher uninflected forms points to some sudden stride in the art of counting. All the facts are readily explained if we conceive that among the Aryans, as among many other races, the counting of low numbers was learnt before the use of the fingers suggested itself, and that so soon as the fingers were seen to be the natural *abacus*, a great advance in arithmetic

---

uech ($=3 \times 6$) for 18 : the Welsh have *dennaw* ($=2 \times 9$) : the old Frisian has *tolftich* ($=$ twelvety) for 120. See further Pott, *Festg.* pp. 33 and 38.

[1] Boethius (p. 395, 3 sqq. ed. Friedlein) says the ancients used to call the units '*digiti*.' The Italian game 'morra' is of very remote antiquity and, like most games, seems to be descended from a serious exercise.

[2] Pott, *Festg.* p. 40, points out that inflexions begin again with the compounds of ten (e.g. *triginta*, a neuter plural, *trecenti*, etc.).

was immediately made. The higher unit-numerals would then be the names of the gestures made in finger-counting or, as among the Algonquins etc., the actual names of the fingers in the order in which they were exhibited in counting.

8. The evidence that 3 or 4 was once the limit of Aryan reckoning has been already adduced. If the fact is so, then the numerals up to that limit probably bear no reference to the fingers, but they are so ancient that it is useless now to inquire into their origin. But the following numerals are neither so ancient nor so curt in form. Their original names appear to have been *pankan* or *kankan* (5), *ksvaks* or *ksvaksva* (6), *saptan* (7), *aktan* (8), *navan* (9) and *dakan* or *dvakan* (10). Some allusion to finger-counting may well underlie these words. Ever since A. von Humboldt first pointed out the resemblance between the Sanskrit *pañk'an* and the Persian *penjeh*, 'the outspread hand,' some connexion between the two has always been admitted. It is possible, indeed, that *penjeh* is derived from *pañk'an* and not *vice versa*, but if we return to the primeval form, *pankan*, as Curtius points out[1], is probably connected with $\pi \acute{v} \xi$, *pugnus* and *fist* or *kankan* with the Germanic *hand*. So also *dvakan* seems to be for *dvakankan*, meaning 'twice five' or 'two hands'[2]' *dakan* points to $\delta \epsilon \xi \iota \acute{o} s$, *dexter*[3], $\delta \acute{\epsilon} \chi o \mu a \iota$ etc. or else to $\delta \acute{a} \kappa \tau v \lambda o s$, *digitus, zehe, toe.* Thus whatever original forms we assume for these two numerals, their roots appear again in some name or other for the hand or fingers. It is intrinsically probable, therefore, that *pankan* means 'hand' and that *dakan* means 'two hands' or 'right hand.' It may be suggested, here, that the intervening numerals are the names of the little, third, middle and fore-fingers of the right hand. Thus the little finger was called by the Greeks $\grave{\omega}\tau\acute{\iota}\tau\eta s$[4], by the Latins *auricularis*. This name is apparently

---

[1] *Griech. Etym.*, Nos. 629 and 384.

[2] The Gothic numerals from 70 to 90 are compounded not with the ordinary *-taihun*, but with *-têhund*, which has been thought (wrongly no doubt) to mean 'two hands' simply.

[3] *Sinister, sem-el, singuli* are curiously analogous to *dexter* (*dec-ister*),

decem, etc. The two hands may possibly have been called the 'one-er' and the 'ten-er.' The ordinary etymology takes *dexter* and *sinister* to mean the 'taker' and 'leaver' respectively.

[4] The finger-names which follow are taken from the appendix to Pott's *Zählmethode*. Comp. also his article

explained by the Germans who call this finger the 'ear-cleaner' (e. g. Dutch *pin, pink* ('poker') *oorvinger*). Now *ksvaks* or *ksvaksva* seems to be a reduplicated form, containing the same root as ξέω, ξαίνω, ξυρέω etc., and meaning 'scraper.' The name *saptan* seems to mean 'follower' (ἕπ-ομαι etc.), and the third finger might very well be so called because it follows and moves with the second, in the manner familiar to all musicians[1]. The name *aktan* seems to contain the common root AK and to mean, therefore, 'projecting,' a good enough name for the middle finger. Lastly, the first finger is known as ἀσπαστικός, *index, salutatorius, demonstratorius* (= 'beckoner,' 'pointer') and this meaning probably underlies *navan*, which will thus be connected with the root of *novus, νέος,* 'new' etc. or that of νεύω, *nuo,* 'nod' etc. or both. Whatever be thought of these suggested etymologies, it must be admitted that there is no evidence whatever that our forefathers counted the fingers of the right hand in the order here assumed. They may have adopted the reverse order, from thumb to little finger, as many savages do and as in fact the Greeks and Romans did with that later and more complicated system of finger-counting which we find in use in the first century of our era and which will be described hereafter in these pages. If this reverse order be

in *Zeitschr.* pp. 164—166. It is curious that this writer should not have attempted to make any use, by comparison, of the facts which he had so industriously collected. Similarly in his article on 'Gender' (*Geschlecht*) in Ersch and Gruber's Encyclopaedia, Vol. LXII., he gives the facts about gender in every language under the sun, but draws no conclusion from them. It is to be observed, however, that both essays were written before the evolution-theory was distinctly formulated. Some other finger-names may be here added. The third finger is generally known either as the 'ring-finger' (δακτυλιώτης, *annularis, golding-er*), or as 'leech-finger' (*medicinalis, arzt.*): in Sanskrit also, *anaman* or

'nameless:' in Greek also ἐπιβάτης or ἐπίβαλος, which may mean 'rider' (ἐπεμβάτης). The first finger is also called 'licker' (λιχανός, Platt D. *pott-licker*). Mr J. O. Halliwell (*Nursery Rhymes and Tales*, p. 206) gives as English finger-names *toucher, longman, lecheman, littleman,* and explains that the third finger is called *lecheman* because people taste with it as doctors try physic. He cites also such names as *Tom Thumbkin, Bess Bumpkin,* etc. with Norse parallels.

[1] So in Odschi or Ashantee the middle finger is called *ensatia hinné,* 'king of the fingers,' and the third is *ensatiasafo-hinné,* 'field-marshal.' (Pott in Ersch and Gr. *sup. cit.*).

assumed, the numerals may still be explained in accordance with other finger-names in common use [1], beside those which have been cited. But after all, the main support of such etymologies is their great *a priori* probability. The theory, on which they are based, brings the history of Aryan counting into accord with the history of counting everywhere else: it explains the Aryan numerals in a way which is certainly correct for nearly all other languages; it explains also the singular discrepancy in the forms of those numerals and some peculiar and very ancient limitations of Aryan counting. It is hardly to be expected that such a theory should be strictly provable at all points.

9. Scanty as is the evidence for the first steps of Aryan calculation, there is none at all for those which follow. It will be conceded, however, that so soon as the fingers were used as regular symbols or a numeral nomenclature was adopted, further progress could not have been difficult. Doubtless at first, as in S. Africa at the present day [2], the numbers from 10 to 100 required two, and those from 100 to 1000, three calculators. But the assistance of coadjutors could be dispensed with, in mere counting, so soon as the memory was trained to remember, without embarrassment, the multiples of 10 or the habit was adopted of making a mark or setting aside a symbol at the completion of each group of 10 [3]. Addition scan be performed with the fingers, but, in the case of high numbers, the process

---

[1] On this plan, *ksvaks* is the thumb, *saptan* the forefinger, *navan*, the third finger. Of these *navan*, 'nodder,' is as good a name for the third finger as *saptan*, for the same reason. *Saptan* may mean 'sucker,' (ὀπός, *sapio*, *saft*, *sap*) pointing to the finger-names λιχανός *pott-licker*, mentioned in the previous note. For this, compare the Zulu names for 7, which are *kombile* 'point,' or *kota* 'lick.' (Pott, *Festg.* p. 48.)

[2] Schrumpf in *Zeitschr. der Deutsch. Morgenl. Gesellsch.* XVI. 463.

[3] Multiples of 10 were expressed by mere compounds, neuter plurals in form, up to 100. This last is supposed to have been named *dakan-dakanta*, of which the last two syllables only survive. But the later word was a neuter *singular*, uninflected (thus the ἑ of ἑκατόν is said to be a relic of ἕν). Multiples of 100 are again compounds and plurals, but in Latin and Greek, curiously enough, they are plural adjectives, with inflexions of gender. The words for 1000 are different in all the great branches of Aryan speech and are all of very obscure origin.

involves a severe tax on the memory. This tax is the more severe with subtraction, because here, to take even the most favourable conditions, the numerals have to be remembered backwards. It is probable therefore that both these operations were very early performed by means of other symbols, such as pebbles ($\psi\tilde\eta\phi o\iota$, *calculi*). The multiplication-table is merely a summarised statement of additions and a division-table would be merely a summarised statement of subtractions. Continual practice, leading to well-remembered inductions, was alone necessary to give considerable facility in the four rules of arithmetic.

10. But division, when it came to be conducted with nicety, introduced a new difficulty. The divisor was not always a whole factor of the dividend and there was then a remainder. What was to be done with this? The question, no doubt, first arose with concrete units, in a case, for instance, where 23 apples were left to be divided among 24 men. Here obviously each man will get a fraction of an apple but there are two ways of ascertaining the fraction. One is to divide each apple into 24 equal parts, and to give to each man 23 such parts. The other is to subdivide 23 into groups, say 12, 8 and 3, and so to give each man first $\frac{1}{2}$, then $\frac{1}{3}$rd, then $\frac{1}{8}$th of an apple. This latter method of treating a remainder (by taking parts of it at a time) is clearly analogous to the way in which the whole dividend has previously been treated, and no doubt it recommended itself, on this account, to the calculators of antiquity. But it had also an especial advantage in this, that the fractions which it produces are more readily represented with primitive symbols. Given only the fingers or pebbles, it would puzzle any man to represent directly that fraction of an apple which we call $\frac{23}{24}$ths, but it would not be so difficult to indicate $\frac{1}{2}+\frac{1}{3}+\frac{1}{8}$[1]. An advantage of the same kind would attend the practice of dividing the unit always into the same fractions (say 12ths or 10ths) and expressing every other fraction, as nearly as

---

[1] It should be mentioned here, also, that fractions with low denominators would naturally be familiar long before those with high denominators were used at all, so that there would be a tendency to express the latter in terms of the former.

possible, in terms of these. As with the former plan the numerator, so with the latter the denominator might be taken for granted and so both the symbols of fractions and calculations with them would be nearly the same as those for whole numbers. And as a matter of fact, the ancient treatment of fractions always did avoid the necessity of handling numerators and denominators together. On the one hand, the astronomical reckoning, introduced into Greece from Babylonia, used only the sexagesimal fractions of the degree, and the Romans used for all purposes the duodecimal fractions of the as[1]; thus on these systems the denominators were implied. On the other hand, and much more commonly, every fraction was reduced to a series of 'submultiples' or fractions with unity for numerator, and thus the consideration of numerators was avoided. This practice was retained in Greek arithmetic to the very last. The Greeks had long since abandoned the old symbolism of numbers but they had adopted another, which, though less clumsy to look at, was even more unmanageable in use. They could state fractions as easily as whole numbers, but calculation of any kind was still so difficult to them that they preferred to get rid of numerators and to reduce denominators to a series of numbers, some of which were so low that they could be handled mechanically and the rest so high that they could often be discarded without materially affecting the result[2].

[1] Each of the Roman fractions had a special name. So we might use *shilling* for $\frac{1}{20}$th, *ounce* for $\frac{1}{16}$th, *inch* for $\frac{1}{12}$th, etc. of any unit whatever. The Aryans, however, do not seem to have had a special name for any merely numerical fraction, except a *half*. The Arabs used to distinguish *expressible* from *inexpressible* fractions. The former are all such as have denomina-tors less than 9 (or compounded of any units), and these had special names: the latter (e.g. $\frac{1}{13}$) had no names. Cantor, *Vorl. über Gesch. der Math.* I. p. 615.

[2] Thus Eutocius, in the 6th century after Christ, reduces $\frac{5}{64}$ to $\frac{1}{6} + \frac{1}{15}$, which is $\frac{1}{960}$ too small. Nesselmann, *Alg. der Griechen*, p. 113.

CHAPTER II.

11. THE preceding pages contain probably all the meagre facts from which it is still possible to discern how the Greeks came by their arithmetical nomenclature, both for whole numbers and for fractions. The subsequent progress of calculation, that is to say, the further use of the elementary processes, depends on many conditions which cannot well be satisfied without a neat and comprehensive visible symbolism. This boon the Greeks never possessed. Yet even without it a retentive memory and a clear logical faculty would suffice for the discovery of many important rules, such for instance as that, in a proportion, the product of the means is equal to the product of the extremes. It is probable, therefore, that much of the Greek arithmetical knowledge dates from a time far anterior to the works in which we find historical evidence of it. It is probable, again, that the Greeks derived from Egypt at an early date as many useful hints on arithmetic as they certainly did on geometry and other branches of learning. It becomes necessary, therefore, to introduce in this place some account of Egyptian arithmetic, both as showing at what date certain arithmetical rules were known to mankind and as providing a fund of knowledge from which the Greeks may have drawn very largely in prehistoric times. The facts to be now stated are in any case of great importance, since they furnish the only compact mass of evidence concerning the difficulties which beset ancient arithmetic and the way in which they were surmounted.

12. Quite recently a hieratic papyrus, included in the Rhind collection of the British Museum, has been deciphered and found to be a mathematical handbook, containing problems in arithmetic and geometry[1]. The latter will be treated on a later page. The book was written by one Ahmes, (Aāhmesu = moon-born), in the reign of Ra-ā-us (Apepa or Apophis of the Hyksos XVIth or XVIIth dynasty), some time before 1700 B.C. but it was founded on and follows, not always correctly, an older work. It is entitled "Directions for obtaining the knowledge of all dark things," but it contains, in fact, hardly any general rules of procedure but chiefly mere statements of results, intended possibly to be explained by a teacher to his pupils. The numbers with which it deals are mostly fractional and it is therefore probable that Ahmes wrote for the *élite* of the mathematicians of his time.

He begins with a series of exercises in reducing fractions, with 2 for numerator, to submultiples. 'Divide 2 by 5' or 'express 2 divided by 7' etc. is his mode of stating the proposition and he gives immediately a table of answers, for all fractions of the form $\dfrac{2}{2n+1}$ up to $\frac{2}{99}$. He does not state, however, why he confines himself to 2 as a numerator or how he obtains, in each case, the series of submultiples which he selects. It is possible that numerators higher than 2 were subdivided[2], but the second question is the more interesting and has been very carefully discussed[3]. It is to be observed that such a fraction as $\frac{2}{29}$, which Ahmes distributes in the form $\frac{1}{24}\ \frac{1}{58}\ \frac{1}{174}\ \frac{1}{232}$ may be expressed also as $\frac{1}{15}\ \frac{1}{435}$, and in various other ways, and

---

[1] Eisenlohr, *Ein mathematisches Handbuch der alten Egypter*, Leipzig, 1877. A short account of the papyrus was given by Mr Birch in Lepsius' *Zeitschrift* for 1868, p. 108. It was then supposed to have been copied, not earlier than 1200 B.C., from an original of about 3400 B.C. The latter was written in the reign of a king whose name is not legible in Ahmes' papyrus, but who is supposed to have been Raenmat or Amenemhat III. The British Museum possesses also an older leather-roll on a mathematical subject, but this apparently is too stiff to be opened.

[2] E.g. $\frac{9}{21} = \frac{7}{21} + \frac{2}{21} = \frac{1}{3} + \frac{1}{14} + \frac{1}{42}$, the last two fractions being copied from the table.

[3] Cantor, *Vorlesungen*, I. pp. 24—28. Eisenlohr, pp. 30—34.

similarly, with all the other fractions, Ahmes has adopted only one of many alternatives. Later on in the book[1] he gives a rule for multiplying a fraction by $\frac{2}{3}$. "When you are asked what is $\frac{2}{3}$ of $\frac{1}{5}$, multiply it by 2 and by 6: that is $\frac{2}{3}$ of it: and similarly for every other fraction." Here it is meant that the denominator must be multiplied by 2 and by 6, and Ahmes' rule is, in effect, that $\frac{2}{3}$ of $\dfrac{1}{a}$ is $\dfrac{1}{2a} + \dfrac{1}{6a}$, and this formula he employs in the table for all fractions of which the denominator is divisible by 3 (e.g. $\frac{2}{9} = \frac{1}{6}\ \frac{1}{18}$ etc.). But the words 'similarly for every other fraction' are of twofold application. They may mean that $\frac{2}{3}$ of any other fraction is to be found by the same method, or that $\frac{2}{5}$, $\frac{2}{7}$ etc. of any fraction may be found by multiplying denominators in a similar manner. The evidence of the table, however, goes to show that Ahmes was ignorant of the latter of these rules[2]. For instance, finding $\frac{2}{5}$ expressed as $\frac{1}{3}\ \frac{1}{15}$, one would expect this formula to be used with all the other fractions of which the denominator is divisible by 5, but it is used, in fact, only for $\frac{2}{25}$, $\frac{2}{65}$, $\frac{2}{75}$. Again, a few of the examples in the table are, as we say, "proved" by being treated backwards. Thus if $\frac{2}{7}$ is $\frac{1}{4} + \frac{1}{28}$, then $\frac{7}{4} + \frac{7}{28}$ should be 2, and this fact (expressed in the form $1\frac{1}{2}\ \frac{1}{4} + \frac{1}{4} = 2$) is what Ahmes points out. It has been suggested therefore that the mode by which the fractions of the table were distributed, was by taking first of all the submultiple which, when multiplied by the original denominator, should be as nearly as possible 2 (e.g. $\frac{1}{4} \times 7 = \frac{7}{4}$), and then adding the remainder. But this process is clearly not employed with most of the distributions (e.g. $\frac{2}{17}$ is given as $\frac{1}{12}\ \frac{1}{51}\ \frac{1}{68}$ instead of $\frac{1}{9}\ \frac{1}{153}$ etc.). This neglect of the most simple and obvious analogies is observable

---

[1] Eisenlohr, p. 150.

[2] The subject is most carefully examined by Cantor. If p be a prime number, then $\dfrac{p+1}{2}$ is a whole number, and $\dfrac{2}{p} = \dfrac{1}{\frac{p+1}{2}} + \dfrac{1}{\frac{p+1}{2} \times p}$, but of 24 prime denominators occurring in the

table, only five are treated on this plan. So again if p and q are odd numbers, then $\dfrac{p+q}{2}$ is a whole number and $\dfrac{2}{p \times q} = \dfrac{1}{q \times \frac{p+q}{2}} + \dfrac{1}{p \times \frac{p+q}{2}}$. Only two denominators, out of the forty-nine, are treated on this principle.

throughout the table and we must conclude that it was compiled empirically, probably by different persons and at different times, certainly without any general theory.

13. Immediately after the table, Ahmes gives six calculations, unfortunately mutilated, showing how to divide 1, 3, 6, 7, 8 and 9 loaves respectively among 10 persons[1], and then follow 17 examples of *seqem* calculation, that is, of raising fractions by addition or multiplication to whole numbers or to other fractions[2]. For this purpose a common denominator is chosen, but not necessarily one which is divisible into a whole number by all the other denominators. Thus, in the problem to increase $\frac{1}{4}$ $\frac{1}{8}$ $\frac{1}{10}$ $\frac{1}{30}$ $\frac{1}{45}$ to 1, the common denominator taken is evidently 45, for the fractions are stated as $11\frac{1}{4}$, $5\frac{1}{2}$ $\frac{1}{8}$, $4\frac{1}{2}$, $1\frac{1}{2}$, 1. The sum of these is $(23\frac{1}{2}$ $\frac{1}{4}$ $\frac{1}{8})$ $(\frac{1}{45})$. Add to this $\frac{1}{9} + \frac{1}{40}$ and the sum is $\frac{2}{3}$. Add $\frac{1}{3}$ and the desired 1 is obtained. From other examples here and elsewhere in the book it is plain that Ahmes did not use direct division. If it was required to raise a by multiplication to b, his plan was to multiply a until he found a product which either was or was nearly b. Thus in the example, numbered by Eisenlohr (32), where $1\frac{1}{3}$ $\frac{1}{4}$ is to be raised by multiplication to 2, he finds on trial that $1\frac{1}{3}$ $\frac{1}{4} \times 1\frac{1}{6}$ $\frac{1}{12}$ produces $\frac{285}{144}$. The difference, $\frac{3}{144}$, between this product and 2 is then separately treated[3].

14. After this preliminary practice with fractions, Ahmes proceeds to the solution of *simple equations with one unknown*[4]. Eleven such are given, expressed, for instance, as follows, (no. 24) 'Heap, its 7th, its whole, it makes 19' (i.e. $\frac{x}{7} + x = 19$).

In this particular case, Ahmes goes on, in effect, to state

---

[1] Eisenlohr, pp. 49—53. In these examples, the denominator is constant, as, in the first table, the numerator.

[2] Eisenlohr, pp. 53—60.

[3] It should be mentioned that Ahmes does not multiply *directly* with a high number but proceeds by many easy stages. In order to multiply by 13, for instance, he multiplies by 2, then (doubling) by 4, then (doubling) by 8

and adds the necessary products. Cantor, *Vorles.* I. pp. 31, 32, and 41.

[4] The unknown quantity is called *hau* or 'heap.' In these examples a pair of legs walking, so to say, with or against the stream of the writing, are used as mathematical symbols of addition and subtraction. Three horizontal arrows indicate 'difference' and a sign ≤ means 'equals.' Cantor, pp. 32, 33. Eisenlohr, pp. 22—26.

$\frac{8x}{7} = 19$: divides 19 by 8 and multiplies the quotient $(2\frac{1}{4} \frac{1}{8})$ by 7 and so finds the desired number $16\frac{1}{2} \frac{1}{8}$, but he has also various other methods of treating the two sides. For instance, in no. 29, where ultimately $\frac{20}{27} x = 10$, he first finds the value of $\frac{27}{20}$ as $1\frac{1}{4} \frac{1}{10}$ and then multiplies this by 10, so as to find $x = 13\frac{1}{2}$[1]. These equations are followed by the table of Egyptian dry measures, and then are added two examples of *Tunnu*- or difference-calculation, i.e. of divisions according to different rates of profit. The examples are 'Divide 100 loaves so that 50 go to 6 and 50 to 4 persons,' and 'divide 100 loaves among 5 persons, so that the first 3 get 7 times as much as the other 2. What is the difference *(tunnu)*?' After this, the writer passes to geometry, but he recurs at the end of the book to these algebraical problems and gives about twenty more examples of the same kind. Most of them are simple, but in at least three Cantor sees evidence that Ahmes was acquainted with the theory of arithmetical and geometrical series. The solution which he gives of the second problem above quoted is as follows: 'the difference is $5\frac{1}{2}$: 23, $17\frac{1}{2}$, 12, $6\frac{1}{2}$, 1. Multiply by $1\frac{2}{3}$: $38\frac{1}{3}$, $29\frac{1}{6}$, 20, $10\frac{2}{3} \frac{1}{6}$, $1\frac{2}{3}$.' The series first given amounts only to 60, and each of its terms must be multiplied by $1\frac{2}{3}$, in order to produce[2] the requisite sum 100. The difference $5\frac{1}{2}$ must have been found from the equation

$$\frac{a + (a - b) + (a - 2b)}{7} = (a - 3b) + (a - 4b),$$

whence $11 (a - 4b) = 2b$ and $b = 5\frac{1}{2} (a - 4b)$. Ahmes then *assumes* $(a - 4b) = 1$, and so by experiment finds its true value. Another example (no. 64) is 'Ten measures of corn for 10 persons. The difference between each person's share and the next's is $\frac{1}{8}$th of a measure.' The solution runs: 'I find the mean, 1 measure. Take 1 from 10: remainder 9. Halve the difference, i.e. $\frac{1}{16}$. Take it 9 times, that gives you $\frac{1}{2} \frac{1}{16}$. Add it to the mean. Deduct $\frac{1}{8}$th of a measure for each person so as to reach the end.'

---

[1] Other examples in Cantor, pp. 32—34.

[2] Upon this Cantor *(Vorles.* p. 36) remarks that it is the first known in-stance of a 'falscher ansatz,' a *falsa positio* or 'tentative assumption,' on which see below § 70. *n.*

These consecutive sentences mean, in modern algebraical form, 'Find $\frac{s}{n}$. Find $(n-1)$. Find $\frac{b}{2}$. Find $\frac{b}{2} \times (n-1)$. Add $\frac{s}{n}$ to $\frac{b}{2} (n-1)$,' i.e. these directions imply a knowledge of the formulæ for finding the sum or the first term of an arithmetical progression. The evidence, however, for Ahmes' knowledge of *geometrical* series is confined to the fact that in one example (no. 79) he states such a series and calls it a 'ladder' (*Sutek*).

15. One might naturally expect that a nation, which at so early a date had acquired so much proficiency in arithmetic, would in another thousand years make much further progress or would at least discover and begin to remove the obstacles which prevented such progress. But the Egyptian intellect, like the Chinese, seems to have been rather shallow, and the ancients themselves, who were indebted to Egypt for the rudiments of many sciences, observed with surprise that no greater advance was made in that country. In geometry, for instance, it is certain that the later Egyptians added nothing whatever to the learning of Ahmes' day, and though as to arithmetic there is little or no direct evidence, yet two facts raise a presumption that Ahmes' book represents the highest attainment of Egypt in that science. First, no improvement was made in Egyptian arithmetical symbolism, and secondly, the Greeks did not derive directly from Egypt any more arithmetical learning than is given by Ahmes. This latter fact renders it unnecessary to pursue further in this place an inquiry into Egyptian arithmetic, but it is probable, nevertheless, as will be seen hereafter, that Egyptians, educated in Greek learning, made some important additions to Greek mathematical methods.

16. The theories suggested and the facts adduced in the foregoing pages may be shortly summarised as follows. Primitive peoples, when they have learnt to generalise, begin to learn to count. They commence counting with groups of two or three things only but soon arrive at counting five. When they reach this limit, they at once begin to use the fingers, or the fingers and toes, as the means and basis of calculation and are hence-

PART II. GREEK ARITHMETIC.

CHAPTER III.

GREEK CALCULATION. *Logistica.*

17. A distinction is drawn, and very naturally and properly drawn, by the later Greek mathematicians between ἀριθμητική and λογιστική, by the former of which they designated the ' science of numbers,' by the latter, the ' art of calculation[1].' An opposition between these terms occurs much earlier and is frequently used by Plato, but though λογιστική can hardly mean anything but ' calculation,' it is not quite clear whether ἀριθμητική then bore the sense which it had undoubtedly acquired by the time of Geminus (say B.C. 50). That it did so, however, is rendered pretty certain by many circumstances. It is probable, in the first place, that the Pythagoreans would have required some variety of terms to distinguish the exercises of schoolboys from their own researches into the *genera* and *species* of numbers[2]. In Aristotle[3] a distinction, analogous to that between the kinds of arithmetic, is drawn between γεωδαισία, the practical art of land-surveying, and the philosophical γεωμετρία. Euclid, who is said to have been a Platonist and who lived not long after Plato, collected a large volume of the theory of numbers, which he calls ἀριθμητική only and in which he uses exactly the same nomenclature and symbolism as we find in those passages where Plato draws a philosophical illustration from arithmetic[4]. It may therefore be assumed that λογιστική and ἀριθμητική covered, respec-

---

[1] See esp. Geminus cited by Proclus, *Comm. Eucl.* (ed. Friedlein), p. 38.

[2] Thus Aristoxenus (*apud* Stob. *Ecl. Phys.* I. 19. c. 2 *ad initium*) says that Pythagoras first raised ἀριθμητική ' above

the needs of merchants,' with which comp. Plato, *Rep.* 525 c.

[3] *Metaph.* II. 2, 26.

[4] Cf. Euclid VII. with Plato, *Theaet.* 147, 148, or *Rep.* 546 c.

forth committed to a quinary or denary or vigesimal scale. The gestures used in finger-counting suggest names for five and the higher units, and with such names and with the use of the fingers it is possible to attain a fair dexterity in calculation with whole numbers. It is not so easy, however, to find names or symbols for fractions, but the difficulty here is very much reduced if a constant numerator or a constant denominator be adopted, and one or the other of these devices was, for more than one reason, employed by all nations which ever got as far as the arithmetic of fractions. It is evident, nevertheless, that fractions were at first and remained a stumblingblock to calculators: for the oldest extant collection of arithmetical examples is chiefly devoted to them and the latest Greek writer on arithmetic still uses the ancient devices for expressing them. Such are the antecedents of Greek arithmetic, so far as they can be discovered from the evidence of the Greek language and of the usages of later Greek calculators. It cannot be doubted, however, that Greece received directly a good deal of arithmetical learning from Egypt, but this, at its best, can hardly have dealt with more abstruse subjects than the solution of simple equations with one unknown and some portions of the theory of arithmetical and geometrical series.

tively, the same subjectmatter in Plato's time, as afterwards and since he uses these terms casually, with no hint that they were novel, we may infer that the distinction between them dates from a very early time in the history of Greek science and philosophy[1].

But though the opposition of ἀριθμητική and λογιστική is as clear as that of theory to practice or science to art, an historical account of either would necessarily involve frequent reference to the other. Just as many of the rules of modern arithmetic are proved by algebra, so with the Greeks the rules of proportion, the rules for finding a greatest common measure and the like were discovered by and belonged to ἀριθμητική, while the discovery of *prime, amicable, polygonal* numbers etc., which are part of the subjectmatter of ἀριθμητική, is obviously due to induction from the operations of λογιστική. It is, however, desirable and even necessary to keep the two apart, for the record of Greek arithmetical theory is far fuller and more exact than that of Greek practice and, besides, the symbolism of the former was entirely distinct from that of the latter. The two departments, therefore, λογιστική and ἀριθμητική, will be kept separate in the following pages, but it is to be premised that probably Greek *logistic*, or calculation, extended to more difficult operations than can be here exhibited and that probably Greek *arithmetic*, or theory of numbers, owed much more to induction than is permitted to appear by its first and chief professors.

---

[1] The Platonic passages may be here mentioned. In *Gorg.* 451 B C ἀρ· and λογ. are opposed, but both are described as τέχναι, dealing with 'odd' and 'even,' the special aim of λογ. being to find out *quantity*, both absolute and relative. In *Euthyd.* 290 B C λογ. is opposed to some *philosophical* use of numbers, not there named. But in *Rep.* 525 C D and *Phileb.* 56 D E, a distinction is drawn between *popular* ἀρ. and λογ. together and the *philosophical* species of both, the basis of the distinction being that the former use unequal and dissimilar units, while the latter use equal units,

contemplated absolutely. The difficulty of course is to perceive what Plato meant by popular ἀρ. and philosophical λογ. It seems to be a satisfactory explanation to suppose that Plato was here thinking of those rules of λογ. which are proved deductively (popular ἀρ.) and those doctrines of ἀρ. which are proved inductively (philosophical λογ.). Thus the proportion 2 *apples* : 1 *obol* :: 6 *apples* : 3 *obols* is a piece of popular ἀριθμητική; the fact that 'all the powers of 5 end in 5' (e.g. 25, 125, 625, etc.), is a piece of philosophical λογιστική.

18. In a historical account of ordinary Greek calculation, the first subject which demands attention is the customary *symbolism.* This also is the subject which ought to be capable of most satisfactory treatment, for here the record, if there is any at all, can hardly be deceptive : and this again is the most important subject, for a good symbolism is itself suggestive, while a bad one stifles the ingenuity, and a nation's arithmetical reputation may be made or marred by the written forms with which it represents numbers.

At the time when the inquiry into the prehistoric development of Greek logistic must perforce be abandoned, we have found the Greeks in possession of a complete numerical nomenclature, with a decimal scale, and accustomed to use the fingers or pebbles (ψῆφοι) as aids to calculation. These symbols were no doubt at first used, and continued always to be used, in the most primitive way, each finger or stone representing a single unit[1]. But the progress of commerce and the increasing adroitness of Greek merchants introduced far more complex conventions into the use of fingers and pebbles, and though it is probable that these improvements were really subsequent to the invention of some sort of written symbols, yet the antiquity of the instruments themselves and the narrow limitations of their use render it desirable that they should be described first, before proceeding to the history of written signs.

19. A mediaeval Greek, one Nicolaus Smyrnaeus (called also Rhabda or Artabasda), in a work entitled ἔκφρασις τοῦ δακτυλικοῦ μέτρου, written probably in the 13th or 14th century[2], describes fully the finger-symbolism which was in use in his time and probably for some fifteen hundred years before. On this system, the operator held up his hands, so that the

---

[1] Herod. vi. 63, 65. Arist. *Problem.* xv.

[2] It is printed in Schneider's *Eclog. Phys.* i. p. 477, also by N. Caussinus in his *Eloquentia Sacra et Humana,* Bk. ix. ch. 8, pp. 565—568 (Paris, 1636), and elsewhere. See Roediger's article in *Jahresb. der Deutsch. Morgenländ. Gesellsch.* for 1845, pp. 111—129.

Roediger, who has been followed by many writers, supposed that Nicolaus Smyrnaeus was of the 7th or 8th century, but Dr Günther (*Vermischte Untersuch. zur Gesch. der Math.* 1876) has lately discovered him to be a contemporary of Manuel Moschopulus, a much later writer. See p. 317.

fingers were erect, the palms facing outwards. The 3rd, 4th and 5th fingers (to use the German description) might be ἐκτει-νόμενοι or straight, συστελλόμενοι 'bent' or 'half-closed,' κλινό-μενοι or 'closed.' The subsequent gestures may be thus described :

(a) *On the left hand :*

for 1, half-close the 5th finger only :

„ 2, „ the 4th and 5th fingers only :

„ 3, „ the 3rd, 4th and 5th fingers only :

„ 4, „ the 3rd and 4th fingers only :

„ 5, „ the 3rd finger only :

„ 6, „ the 4th finger only :

„ 7, close the 5th finger only :

„ 8, „ the 4th and 5th fingers only :

„ 9, „ the 3rd, 4th and 5th fingers only.

(b) The same operations on the *right hand* gave the *thousands*, from 1000 to 9000.

(c) *On the left hand :*

for 10, apply the tip of the forefinger to the bottom of the thumb, so that the resulting figure resembles δ :

„ 20, the forefinger is straight and is separated by the thumb from the remaining fingers, which are slightly bent :

„ 30, join the tips of the forefinger and thumb :

„ 40, place the thumb behind (on the knuckle of) the forefinger :

„ 50, place the thumb in front (on the ball) of the fore-finger :

„ 60, place the thumb as for 50 and bend the forefinger over it, so as to touch the ball of the thumb :

„ 70, rest the forefinger on the tip of the thumb :

„ 80, lay the thumb on the palm, bend the forefinger close over the first joint of the thumb and slightly bend the remaining fingers :

„ 90, close the forefinger only as completely as possible.

(d) The same operations on the *right hand* gave the hundreds, from 100 to 900.

Nicolaus himself does not give signs for numbers above 9000, but Martianus Capella, a writer of the 5th century, says (*De Nuptiis,* Lib. VII. p. 244 of Grotius edn. 1599) 'nonnulli Graeci etiam μυρία adjecisse videntur' by means, apparently, of 'quaedam brachiorum contorta saltatio' of which he does not approve. The motions were probably the same as those described by Bede in his tract 'De loquela per gestum digitorum[1].' Different positions of the left hand on the left breast and hip gave the numbers from 10,000 to 90,000 : the same motions with the right hand gave the hundred thousands and the hands folded together represented a million.

20. The finger-symbolism here described was in use, in practically the same form, in Greece and Italy and throughout the East certainly from the beginning of our era[2], but there is unfortunately no evidence as to where or when it was invented. By far the oldest passage in which any reference to it may be supposed to occur is Aristophanes, *Vespae,* ll. 656—664, where Bdelycleon tells his father to do an easy sum, οὐ ψήφοις ἀλλ᾽ ἀπὸ χειρός. "The income of the state," says he, "is nearly 2000 talents: the yearly payment to the 6000 dicasts is only 150 talents." "Why," answers the old man, "we don't get *a tenth* of the revenue." It is clear, from this reply, that the 'easy sum' in question amounted only to dividing 2000 by 10 or multiplying 150 by 10, an operation which does not require the more elaborate finger-signs. Failing this passage, there is

---

[1] *Opera,* Basileae, col. 171—173. The material part is given by Roediger. The finger positions described by Bede differ slightly, in one or two cases, from those of Nicolaus Smyrnaeus, and both again vary slightly from those used in the East, where the units and tens were represented (*not always,* v. the Arabic poem in *Bulletino Boncompagni,* 1863, I. pp. 236, 237) on the *right* hand and not on the left. The reader is referred to Roediger's article above mentioned. Plates will be found in *Journal of Philology,* Vol. II. p. 247, in Stoy's pamphlet *Zur*

Gesch. des Rechenunterrichts (Jena, 1876), in *Neue Jahrb. für Phil. u. Päd.* 15th supplbd. p. 511, and in many other places. A large collection of references is given by Prof. Mayor in his note to Juvenal, *Sat.* x. 248. More, esp. to late Jewish and Arabic writers, in Steinschneider's *Bibliogr. Hebr.* Vol. XXI. pp. 39, 40.

[2] The same or something like it is still used by Persian merchants. See De Sacy in *Journ. Asiatique,* Vol. 2, and Tylor, *Primitive Culture,* I. p. 246, n.

another possible reference, equally doubtful, to this system of finger-symbolism in Plautus, *Miles Gloriosus*, II. 3, but the first clear references to it occur in Plutarch and authors of his time[1]. Pliny, indeed, says that there was, in his time, a statue of Janus, erected by Numa, of which the fingers indicated 365 or 355 (the reading is doubtful, cf. also Macrobius, *Conv. Sat.* I. 9), the number of days in the year, but no importance can be attached to such a statement. All that we can allege of the system is that it is mentioned only in later classical literature, that it then appears to be of universal diffusion and that it was far more persistent in the East than in the West[2]. If we consider that such a system can have been of no use in calculation, save as a *memoria technica* for some number with which the mind of the reckoner was not immediately engaged—if, in other words, we consider that such a system was useful to *represent* numbers but not to calculate with them, then it becomes probable that it was invented in the first instance as a secret means of communication between merchants[3] or as a numerical gesture-language between persons who were ignorant of one another's tongues. Phoenician and Greek commerce would make it widely known: the later diffusion of Latin and Greek and the larger use of writing would ensure its gradual extinction in the West, but it would still preserve its original utility in the motley and ignorant crowds of the Eastern bazaars.

21. In reckoning with pebbles, no doubt at first each pebble represented one of the objects to be counted, the advantage of course being that space was saved and the memory relieved by a good *coup d'œil*, for it will be conceded that it is easier to count 100 pebbles than 100 cows or to find 10 times

---

[1] Plut. *Apophth.* 174 b. Pliny, *Hist. Nat.* XXXIV. s. 33. For other reff. see Prof. Mayor's note on *Juv.* x. 249, above referred to, or Dean Peacock's article *Arithmetic* in *Encycl. Metropolitana.*

[2] Erasmus, in his ed. of Jerome (III. 25 B c) published in 1516, confessed his ignorance of the finger-symbolism referred to by the saint. He under-

stood it afterwards (see *ibid.* p. 313).

[3] The Persian system mentioned by De Sacy and Tylor (see *note* above) is used only in secret, when for instance a dragoman wishes to have one price with the seller and the other with his master. See also the opening words of Roediger's article. Another suggestion as to the origin of this symbolism will be made below, § 25.

10 in pebbles than in sacks or such other articles of commerce. So soon as the heap contained one pebble for each object, the calculator would begin afresh and by arranging the pebbles in groups of 10, arrive at the total and the name of the total, without having his attention embarrassed by petty circumstances[1]. This use of pebbles in mere counting, where each represents a real object, would naturally precede their use in calculations where some pebbles would represent imaginary objects. A great number of pebbles could be dispensed with if the operator, on completing a group of 10, laid aside a large pebble or a white one and then began again with the pebbles of the original group. He would soon find that there would be no need for a variety of pebbles, if he always laid pebbles representing 10 in a separate place from those representing units. In this way, he would arrive at a neat visible symbolism for a high number, which would greatly facilitate operations in the four rules of arithmetic. Such an advanced pebble-symbolism the Egyptians and the Chinese had from a time 'whereof the memory of man runneth not to the contrary.' It can hardly be doubted that they invented it independently and imparted it to the nations around. Wherever and whenever invented or borrowed, the Greeks and Italians had it also and used it by preference for all ordinary calculations down to the 15th century of our era. The evidence for its use, however, is singularly late. Homer and Pindar do not allude to it, but it is plain that it was in regular use by the 5th century B.C., though the authorities even of that time do not state explicitly how the calculation with pebbles was conducted[2]. It cannot be doubted,

---

[1] In a London night-school I have often seen a boy, in order to multiply say 12 by 10, make 120 dots on his slate and then count these. What he wanted was *the name* of the total and he did not always get this right. With primitive man, I imagine, the use of pebbles would not arise till numeral names had partly superseded finger-counting. If, for instance, a savage sold something for 50 cows, he would indicate his price by naming it, and would then, with the aid of pebbles, ascertain whether he had got the price he bargained for. Thus the Mexicans acquired a set of numerals, used in counting animals and things, which runs *centetl, ontetl,* etc. or 'one-stone,' 'two-stone,' etc. Other similar examples are cited in Tylor, *Early Hist.* p. 163.

[2] Diogenes Laertius (I. 59) ascribes to Solon a saying that courtiers were like the pebbles on a reckoning-board, for they sometimes stood for more,

however, that the pebbles were arranged in lines, either horizontal or perpendicular, and that the pebbles on the first line
represented units, those on the second tens, those on the third
hundreds and so on. How many lines there were and how
many pebbles might be placed on each there is no evidence to
show. It may be added that fractions in the form of 'submultiples' would not present any difficulty when the system of
local values for the pebbles was once introduced. If for instance
a line were appropriated to pebbles of the value of $\frac{1}{12}$, it would
be as easy to discern that 12 pebbles on that line are equal in
value to 1 on the units line, as to perceive that 10 pebbles on
the unit line may be replaced by 1 on the tens line. But since
a great many lines devoted to fractions would have been inconvenient, probably a few lines only were devoted to certain
selected fractions, and all other fractions were reduced as nearly
as possible to terms of these.

22. The surface on which such lines were drawn, or the
frame on which strings or wires were stretched, for the purpose
of pebble-reckoning, was called by the Greeks ἄβαξ or ἀβάκιον.
This name seems to point to the common Semitic word *abaq*
meaning 'sand,' and it is said that a board strewn with sand,
on which lines might be drawn with a stick, was and still is a
common instrument for calculation in the East. It is the more
desirable also that some Oriental origin for the ἄβαξ should
be found because, in late Greek writers, we find a general
tradition that Pythagoras, who certainly studied out of
Greece, was the inventor or introducer of the instrument. It
cannot, however, be considered that the Semitic origin of
ἄβαξ is rendered at all probable by such considerations. The

---

sometimes for less. This, if genuine
(but cf. Polyb. v. 26, 13), is the first and
also one of the most explicit references
to the pebble-symbolism. If this be
doubted, then the earliest authentic
reference is probably a fragment of
Epicharmus (ed. Ahrens, 94, 8): then
Aeschylus (*Agam.* 570), then perhaps
Herodotus (II. 36), who says that, in
pebble-reckoning, the Egyptians counted as they wrote from right to left, the
Greeks from left to right. It may be
that the *abacus* with the Greeks was
not so old as writing, for the Greeks
did not originally write from left to
right, but either from right to left or
βουστροφηδόν. They may have counted
from right to left, but can hardly have
counted βουστροφηδόν.

word itself in the sense of 'reckoning-table' is not used for certain in any writer before Polybius (ἀβάκιον in v. 26, 13) who belongs only to the 2nd century B. C. It is, however, used in the sense of plain 'board' in many different connexions[1].

Assuming it to be true, also, that the Semites did generally use a sanded board for their calculations[2], it does not appear that this was called *abaq,* and the step from Semitic *abaq* 'sand' to Greek ἄβαξ a 'board' remains practically as wide as before. Lastly, the tradition which connects the ἄβαξ with Pythagoras as well as that which connects him with a Semitic people, is so late and belongs to so imaginative authors[3] that no reliance can be placed upon it. Of course, a few lines drawn with a stick in the dust and a handful of stones were as efficient an instrument for calculation as was needed and must always have been used by Greeks upon occasion. Such an *impromptu* ledger would indeed frequently be preferable to a more elaborate device, since it could be adapted to different fractions, different monetary scales etc., while a permanent machine would probably be restricted to one scale and a few selected fractions. But whether such a scheme of lines drawn on the ground could ever in Greek have been called ἄβαξ there is no evidence to show.

23. It must be admitted, also, that hardly anything is known of the normal Greek ἄβαξ, using that word in the sense of a reckoning-board with permanent lines drawn on it and possibly permanent balls or pebbles attached to it. Three types

---

[1] The word seems first to occur in the sense of 'trencher' in Cratinus, Κλεοβ. 2 (*cit.* Poll. x. 105). Hesychius says it was a synonym for μάκτρα 'trough.' Pollux also cites ἀβάκιον from Lysias, without stating its meaning. It is oddly accented.

[2] The evidence adduced by Cantor, *Math. Beiträge,* p. 141, is not satisfactory on this point, but the fact is hardly worth disputing. A sanded board was certainly used by Greek *geometers,* but is nowhere attributed to arithmeticians. Cf. Cic. *Nat. Deor.*

2, 18, 48. *Tusc.* 5, 23, 64, and other quotations collected by Friedlein, *Zahlz.* § 76, pp. 52, 3. See also Cantor, *Vorl.* pp. 109—111. It seems to me not unlikely that ἄβαξ was a childish name for the board on which the alphabet was written and from which the children read their βῆτα ἄλφα βα, βῆτα εἶ βε, etc. (Athenaeus, x. 453). Ἄβαξ would be the 'ABC board,' the termination being chosen by analogy from πίναξ.

[3] Iamblichus, for instance, and the pseudo-Boethius, cited *post.*

at least of such a machine are well known. One of these is the Russian *tschotü*, in which each wire carries 10 balls[1]. Some advance is shown in the Chinese *suan-pan*[2], where the whole field of the frame is divided by a transverse string: each wire on that part of it which is below this string carries 5 balls: and on the part which is above 2 balls, each of which is worth 5 of those below. On both these machines, apparently, it is possible and usual to remove balls from one wire to another as the case may require. But the third type is the Roman *abacus*, which, at any rate in its highest development, was closed, so that balls or buttons could not be removed from the wire or groove in which they were originally placed. A few specimens of this sort, constructed with grooves in which buttons (*claviculi*) slide, are still preserved. One of them which is figured in Daremberg's *Dictionnaire des Antiquités* (*s.v.* abacus) and is in the Kircher Museum at Rome, may be roughly represented thus:

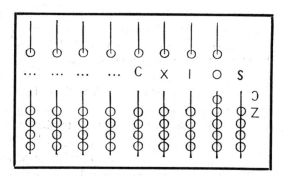

Leaving out of consideration, for a moment, the two grooves on the extreme right, it will be seen that the remaining 7 contain buttons representing units, tens, etc. up to millions. The lower

---

[1] The balls are differently coloured, some of the 10 being white and some black. The instrument was introduced into the schools of Eastern France after the great Russian campaign. It is common enough in Pestalozzian schools. See further Cantor, *Math. Beitr.* pp. 129, 130.

[2] *Suán* = reckon : *p'huán* = board. Goschkewitsch, an authority quoted by Hankel, *Zur Gesch. der Math.* p. 54, says that "the practised Chinese reckoner plays with the fingers of the right hand on the *suan pan* as on a musical instrument and grasps whole numerical chords."

grooves contain 4 buttons each, the higher 1 each, which represents 5 of the same value as those in the lower corresponding groove. The letters indicating the values of the buttons are obscure above C, but are plain enough on another specimen, which once belonged to one Welser, in whose works published at Nuremberg in 1682 there was given a drawing of his *abacus* (pp. 442 and 819)[1]. The sign O which distinguishes the penultimate groove on the right, stands for *uncia*, and as there are 12 *unciae* to the *as*, here the lower portion of the groove has 5 buttons for 5 *unciae*, the upper 1 button for 6 *unciae*. The signs appended to the last groove on the right are S for *semuncia* ($\frac{1}{24}$th of an *as*) : $\mathfrak{I}$ for *sicilicus* ($\frac{1}{48}$th of an *as*) : and Z for *sextula* ($\frac{1}{72}$nd of an *as*). It is not, however, very clear why there should be 4 buttons in this groove or what was the value of each and how, if of different values, they were distinguished from each other. Welser's *abacus*, which in other respects is exactly similar to this, had three separate grooves for these fractions, the first containing 1 button for the *semuncia* ($\frac{1}{24}$th) : the second 1 button for the *sicilicus* ($\frac{1}{48}$th) : the third 2 buttons, each representing a *sextula* ($\frac{1}{72}$nd). These grooves therefore together (and no doubt the last groove of the Kircher *abacus*) represent $\frac{18}{12}$ths of an *uncia*[2]. Both *abaci* are capable of representing all whole numbers from 1 to 9,999,999 and the duodecimal fractions of the *as* in common use. Since such an *abacus* could seldom represent more than one number at a time, it is probable that, in calculating with it, the larger of the two numbers to be dealt with would be represented on the table. The smaller would be mentally added or subtracted

---

[1] Reproduced by Friedlein in *Zeitschr. für Math. u. Physik.* Vol. IX. 1864. Plate 5. See also p. 299. A description of this *abacus* is given also in Friedlein's *Zahlzeichen*, p. 22, § 32. A figure of it is given in Daremberg, *Dict. des Ant.* s.v. *arithmetica*. M. Ruelle, the writer of the article, says that 4 Roman *abaci* (which he names) are known, but it does not appear that they are all now in existence.

[2] This statement, which is taken from Friedlein, seems unlikely. On the analogy of all the preceding grooves, we should expect the table to conclude with $\frac{14}{12}$ths of the *uncia*, and not $\frac{18}{12}$ths. It will do so if the last two buttons be taken to represent, not sextulae, but *dimidiae sextulae*, the ordinary sign of which is easily to be confused with that of the sextula. See Friedlein, *Zahlz.* Plate to § 48.

as the case may be, and the buttons would be successively altered so as to represent the sum or remainder. Multiplication can only have been performed by repeated additions, and division by repeated subtractions[1].

24. It will be seen that the types of *abacus* now known are not very diverse from one another, and there is no cause to be greatly distressed by our ignorance of what the Greek ἄβαξ was. A certain table, however, which may be an ἄβαξ, was discovered in 1846 in the island of Salamis and this, which can be partly explained by reference to the Roman instruments, must serve to assure us that there cannot have been any great superiority in the Greek ἄβαξ at any time. This "Salaminian table" may be figured thus[2]:

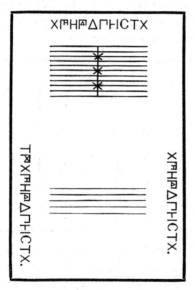

---

[1] Unless indeed the *abacus* is used merely as a *memoria technica*. Thus the Chinese, in dividing, first represent the dividend, then, breaking it up as the remainders successively are obtained, place, on the wires from time to time vacated, balls to represent the successive ciphers of the quotient. The actual division is done in the mind by use of the multiplication-table. Thus Goschkewitsch (cited by Hankel *uti sup.*) says many modes of division have been proposed for the Russian *tschotü*, but they all involve the use of a second board or of a board and paper.

[2] There is a drawing of it in Daremberg s.v. *Abacus:* also in *Revue*

It is made of marble and is very large, being about 5 ft. (1·5 metres) long by 2½ ft. (·75 metre) wide. The letters upon the margin are easily explained. ⊢ is the customary Attic sign for a *drachma.* The letters which, in the table, stand on the left of this sign are Π for 5 (πέντε), Δ for 10 (δέκα), ⋈ for 50, H for 100 (ἑκατόν), ⋈ for 500 and X for 1000 (χίλιοι) in the ordinary Attic style. To these are added, in one row, the signs ⋈ for 5000 and T for τάλαντον or 6000 *drachmae.* The signs which stand to the right of ⊢ in the table are the fractions of the *drachma,* viz. ⏐ for ⅙th (obol), C for $\frac{1}{12}$th (½ obol), T¹ for $\frac{1}{24}$th (τεταρτημόριον of the obol) and X for χαλκοῦς (⅛th of the obol, $\frac{1}{48}$th of the drachma). The last three fractions, it will be observed, when added together make ⅞ths of an obol, which is the real unit of the table. On the principle of a Roman *abacus,* this scale would be thus distributed :

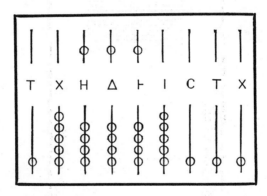

But it will be seen that the lines of the Salaminian table do not fall in with this arrangement. Here we have 11 lines, with 10 intervals, in one place : and 5 lines, with 4 intervals, in another. If the table be really an ἄβαξ, the simplest explanation is that

Archéolog. 1846, p. 296, where a very minute description of the stone is given by M. Rangabé. Another Greek *abax* is also figured on the Darius-vase at Naples. The numerals on it are of the same kind as those on the Salaminian table and it is held by the

reckoner so that the columns are perpendicular to his body. But it is too small and roughly drawn to furnish important information.

[1] τεταρτημόριον is Böckh's explanation : M. Vincent proposed τριτημόριον.

the 10 spaces at one part of the board contained stones representing values from a talent to an obol, in the order ΤΡΧΡΗΡ ΔΓΗΙ: and the 5 spaces at the other part of the board contained stones representing values from an obol to a chalcus, in the order Ι, C, Τ, Χ. The transverse line would serve to distinguish two sums which were to be added together or subtracted one from the other: or again, as in the Roman system, the numbers compounded with Γ might have been placed above this line. The crosses on the line are merely aids to the eye in keeping the various rows distinct. Operations with fractions of the obol would be separately conducted at the lower end of the table. The table also being very large, perhaps two people would work at it at once, or because it was heavy, it might be desirable to use it from either side and therefore a table of customary values would be repeated in various parts of the table. The received explanation, however, of the use of the table is different from this. The well-known archaeologist, M. Vincent[1], considered that the table served two purposes: that it was an ἄβαξ, and also a scoring board for a game something like *tric-trac* or backgammon. When it was used as an ἄβαξ, fractions of the drachma were calculated at the end of the table on the 4 spaces there reserved: sums from the drachma to the talent were calculated on five of the other ten spaces, and the remaining five were used for calculations from one talent to 10,000 talents. It is an objection to this theory that a Greek merchant or tax-gatherer can seldom have had occasion to calculate above a few talents, since the whole revenue of Athens in her prime was not 2000 talents[2]. But the suggestion which M. Vincent adopts from M. Rangabé[3], that the Salaminian table was also a scoring-board for some kind of πεττεία is extremely attractive. Pollux, who vaguely describes two kinds of this game (IX. 97), says that each player had 5 lines and 5 counters, and that the middle line was 'sacred' (ἱερὰ γραμμή). M. Rangabé therefore sug-

---

[1] *Revue Archéol.* 1846, p. 401 sqq.

[2] See, however, Theophrastus, *Char.* VI. (ed. Jebb), where the boastful man reckons, on just such an abacus, that he must have spent 10 talents in charity.

[3] *Ibid.* p. 295 sqq.

gested that the lines marked with a cross on the table are really
the ἱεραὶ γραμμαί, and that two players, sitting opposite to one
another, would play at some kind of backgammon, each player
confining his counters to his own side of the transverse line.
The counters were moved according to throws with dice, as in
backgammon. Anyone, who is acquainted with the latter
game, will be able to suggest two or three very good forms of it
which might be played on this table and in which the lines
marked with a cross should be truly ἱεραί, either because no
'blot' might be left there or because they should be an asylum
where no solitary wanderer could be 'taken up.' The 5 lines
at the opposite end of the table would serve for some less
elaborate πεττεία or for a third player or might, in some way,
have been used to determine the values of the throws.

25. It will be seen, from the preceding observations, that our
knowledge of the *abacus* of antiquity is derived entirely from
Roman sources, and that the mode in which it was used must
be inferred simply from the appearance of extant instruments
and the practice of modern nations. It would seem, however,
that the use of the *abacus* was combined with the more
advanced finger-symbolism above described. Thus the Emperor
Frederick the Second (Imp. A.D. 1210—1250) in a treatise on
the art of hawking[1], says that the hands must, on various
occasions, be held in certain positions, such as *abacistae* use for
representing certain numbers in accordance with Bede's or
Nicolaus' instructions. Now since only one number could, as a
rule, be represented at a time on the *abacus*, a calculator who
was operating with two high numbers, would require a *memoria
technica* of both, and it would be very convenient to represent
one on the *abacus*, the other on the hands. It is indeed

---

[1] *Reliqua Libr. Frederici II.* (ed.
Schneider I. p. 102) quoted by Roediger : "Replicet indicem ad extremi-
tatem pollicis et erit modus secundum
quem abacistae tenent septuaginta cum
manu," with more directions of the
same kind. Thus also Leonardo of
Pisa (about A.D. 1200) in his *Liber
Abbaci* (ed. Boncompagni p. 5) says
that after mastering the *apices* (*i.e.*
the numerical signs used with the
abacus) the pupil must learn 'com-
putum per figuram manuum secundum magistrorum abbaci usum anti-
quitus sapientissime inventam.' He
then gives the scheme after Bede.
Vide Friedlein, *Zahlz.* p. 56.

possible, considering the lateness of all allusions to this finger-symbolism, that it was originally invented as a companion to the abacus.

26. A few words only remain to be added on this branch of Greek practical arithmetic. Western *abacistae* had introduced, certainly by the 10th century, a considerable improvement in the use of their instrument, which consisted in discarding pebbles and substituting for them the Roman numeral signs or the letters of the alphabet in order, so that thenceforth 525, for instance, was represented by V. II. V. or EBE in the last three columns of the *abacus*[1]. At the end of the first book of the *Geometria,* attributed to Boethius (who died A.D. 524) the author states (Friedlein's ed. pp. 395—397) that Pythagoreans used with the abacus certain nine signs which he calls *apices* of which he gives the forms. (The names are added apparently by a later hand.) The forms are obviously the parents of our own so-called Arabic numerals (except 0, which is not mentioned in Boethius)[2], and some of the names are also pure or nearly pure Arabic: the forms are also practically identical with the Gobar-numerals used by the Arabs of N. Africa in the 9th century, which again are admittedly of Indian origin. Upon these facts an endless controversy has arisen among historians, the questions in dispute being whether Pythagoras or any Pythagoreans might not have procured these signs from India and used them secretly for their quasi-theosophical arithmetic[3]: whether the later Alexandrians might not have obtained the

---

[1] Gerbert (ob. 1003) sometimes uses the Roman numerals, but generally the *apices.* Boethius or rather a pseudo-Boethius (Friedlein's ed. pp. 426—429) of a much later date mentions the use of the alphabet (cf. Friedlein *Zahlz.* pp. 54, 55, §§ 78—80). It will be observed that with an abacus on which numbers were represented by signs in appropriate columns, the four rules of arithmetic could be performed precisely as we perform them. Addition, subtraction and multiplication were so in fact. Division, however, was not.

See Hankel, *Zur Gesch. der Math.* pp. 317—323.

[2] It is doubtful whether the cipher was at first used by the Western Arabs among the Gobar-signs. It was introduced to Europeans first apparently in the book *Liber Algorismi,* a translation of the work of Mohammed ben Musa Alkhârismî, made in the 12th century. *Gobar* or *gubâr* means 'dust.'

[3] This was Cantor's opinion, *Math. Beiträge,* p. 221 sqq., but in *Vorlesungen,* p. 610 and elsewhere he follows Woepcke (see next note).

signs from India or elsewhere and given them to the Italians on
the one hand, the Arabs on the other[1]: and lastly whether the
passage in Boethius is not a forgery[2]. It is sufficient here to
repeat, what is admitted by all parties that there is no evidence
in any Greek author that these *apices* were known to the Greeks :
that there is also no evidence whatever that the Greeks ever
used any written numerical signs with the abacus: that the
MSS. of Boethius containing the *apices* are certainly not older
than the 11th century: that no trace of such signs is to be
found elsewhere in any European writer before the end of the
10th century or thereabouts: that the Indian signs, from which
the *apices* are derived, seem to be not older than the 2nd or 3rd
century: that the Arabs themselves did not obtain the Indian
arithmetic and Indian numerals till the time of Alkhârizmî
(cir. A. D. 800) and that Arabian mathematics did not begin to
pass from Spain to other European countries till about the
time of Gerbert (ob. A.D. 1003). The mere statement of these
facts is surely sufficient to assure any reader that the connexion
of the Greeks with the *apices*, if not absurd, is purely conjectural and need not be discussed at length in a short history
of Greek mathematics[3]. If it were admitted that the Greeks
knew of the *apices* at all, there would still be no reason whatever to think that they ever used them in calculation.

27. The *apices*, it must be remembered, were used only
with the abacus. No writer, even of the middle ages, ever in
the course of his text exhibits a number in these symbols. If
he purposes to illustrate the method of division, he states his
example with Roman numerals, then draws an abacus and

---

[1] Woepcke (*Journal Asiatique*, 1863,
I. p. 54 sqq.) suggests that the later
Alexandrians got these signs from
India: Theod. Henri Martin (*Annali
di matem.* Rome 1863, p. 350 sqq.)
suggests that they got them partly
from Egyptian, partly from Semitic
sources.

[2] This is Friedlein's opinion, maintained in many articles. The most
convenient reference is to *Zahlz.* pp.
15—19, 23—26, 51—54, 66, 67. Still

later and on the same side is Weissen-
born, in *Zeitschr. Math. Phys.* XXIV.
(1879), *Hist. Lit. Abth. Supplement-
heft*, published also in Teubner's *Ab-
handlungen zur Gesch. der Math.* Part
II. 1879. This latter writer rejects the
whole *Geometria*, not merely the arithmetical passages at the end of Bk. I.
and Bk. II.

[3] Another brief discussion of the
apices question is given by Hankel,
pp. 323—328.

inserts in it the necessary numbers with *apices.* Hence, closely as *apices* upon an abacus resemble, and serve all the purposes of our modern numerals, there is still a great gulf between the two. The cipher is yet to be invented before the abacus can be discarded. It follows again, from the same fact, that whatever be thought of the Greek acquaintance with *apices*, there can be no doubt at all that these were never entitled to be described as ordinary Greek characters for the numerals. They were not, and could not have been, used in inscriptions or other writings. It remains to consider, in this place, what characters were used in such documents.

28. It has been suggested above that probably, when the use of the fingers in counting was first discovered, it required, as in S. Africa at the present day, two men to count the higher tens, three to count the higher hundreds and so on. A single man, in counting say 40 or 60, would be apt to forget how many times he had counted his fingers through and would take an assistant to record them. But he would soon find that he could count high numbers by himself, if he kept some visible record, to which he could afterwards return, of each group of 10. Suppose, for instance, that each time he had counted through both hands, he pressed them on the ground, so as to leave an imprint of his fingers. He would thus have a *written* record, in groups of 10 perpendicular strokes[1]. Any other marks would, of course, serve his purpose, but it is a curious fact that in all

---

[1] In order clearly to represent the arithmetical resources of primitive man, I may as well state here what I conceive to be a very early method of counting. Suppose a man, who has names for his fingers and knows that all human beings have the same number of fingers, desires to count by himself 96 cows or other large unmanageable objects. I suppose he would first take a pebble for each cow and seat himself before the heap of pebbles. He would then take a pebble for each finger up to 10, then press his fingers on the ground: and would repeat this process till he had exhausted the pebbles. The imprints of his fingers would then show nineteen hands and one finger over, or 9 men + 1 hand + 1 finger. This he would *call* 'ring finger-men and *right*-thumb' or by some such name. The pebbles, the fingers and the written marks are used concurrently, but in time as his nomenclature became settled and his memory improved, he would omit first one, then two of the three symbols, and would finally dispense with them all and trust to his nomenclature alone.

the most ancient specimens of any sort of writing, the units at least are represented, not by dots or crosses or any other marks, but by perpendicular strokes only. This want of variety suggests that such strokes represent the fingers. The invention of separate symbols for 10 and 100 would follow at a far later time. The oldest known writings of the Egyptians and Phoenicians have such signs, but have no intermediate signs (e.g. for 50 or 500). They repeat the unit-strokes up to 9: they repeat the signs for 10 and 100 up to 9 times[1]. The ancient Greeks, according to Iamblichus[2], did the same. It is probable enough that such was the case, since an arithmetical written symbolism may well suggest itself long before any other kind of writing; but on the other hand, as some kind of writing is necessary to explain to us the purport of arithmetical symbols, and as the oldest Greek writings are of very late date and of the most advanced art, we can hardly expect to find evidence in support of Iamblichus' statement.

29. It goes without saying that, in a very large proportion of Greek inscriptions, the names of such numbers as occur are written in full. The oldest known compendious numerical *symbols* are those which used to be called *Herodianic signs.* The attention of modern students was first called to them by one Herodianus, a Byzantine grammarian of the 3rd century, who, in a passage printed by Stephanus in the *Appendix Glossariorum* to his Thesaurus, declared that he had frequently seen these signs in Solonic laws and other ancient documents, coins and inscriptions. While Greek epigraphy was an unknown science, this statement excited little interest, but it has since been abundantly confirmed by the enormous mass of inscriptions which the industry of scholars has, of late years, collected. In this sort of numerals, a stroke I repeated not more than four times, is the unit-sign *par excellence* and the other symbols are

---

[1] See Pihan, *Exposé des Signes de Numération*, etc. pp. 25—41, 162—168.

[2] *In Nicom. Arithm.* ed. Tennulius, p. 80. An inscription from Tralles has ετεος ||||||| but Böckh suspects this to be a forgery of late imperial times.

(See Franz, *Epigr. Graeca*, p. 347. Böckh, *C. I. G.* no. 2919, Vol. ii. p. 584.) Such forgeries were, of course, not unusual when a city wished to produce a documentary title to some ancient privilege.

merely initial letters of numeral names. Γ (πέντε) stands for
5: Δ (δέκα) for 10: Η (ἑκατόν) for 100: Χ (χίλιοι) for 1000 :
Μ (μυρίοι) for 10,000, and there are further *compendia* ☪, ☍ etc.
for 50, 500 etc. as may be seen on the Salaminian table figured
above. From the frequency with which these signs occur in
Athenian inscriptions, they are now generally called *Attic*. As
a matter of fact, no others are used in any known Attic inscription of any date B.C.[1] But they are by no means exclu-
sively Attic. They were used for instance in Boeotia, at first in
the forms of the local alphabet (thus Μ, ⱴ, ↓, ⌐Ε, ⊦Ε, ◁, ▷, I)
and afterwards, down to a late date, in the Attic forms[2]. It is
probable, in fact, that these numerals were once universally
used in Greece but at present there is not enough evidence on
this point. They were at any rate known and used outside
Attica long after the alphabet came to be used for numerical
purposes. A great number of papyrus-rolls preserved at Hercu-
laneum, state on the title-page, after the name of the author,
the number of *books* in his work, given in alphabetic numerals,
and the number of *lines* in Attic numerals: *e.g.* Ἐπικούρου περὶ
φύσεως ΙΕ (ἀριθ.) ΧΧΧΗΗ. We might in the same way use
Roman numerals for the one division, Arabic for the other.
One author, who is presented with such a title page in these
rolls, is a certain rhetorician called Philodemus, of Cicero's time.
The papyri therefore cannot be older than 40 or 50 B.C. and
may be much later[3].

---

[1] In other words, no others occur in
Vols. I. and II. of the *Corpus Inscr.
Atticarum.*

[2] See Franz, *Epigr. Graeca*, App. II.
ch. 1, p. 348, Böckh, *C. I. G.* Vol. I.
no. 1569 (p. 740 sqq.) and no. 1570
(p. 750 sqq.) The latter inscription
Böckh dates about 70 or 100 B.C. A
large majority of Greek Inss. (including all the oldest) do not contain
numerals at all. Inss. from places
outside Attica are very seldom older
than the 2nd century B.C. and are
mostly of imperial times. The monu-

mental evidence, therefore, as to the
early numeral signs, is very scanty.
The Herodianic signs are found, beside
Boeotia, in Arcadia with local peculiarities (Le Bas and Foucart, *Inss. de
Peloponn.* no. 341e): in Erythrae near
Halicarnassus about 250 B.C. (Rayet
in *Revue Archéol.* 1877, Vol. 33, p. 107
sqq.): and in Rhodes about B.C. 180
(Brit. Mus. Inss.), cf. Curtius in Bur-
sian's *Jahresb.* for 1878.

[3] See Ritschl, *Die Alexandrinischen
Bibliotheken*, pp. 99, 100, 123 *note.*

30. But at some time which cannot now be certainly determined, the Greeks adopted the practice of using the letters of the alphabet in order as their numeral symbols, and this style ultimately superseded the Attic in Attica itself and became universal among Greek speaking peoples. The alphabet, however, as used for numbers, was not the same as that used for literary purposes, but contained some additions. The following table will show clearly enough what the numerical alphabet was:

$$\alpha', \beta', \gamma', \delta', \epsilon' = 1, 2, 3, 4, 5.$$
$$*\varsigma = 6.$$
$$\zeta', \eta', \theta', \iota' = 7, 8, 9, 10.$$
$$(\iota\alpha', \iota\beta' \ldots\ldots\iota\theta' = 11, 12\ldots\ldots 19.)$$
$$\kappa', \lambda', \mu', \nu', \xi', o', \pi' = 20, 30, 40, 50, 60, 70, 80.$$
$$(\kappa\alpha', \kappa\beta', \lambda\alpha', \lambda\beta' \text{ etc.} = 21, 22, 31, 32 \text{ etc.})$$
$$* \wp = 90.$$
$$\rho', \sigma', \tau', \upsilon', \phi', \chi', \psi', \omega' = 100, 200, 300\ldots\ldots 800.$$
$$* \lambda = 900.$$
$$(\rho\iota\alpha', \rho\kappa\beta' \text{ etc.} = 111, 122 \text{ etc.})$$
$$,\alpha, ,\beta, ,\gamma, ,\delta, ,\epsilon \text{ etc.} = 1000, 2000, 3000 \text{ etc.}$$
$$\text{M}\upsilon \text{ or M, } \overset{\beta}{\text{M}}, \overset{\gamma}{\text{M}} \text{ etc.} = 10,000, 20,000, 30,000 \text{ etc.}$$

It will be seen that an alphabet of 27 letters[1] (including 3 strange letters, the so-called ἐπίσημα ϛ, ϙ, and λ) represents all the numbers from 1 to 999 and that numbers under this limit are marked with an acute accent, placed immediately behind the last letter. At 1000, the alphabet recommences, but a stroke is now placed before the letter and usually, but not always, somewhat below it. For 10,000 Mυ, or M, the initial of μυρίοι was generally used, and the *coefficient* of the myriad, to use an algebraical expression, was usually written over (but sometimes before or behind)[2] this M. Sometimes

---

[1] The 24 letters, exclusive of the ἐπίσημα, are those of the Ionic alphabet, introduced formally at Athens in 403 B.C. It was in use in Asia as early as 470 B.C.

[2] If the coefficient was written first, M was often omitted and a dot substituted (e.g. β.͵θτμβ' = 29,342). In MSS. again the myriads are sometimes represented by ͆α, ͆β, ͆γ etc. hundreds of thousands (μυριάκις μυρίοι) by ͆α, ͆β, ͆γ etc. *Vide*, for authorities, Hultsch, *Metrologicorum Scriptorum Relliquiae*, Vol. I. pp. 172, 173. Ritschl, *Die Alex. Bibl.* p. 120. *Nicomachus* (ed. Hoche) *Introd.* p. x. Friedlein *Zahlz.*

also (*e.g.* in MSS. of Geminus) $,\iota$, $,\kappa$, $,\lambda$ etc. are used for 10,000 20,000, 30,000 etc. in the ordinary sequence of the alphabet.

Thus the number 29,342, would be written $\overset{\beta}{\mathsf{M}}.,\theta\tau\mu\beta'$ or $,\kappa\theta.\tau\mu\beta'$. But in a high number, since the digits were always arranged in the same order, from the highest multiple of 10 on the left to the units on the right, the strokes or accents which distinguish thousands and units were often omitted and a stroke drawn over the whole number. The left-hand letter would then have a *local* value (*e.g.* $\overline{\theta\tau\mu\theta} = 9349$)[1]. The symbolism for fractions will be mentioned later.

31. It has been commonly assumed, since the use of the alphabet for numerals was undoubtedly a Semitic practice and since the Greek alphabet was undoubtedly derived from Semitic sources, that therefore the Greeks derived from the Semites the numerical use of the alphabet with the alphabet itself[2]. And this theory derives further colour from the fact that the Greek numerical alphabet contains three Semitic letters which were, within historical times, discarded from the literary alphabet. Yet this evidence is in all probability wholly illusory. The Greek alphabet was derived from the Phoenicians but the Phoenicians never used the alphabet for numerical purposes at all[3]. The Jews and Arabs did, but the earliest documentary evidence for the practice, even among them, is not older than 141—137 B. C. when dates, given in alphabetic numerals, appear on shekels of Simon Maccabaeus[4]. The Greek evidence goes a good deal further back than this.

---

pp. 9—11, §§ 12—17. Nesselmann, *Algebra der Griechen*, pp. 74—79.

[1] Cf. for instance *C. I. A.* Vol. III. nos. 60 and 77.

[2] See for instance Nesselmann, *Alg. der Griechen*, pp. 74—79. Cantor, *Math. Beiträge*, pp. 115—118. *Vorles.* pp. 101—107. Friedlein, *Zahlz.* p. 9, § 12, etc.

[3] The ordinary forms of Phoenician numerals are *upright strokes* for units: a *horizontal stroke* for 10: $\wedge$ for 20, and $|<|$ for 100.

[4] See Schröder *Phönikische Sprache.*

Madden, *Coins of the Jews*, p. 67. Also Dr Euting's letter quoted by Hankel, *Zur Gesch. der Math.* p. 34. It was Hankel who first proclaimed the relevancy of these facts to the history of the Greek numerals. But Ewald and Nordheimer had, long before, stated that the Hebrew numerals were used "after the Greek fashion" and that they do not appear till a late time. Hankel abides by the common opinion that the Greek numerical alphabet dates from the 5th century B. C.

Against these facts it may be urged (1) that the Jewish practice of *Gematria*,[5] adopted by the later Kabbalists, is said by them to be very early, and is perhaps as old as the 7th century B. C. This was a curious system of Biblical interpretation, whereby two words were treated as interchangeable, if their letters, considered as numerals amount, when added together, to the same sum[1]. And again (2) both the Hebrew and the Greek literary alphabets are too short for a good arithmetical symbolism and both are supplemented up to the same limit (the 27th letter in each standing for 900). But as to (1), it must be observed that the supposed antiquity of *gematria* depends solely on a merely conjectural and improbable comment on Zechariah xii. 10[2]. There is in fact no clear instance of *gematria* before Philo or Christian writers strongly under Philonic influence (e. g. Rev. xiii. 18; *Ep. Barn.* c. 9)[3]. The practice belongs to Hellenistic Jews; its name is Greek and it is closely connected with Alexandria, where, we shall see, alphabetic numerals are first found. And as to (2), it seems more likely that the Jews took the idea of alphabetic numerals from the Greeks than *vice versa*. The Greeks could, by hook or by crook, furnish the necessary 27 alphabetic symbols. The Jews could not. Their alphabet is only 22 letters, and the numbers, 500 to 900, must be represented by the digraphs תק‎, תר‎ etc. compounded of 100-400, 200-400, etc.[4] There is in

---

[1] See Cantor, *Vorles.* I. pp. 87, 104—5, quoting Lenormant, *La Magie chez les Chaldéens* p. 24. Also Dr Ginsburg's monograph, *Kabbalah* p. 49 and the same writer's article *Kabbalah* in *Ency. Brit.* 9th ed. Vol. xiii. The *Gematria* is employed in Rev. xiii. 18, where 666, the number of the beast, is the sum of the Hebrew letters in *Nerun Kesar.* So in Gen. xviii. 2 'Lo! three men' is by *gematria* found equivalent to 'These are Michael, Gabriel and Raphael.' *Gematria* is by metathesis from γραμματεία.

[2] Hitzig, *Die* xii. *kleinen Propheten,* p. 378 sqq. cited by Cantor *Vorl.* p. 87.

[3] Cf. Siegfried's *Philo,* p. 330.

[4] The later Hebrew alphabet has five *final* forms ך‎, ם‎, ן‎, ף‎, ץ‎ (cf. Greek σ and ς), which are sometimes used to represent the numbers 500—900. But this cannot be an ancient practice. The square Hebrew characters, which alone have finals, did not come into use till the 1st or 2nd century B. C., and these five finals were not definitely fixed for many centuries afterwards. Vide the Table of alphabets in Madden 'Coins of the Jews' or Dr Euting's, appended to Bickell's *Outlines of Hebrew Gram.* 1877.

[5] See p. 317.

fact no evidence against, and a good deal for, the supposition
that the Jews derived alphabetic numerals from the Greeks.
The contrary belief is perhaps only a relic of the old superstition
which counted it profane to question the priority of the Hebrews
in all arts.

32. But the date at which the Greeks adopted the alphabetic numerals is not easily to be determined. The alphabet
was indeed, at an early date, used *quasi* numerically but not in
the manner now under discussion. The tickets of the ten panels
of Athenian jurymen (*heliastae*) were marked with the letters of
the alphabet from *a* to *κ*, *ϛ* being omitted[1]. So also the books
of Homer, as divided by Zenodotus (flor. c. B.C. 280) were
numbered by the 24 letters of the ordinary Ionic alphabet,
ϛ and *Ϙ* being omitted: and the works of Aristotle were also
at some ancient time divided into books, numbered on the same
principle[2]. It seems unlikely that the regular numerical alphabet (with *ϛ*, *Ϙ*, *ʒ*) was in common use at the time when these
divisions were made. Secondly, in the numerical alphabet *ϛ* is
undoubtedly the *digamma* and this and *Ϙ* occur at their proper
(*i.e.* original) places in the alphabet. But the evidence at
present forthcoming shows that there never was, in any Greek
country, a literary alphabet which contained both *ϛ* and *Ϙ* along
with both *ψ* and *ω*. One or other of the first had dropped out
before one or other of the second had been introduced[3]. The
last numeral *ʒ*, whether it represents the Phoenician *shin*[4] or
tsadé, occurs in either case out of its place and is clearly
resumed into the alphabet for numerical purposes only. These
facts surely raise a presumption that the numerical alphabet
was settled not casually and by local custom, but deliberately

[1] Schol. to Aristophanes *Plut.* 277.
Hicks, *Greek Hist. Inss.* no. 119, p.
202. Franz, *Epigr. Gr.* p. 349.

[2] This appears from Alexander Aphrodisiensis, who (in *Metaph.* 9, 81, b.
25) quotes from *ϛ′ τῶν Νικομ.* a series
of definitions which belong to the
sixth book. The Aristotelian books so
numbered are the *Ethics*, *Politics* and
Topics.

[3] See the charts appended to Kirch-
hoff *Zur Gesch. des Griech. Alphabets*
3rd ed. and pp. 157—160 of the text.
Such transcripts as that in Hicks *Gr.
Inss.* no. 63, p. 117 sqq. are misleading. The original of this (see
Rhein. Museum, 1871 p. 39 sqq.) contains neither *η* nor *ω.*

[4] The Greek *σάν.* Herod. I. 139.
Franz, *Epigr. Graeca*, p. 19.

and by some man of learning[1]. Further, since no antiquarian could of his own motion persuade a people to revive, and to revive in their right places, letters which they had long since discarded, it is probable that this particular antiquarian was supported by some paramount political authority. It is plain also that this authority did not reside at Athens or near thereto, for the Athenians and Boeotians continued to use the Herodianic signs for two or three centuries at least after the alphabetic numerals appear elsewhere. It may be conceded, indeed, that public inscriptions would be the last place in which the new numerals would appear, but it is incredible that the old signs should have been retained by mere custom so long if the new had meanwhile been in common use. Lastly, it must be mentioned that the alphabetic numerals were a fatal mistake and hopelessly confined such nascent arithmetical faculty as the Greeks may have possessed. The Herodianic signs were clumsy but they did not conceal those analogies which ought to be obvious to the tiro in arithmetic. An Athenian boy who had been taught that III multiplied by III amounted to ⌐IIII would very soon have learnt that ΔΔΔ multiplied by ΔΔΔ would amount to ⌐ΗΗΗΗ and he might have guessed that, if ⌐I added to ⌐ amounts to ΔI, then ⌐Δ added to ⌐ would amount to ΗΔ. And these are really the severest difficulties which can occur with Herodianic signs. But, with alphabetic signs, $\gamma' \times \gamma' = \theta'$ is no clue to $\lambda' \times \lambda' = \text{ᾳ}$ or $\varsigma + \epsilon = \iota\alpha'$ to $\xi' + \nu' = \rho\iota'$. Such signs as these are no assistance to calculation and involve in themselves, a most annoying tax on the memory. Their advantage lies in their *brevity* alone, and it is to be suspected

---

[1] It should be mentioned here that we know of no fluctuation in the value of the Greek letters. Ϙ for instance might occasionally have its Semitic value 100, instead of 90, or Σ might occasionally (Ϙ or ϛ being omitted) represent 100, instead of P. But there is no known case in which any such doubt arises. It is, no doubt, only an accident that ᾳ does not occur, on any inscription earlier than the 13th or 14th century. Similar accidents may have affected the record in many other particulars, but we must of course use the record as we find it. As it stands, it points to Alexandria as the place where the numerical alphabet was invented and there never was any reason to doubt this.

that they were invented first for some purpose to which brevity was essential or desirable.

33. It curiously confirms all the inferences which have here been made, to find that the earliest evidence of these alphabetic numerals is found on coins of Ptolemy II. (Philadelphus) assigned to 266 B.C. The lateness of this date accounts for the later persistence of Herodianic signs. Alexandria, if anywhere, was the place where an antiquarian might have formed the numerical alphabet, and a king have published it, with effect. Coins are precisely the documents on which it is desirable to state numbers as concisely as possible[1]. Other evidence begins also soon after the date of these coins and in the same place. The oldest Graeco-Egyptian papyrus, which is ascribed to 257 B.C.[2], contains the numerals $\kappa\theta'$ ($= 29$), and after this alphabetic numerals are common enough on Ptolemaic coins and papyri[3]. They do not occur, however, on stone-inscriptions, as might be expected, till somewhat later. The earliest instance is probably one of uncertain place (though certainly from the Levant) ascribed to about 180 B.C.[4] or another of Halicarnassus[5] of about the same date. A Rhodian inscription of the same time still uses the Herodianic signs[6] but soon afterwards, say from 150 B.C. the alphabetic numerals are used invariably on all Asiatic-Greek monuments[7].

The cumulative evidence is surely very strong that the

---

[1] It will be remembered that the earliest Jewish evidence is found on coins.

[2] Now at Leyden, no. 379. See Robiou, quoting Lepsius, in *Acad. des Inscr. Suj. div.* 1878, Vol. 9.

[3] The K on some coins of Ptolemy I. (Soter) and the double signs AA, BB etc. on those of Arsinoe Philadelphi are of doubtful signification.

[4] *C. I. G.* Vol. IV. pt. XXXIX. no. 6819. No. 6804 was clearly not written at the dates which it mentions.

[5] *C. I. G.* Vol. II. no. 2655. Franz, *Epigr. Gr.* p. 349.

[6] In British Museum, not yet published.

[7] In the Asiatic inscription no. 6819, above cited, and many more, the numbers are arranged in their *alphabetic* order, e.g. $\overline{\eta\kappa}$, $\overline{\varsigma\kappa}$. The coins of Ptolemy Philadelphus above-cited were struck at Tyre. These two facts may perhaps suggest some Semitic influence in the use of alphabetic numerals, but I cannot attach any weight to them. The practice of writing numerals in their alphabetic order survived in Macedonia and N. Greece till the 2nd cent. See *C. I. G.* II. nos. 1965, 1970, 1971.

alphabetic numerals were first employed in Alexandria early
in the 3rd century B. C.   It remains to be added that two of the
foremost Greek mathematicians were during this century very
much interested in the further *abbreviation* of Greek numerals.
Archimedes (B C. 287—212) and Apollonius of Perga (flor.
temp. Ptol. Euergetes B. C. 247—222) both suggested new
modes of stating extremely high numbers, the former in his
ψαμμίτης, the latter probably in his ὠκυτόκιον. These will
be described later on but are mentioned here to show that
probably arithmetical symbolism was one of the Alexandrian
subjects of inquiry at precisely the time when the new symbolism first appears on Alexandrian records.

34. But it is time to return to the alphabetic numerals as used
in calculation.   Fractions (λεπτά) do not appear on inscriptions
but are represented in MSS. in various ways.   The most common
methods are either to write the denominator *over* the numerator
or to write the numerator with one accent and the denominator
twice with two accents each time (e.g. $\overset{\kappa a}{\iota\zeta}$ or $\overset{-\kappa a'}{\iota\zeta}$ or $\iota\zeta'\ \kappa a''\ \kappa a''$).
Submultiples, or fractions of which the numerator is unity, are
the most common.   With these, the numerator is omitted, and
the denominator is written above the line or is written once
with two accents, (e.g. $\lambda\beta'$ or $\lambda\beta'' = \frac{1}{32}$)[1].   Some special signs are
found, viz. signs similar to L, C' and S for $\frac{1}{2}$ and w'' for $\frac{2}{3}$.
Brugsch gives, on the authority of Greek papyri[2], the signs | for
addition, $\supset$ for subtraction, and $\frown\!\!\!1$ for a total.   Another common compendium is the form $\times$ for ἐλάττων and its inflexions[3].

---

[1] For some more minute details see
Nesselmann, *Alg. der Gr.* pp. 112—115.
Hultsch, *Metrol. Scriptt.* I. pp. 172—
175.   Friedlein *Zahlz.* pp. 13—14.

It is to be remembered that though
fractions with high numerators occur
in Greek writers, yet they represented
only the ratio between the numerator
and denominator.   In calculation, they
were reduced, as among the Egyptians,
to a series with unity for numerator
and these two conceptions of a fraction,
as a ratio and as a portion of the unit,

were alone permissible in Greek arithmetic.   See Cantor *Vorl.* pp. 107, 174,
405.   Hankel, p. 62, and Hultsch, *loc.
cit.*

[2] *Numerorum Demoticorum Doctrina*,
1849, p. 31.   See plate I. appended
to Friedlein *Zahlz.* and reff. there
given.

[3] In Heron's *Dioptra* (ed. Vincent
p. 173) and the scholia to the Vatican Pappus (ed. Hultsch, Vol. III. p.
128).   Nesselmann, *Alg. Gr.* p. 305 and
n. 17.

It remains to be mentioned only that the Greeks had no cipher. The $\bar{o}$ which Delambre found in the Almagest is a contraction of οὐδέν, and occurs only in the measurements of angles, which happen to contain no degrees or no minutes[1]. It stands therefore always alone and is not used as a digit of a high number. The stroke which Ottfried Müller found on an Athenian inscription, and which Böckh thought to be a cipher, is clearly explained by Cantor[2] as the *iota*, the alphabetical symbol for 10.

35. Of calculation with these alphabetic numerals very little mention is made in any Greek literature. It would seem from the technical names for addition and subtraction (viz. συντιθέναι and ἀφαιρεῖν, ὑπεξαιρεῖν) and from some passages of classical authors that these operations were generally performed on the ἄβαξ[3]. Multiplication, also, was, if possible, performed by addition[4], but it cannot be doubted that an expert reckoner would master a multiplication-table and have the alphabetic signs at his finger-ends. For such a person, for a mathematician, that is, who was competent to read Archimedes, Eutocius, a commentator of the 6th century after Christ, performs a great number of multiplications with alphabetical numerals[5]. The date of the writer and the work to which they are appended alike show that these are masterpieces of Greek arithmetic. A specimen or two, with modern signs added for more convenient explanation, may be here inserted :

---

[1] *Astronomie Ancienne*, I. p. 547, II. pp. 14 and 15. Theon in his commentary says nothing of this $\bar{o}$ which indeed may be only the introduction of late transcribers who knew the Arabic signs (v. Nesselmann, *Alg. Gr.* p. 138, and note 25. Friedlein, *Zahlz.* p. 82).

[2] See *Math. Beitr.* p. 121 sqq. and plate 28. Hultsch, *Scriptt. Metrol. Graeci*, Praef., pp. v. vi., Friedlein, *Zahlz.* p. 74.

[3] Cf. Theophrastus *Char.* (ed. Jebb) IV. n. 10 and XIII. n. 2.

[4] Lucian, Ἑρμότιμος, 48. Friedlein, *Zahlz.* p. 75.

[5] Torelli's ed. of Archimedes (Oxford, 1792), *Circuli Dimensio*, p. 208 sqq. The forms as they stand in MSS. are given p. 216. See also Nesselmann, *Alg. Gr.* pp. 116—118. Hankel, p. 56. Friedlein, *Zahlz.* p. 76, where many misprints in Nesselmann are corrected.

$$\frac{\begin{array}{c}\overline{\sigma\xi\epsilon}\\ \overline{\sigma\xi\epsilon}\end{array}}{\begin{array}{c}\overset{\delta}{}\quad\overset{a}{}\\ \mathsf{M}\ \mathsf{M}{,}\beta\ {,}\alpha\\ \overset{a}{\mathsf{M}{,}\beta}\ \overline{{,}\gamma\chi}\ \overline{\tau}\\ \underline{{,}\alpha\ \overline{\tau}\ \overline{\kappa\epsilon}}\\ \overset{\zeta}{\mathsf{M}}\ \overline{\sigma\kappa\epsilon}\end{array}}$$

$$\begin{array}{l}
265\\
265\\
\hline
40000,\ 12000,\ 1000\\
12000,\ 3600,\ \ 300\\
1000,\ 300,\ \ 25\\
\hline
70225.
\end{array}$$

The mode of proceeding is apparent on the face of this example. Each digit of the multiplier, beginning with the highest, is applied successively to each digit of the multiplicand beginning with the highest. Examples of multiplication, where fractions are involved, are also given by Eutocius. One of them is as follows [1]:

$$\frac{\begin{array}{c}\overline{{,}\gamma\iota\gamma}\ L\ \delta'\\ \overline{{,}\gamma\iota\gamma}\ L\ \delta'\end{array}}{\begin{array}{c}\overset{\overline{\pi}}{}\quad\overset{\gamma}{}\\ \mathsf{M}\ \mathsf{M}{,}\theta\ \overline{{,}\alpha\phi}\ \overline{\psi\nu}\\ \overset{\gamma}{\mathsf{M}}\ \overline{\rho\lambda}\ \overline{\epsilon}\ \overline{\beta L}\\ {,}\theta\ \overline{\lambda\theta}\ \overline{\alpha L}\ \overline{L\delta'}\\ \overline{{,}\alpha\phi}\ \overline{\varsigma L}\ \overline{\delta'}\ \overline{\eta'}\\ \overline{\psi\nu}\ \overline{\gamma\delta'}\ \overline{\eta'}\ \overline{\iota\varsigma'}\end{array}}$$

$$\overset{\pi\eta}{\mathsf{M}}\ {,}\beta\ \overline{\chi\pi\theta}\ \overline{\iota\varsigma'}.$$

$$\begin{array}{l}
3013\ \tfrac{1}{2}\ \tfrac{1}{4}\\
3013\ \tfrac{1}{2}\ \tfrac{1}{4}\\
\hline
9000000,\ 39000^2,\ 1500,\ 750.\\
30000,\ 130,\ 5,\ 2\tfrac{1}{2}.\\
9000,\ 39,1\tfrac{1}{2},\ \tfrac{1}{2}\ \ \tfrac{1}{4}.\\
1500,6\tfrac{1}{2},\tfrac{1}{4},\tfrac{1}{8}\\
750,3\tfrac{1}{4},\tfrac{1}{8},\tfrac{1}{16}\\
\hline
9082689\ \tfrac{1}{16}.
\end{array}$$

Multiplications are given also by Heron of Alexandria (*flor.* B.C. 100) and are conducted in precisely the same way as those of Eutocius [3]. In other words, for 700 years after the introduction

---

[1] In this specimen, the letter L represents the Greek sign for $\tfrac{1}{2}$. See above, p. 48.

[2] It will be observed here that Eutocius treats 13 as a single digit. He knew the multiplication table for 13.

[3] *Geometria*, ed. Hultsch, 36 and 83, pp. 81 and 110. They are printed also by Friedlein, *Zahlz.* pp. 76, 77. The second of them begins: μονάδες

ιδ' καὶ λεπτὰ τριακοστότριτα κγ· ὧν ὁ πολυπλασιασμὸς γίνεται οὕτως· ιδ' ιδ' ρϘϛ'· καὶ ιδ' τὰ κγ λγ'' λγ'' τκβ' λγ'' λγ'' κ.τ.λ. In modern figures, the problem is $14\tfrac{23}{33} \times 14\tfrac{23}{33}$.

It is worked out as follows:
$14 \times 14 = 196 : 14 \times \tfrac{23}{33} = \tfrac{322}{33} : \tfrac{23}{33} \times 14 = \tfrac{322}{33}$ and $\tfrac{23}{33} \times \tfrac{23}{33} (= \tfrac{529}{33} \cdot \tfrac{1}{33}) = \tfrac{16}{33} + \tfrac{1}{33} \cdot \tfrac{1}{33}$. The sum (ὁμοῦ) is $196\tfrac{660}{33} + \tfrac{1}{33} \cdot \tfrac{1}{33} = 216 + \tfrac{1}{33} \cdot \tfrac{1}{33}$.

of the alphabetic numerals, no improvement was made in the style of Greek calculation. And if such were the performances of professional calculators, it may be conceived that those of the unlearned were yet more clumsy. Thus Hankel[1] quotes from a work written as late as 944 A.D., some multiplications in which the writer finds by *addition* that 5 times 400 is 2000 and that 5 times 9 are 45! It can hardly be doubted that some Greek compiled a multiplication table and that children at school were practised in the use of it, as Roman children were, but no trace of such a table survives nor is any clear mention of it made in any Greek writer.

36. No example of simple division nor any rules for division are found in Greek arithmetical literature. The operation must have been performed by subtracting the divisor or some easily ascertained multiple of the divisor from the dividend and repeating this process with the successive remainders. The several quotients were then added together[2]. But the Greeks had no name for a *quotient* and did not conceive the result of a division as we do. To a Greek 5 was not the *quotient* of $\frac{35}{7}$. The operation did not discover the fact that 5 times 7 is 35 but that a seventh part of 35 contains 5, and so generally in Greek a division sum is not stated in the form "Divide *a* by *b*," but in the form "Find the *b*th part of *a*." This is the sort of nomenclature which would naturally be expected among a people who were constantly compelled to resort to the ἄβαξ with its concrete symbols.

But though there is no instance of a simple division, there is more than one of what, in our schools, is called ' compound ' division, where the dividend and the divisor both consist of a

[1] p. 55, citing *De argumentis lunae,* wrongly attributed to Bede. (*Patrologia,* ed. Migne, Vol. 90, p. 702.) On p. 56 Hankel gives a division from the same book. To divide 6152 by 15, multiples of 15 are first tried in order up to 6000. The remainder is 152. Then 15, 30, 60, 90, 120, 150. Remainder 2. The answer is 400+10 and 2 over.

[2] The process with whole numbers may be inferred from that with fractions. Heron (*Geometria,* ed. Hultsch, 12. 4, p. 56) divides 25 by 13, finds a quotient $1 + \frac{1}{2} + \frac{1}{3} + \frac{1}{13} + \frac{1}{78}$ and adds these terms together to $1\frac{12}{13}$. Obviously the intermediate stages were $\frac{12}{13} = \frac{24}{26}$ $= \frac{1}{2} + \frac{11}{26} = \frac{1}{2} + \frac{33}{78}$ etc. See Friedlein, *Zahlz.* p. 79.

whole number with fractions. These occur in Theon's commentary on Ptolemy's μεγάλη σύνταξις (the Almagest). Here for astronomical purposes it is frequently necessary to conduct operations with degrees and the sexagesimal fractions, minutes, seconds etc. (πρῶτα ἑξηκοστά, δεύτερα ἑξηκοστά, etc.)[1]. The rules for such operations are easy to perceive, if it be remembered that degrees are the units, minutes $\frac{1}{60}$ths and seconds $\frac{1}{3600}$ths of the unit. Hence Theon rightly premises that where a dividend consists of degrees, minutes, seconds, etc., division by degrees produces a quotient of the same denomination as the dividend : division by minutes produces a quotient of the next higher denomination to the dividend : division by seconds a quotient of two denominations higher than the dividend etc. And in multiplication, of course, the denominations are similarly lowered. There is no occasion here to give a specimen of Theon's multiplication, for it follows precisely the same lines as that of Eutocius, exhibited above, p. 50. But it is desirable to show his method of division, since no other specimen of the process is procurable. He divides $\overline{,α φιε}$ κ″ ιε‴ (i.e. 1515° 20′ 15″) by $\overline{κε}$ ιβ″ ι‴ (i.e. 25° 12′ 10″) in the following manner[2]:

---

[1] The Latin for these was *partes minutae, partes minutae secundae.* The sexagesimal system is beyond question of Babylonian origin. In Greek mathematical literature, the circle is divided into 360 parts (τμήματα or μοῖραι) first in the Ἀναφορικός of Hypsicles (cir. B.C. 180). The division of the diameter into 120 parts with sexagesimal fractions appears first in Ptolemy (cir. A.D. 140), but was probably introduced by Hipparchus (cir. B.C. 130). This trigonometrical reckoning was never used save by astronomers. See Cantor, *Vorles.* pp. 70, 76, 274, 311, 336, 351. Hankel, p. 65. Friedlein, *Zahlz.* pp. 81—82. Nesselmann, *Alg. der Griech.* pp. 139—147. Theon's *Commentary* (ed. Halma) pp. 110—119. 185—6. A summary of Hypsicles' book is given by Delambre, *Astron. Anc.* I. See also *post* §§ 55, 140.

[2] Theon does not himself give a scheme of a division, as he does of a multiplication. He merely describes the process. The scheme in the text with modern figures is from Delambre, *Astron. Anc.* II. p. 25. A translation of Theon's words is given by Nesselmann, p. 142.

$$25^\circ \ 12' \ 10'' \ )1515^\circ \ 20' \ 15'' \ (60^\circ \ 7' \ 33''$$
$$25^\circ \times 60^\circ = 1500^\circ$$

$$\text{Remr. } 15^\circ = 900'$$

$$920' \quad \text{(bringing down 20' from dividend.)}$$
$$12' \times 60^\circ = 720'$$

$$\text{Remr.} \qquad 200'$$
$$10'' \times 60^\circ = \quad 10'$$

$$\text{Remr.} \qquad 190'$$
$$25^\circ \times 7' = 175'$$

$$\text{Remr.} \qquad\quad 15' = 900''$$

$$915'' \quad \text{(bringing down 15''.)}$$
$$12' \times 7' = \quad 84''$$

$$\text{Remr.} \qquad\quad 831''$$
$$10'' \times 7' = \quad 1'' \ 10'''$$

$$\text{Remr.} \qquad 829'' \ 50'''$$
$$25^\circ \times 33'' = 825''$$

$$\text{Remr.} \qquad\qquad 4'' \ 50''' = 290'''$$
$$12' \times 33'' \qquad\qquad = 396'''$$

The quotient therefore $60^\circ \ 7' \ 33''$ is a little too high, but here Theon leaves it. The length and timidity of the operation sufficiently show with what difficulty it was performed[1].

37. There is another operation, the *Extraction of a square root*, which—though indeed no specimen of it with ordinary numbers occurs in any Greek writer,—was so frequently performed, and at such an early date, by Greek arithmeticians that some mention of it must be made in this place. Archimedes in his *Circuli Dimensio*[2] gives a great number of approximate square-roots. He states, for instance, that $\frac{1351}{780}$ is

---

[1] There is extant a meagre tract on Multiplication and Division with Sexagesimal Fractions, attributed either to Pappus or to Diophantus. It was edited by C. Henry, *Opusculum de Multiplic.* etc. Halle, 1879. See also the preface, pp. XII. XVI. of Hultsch's III. Vol. of Pappus.

[2] Prop. III. pp. 206—208 (ed. Torelli). The whole work is given *post* § 126.

greater than $\sqrt{3}$ which is greater than $\frac{265}{153}$: so also

$$\sqrt{349,450} > 591\tfrac{1}{8} \text{ and } \sqrt{1,373,943\tfrac{33}{64}} > 1172\tfrac{1}{8} \text{ and}$$

$$\sqrt{5,472,132\tfrac{1}{16}} > 2339\tfrac{1}{4}{}^1.$$

He does not, however, give any clue to the mode by which he obtained these approximations. Nor does his commentator Eutocius, but the latter states that the rule for finding an approximate square-root (ὅπως δεῖ σύνεγγυς τὴν δυναμένην πλευρὰν εὑρεῖν) was given by Heron in his *Metrica*, by Pappus, Theon and several other commentators on Ptolemy. Only one of the works, to which Eutocius here alludes, is now extant. Theon, in his commentary on the Almagest, gives the rule, and an explanation of the rule, and some examples, of extracting a square-root with sexagesimal fractions. It is clear that Archimedes did not use Theon's method, and no other is

---

[1] The approximations might still be improved. $591\tfrac{1}{7}$, $1172\tfrac{1}{4}$, and $2339\tfrac{1}{2}$ are nearer to, and also smaller than, the roots of the numbers in question. Other roots which Archimedes gives, are too large. Nesselmann, *Alg. Gr.* p. 108—110. From the fact that Archimedes gives both too small and too large approximations, it has been supposed that he used continued fractions, but (apart from the difficulty of suggesting a Greek symbolism for these) it is objected to this theory that Archimedes' approximations are not so close as those which continued fractions would produce. Many other modes, by which he might have found his values for $\sqrt{3}$, have been suggested. The simplest (De Lagny's) is as follows. Archimedes selected fractions such that the square of the numerator is nearly 3 times the square of the denominator. He would in this way find two series:

$\frac{2}{1}, \frac{7}{4}, \frac{26}{15}, \frac{97}{56}, \frac{362}{209}, \frac{1351}{780}$ all $> \sqrt{3}$: and

$\frac{5}{3}, \frac{19}{11}, \frac{71}{41}, \frac{265}{153}, \frac{989}{571}, \frac{3691}{2131}$ all $< \sqrt{3}$.

Both these series are constructed on the

same principle, each numerator being twice the preceding numerator + thrice the preceding denominator, and each denominator being twice the preceding denominator + the preceding numerator. This is closely similar to the procedure of Diophantus. Archimedes, however, takes the 6th term of the first series and only the 4th of the second. It is therefore essential to this, as to every other explanation of the same kind, that Archimedes be supposed to have been less careful with one approximation than with the other. (For more theories and criticisms thereon see Heiberg, *Quaestiones Archimedeae,* 1879, pp. 60—66.) As it is unlikely that Archimedes, if he had a scientific method, would have failed to use it rigorously, some writers (e.g. Nesselmann, *loc. cit.* and Friedlein, p. 81) are of opinion that he found his approximations only by repeated trial: others however (e.g. Cantor, *Vorl.* pp. 272—4, and Heiberg *sup. cit.*) believe that he had a method which we cannot discover.

forthcoming in any Greek writer. It would seem also, from Theon's language, that his method was by no means old or familiar, and we must conclude, therefore, in default of evidence, that the earlier Greeks found square-roots by experiment only. The process would certainly take a long time, but we have no reason to suppose that the Greeks were unwilling to spend a long time on a simple arithmetical problem. They may, of course (without going so far as Theon's method), have derived many useful hints from geometry: e. g. the square-root of a number is twice that of one quarter of the number: or 4 times that of $\frac{1}{16}$ or 9 times that of $\frac{1}{81}$, etc., and in this way, they may easily have reduced the number to be experimented on down to some reasonable limit. It is useless, however, to expend conjecture on a subject on which there is not a particle of evidence.

38. Theon's method of extracting a square-root may be best explained by a paraphrase of his own words. " I ought to mention " he says " how we extract the approximate root of a quadratic which has only an irrational root. We learn the process from Euclid II. 4, where it is stated: ' If a straight line be divided at any point, the square of the whole line is equal to the squares of both the segments together with twice the rectangle contained by the segments.' So, with a number like 144, which has a rational root, as the line $\alpha\beta$, we take a lesser square, say 100, of which the root is 10, as $\alpha\gamma$. We multiply 10 by 2, because there are two rectangles, and divide 44 by 20. The remainder 4 is the square of $\beta\gamma$, which must be 2. Let us now try the number 4500, of which the root is 67° $4'$ $55''$. Take a square $\alpha\beta\gamma\delta$, containing 4500 degrees ($\mu o \hat{\iota} \rho \alpha \iota$). The nearest square number is 4489, of which the side (root) is 67°. Take $\alpha\eta = 67^\circ$, and $\alpha\epsilon\zeta\eta$ the square of $\alpha\eta$. The remaining gnomon $\beta\zeta\delta$ contains 11°, or $660'$. Now divide $660'$ by $2\alpha\eta$, i.e. 134. The quotient is $4'$. Take $\epsilon\theta$, $\eta\kappa = 4'$, and complete the rectangles $\theta\zeta$, $\zeta\kappa$. Both these rectangles contain $536'$ ($268'$ each). There remain $124'$, $= 7440''$. From this we must subtract the square $\zeta\lambda$, containing $16''$. The remaining gnomon $\beta\lambda\delta$ contains $7424''$. Divide this by $2\alpha\kappa$

(= 134° 8′). The quotient is 55″. The remainder is 46″ 40‴, which is the square λγ, of which the side is nearly enough 55″."

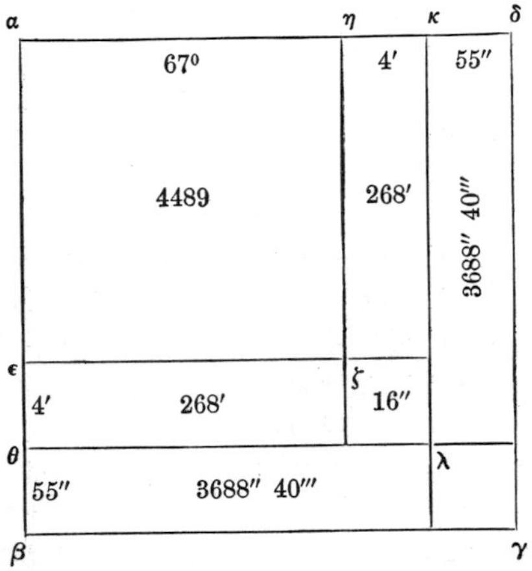
α
η
κ
δ
67⁰
4′
55″
4489
268′
3688″ 40‴
ε
ζ
4′
268′
16″
θ
λ
55″
3688″ 40‴
β
γ

So in general Theon concludes "when we seek a square-root, we take first the root of the nearest square-number. We then double this and divide with it the remainder reduced to minutes and subtract the square of the quotient, then we reduce the remainder to seconds and divide by twice the degrees and minutes (of the whole quotient). We thus obtain nearly the root of the quadratic[1]." In this procedure, with its continual references to a geometrical figure, we have a conspicuous instance of the fact, stated at the beginning of this chapter, that Greek λογιστική must often have sought its rules in the discoveries of the scientific ἀριθμητική. No doubt it was so in many other cases. It is hardly to be believed that while philosophers were aware of the modes of finding a Greatest Common Measure and a Least Common Multiple and well versed in the treatment of series and proportions, the common people should have been unable to adapt these results of ἀριθμητική to the needs of their own daily calculation. The meagre records of Greek *logistic*, however, contain no mention

[1] Theon gives another example, also with a figure. In this case he finds the square root of 2° 28′ and finds 1° 34′ 15″ as the result. The procedure is exactly the same as in the preceding example.

of any of these subjects. The theoretical treatment of them is alone known and this belongs to ἀριθμητική the subject of the next chapter.

39. Before closing this account of λογιστική, it remains to add a few facts, isolated here either because they did not have, or do not seem to have had, any real influence on the methods of Greek calculation or because the original report of them is so meagre or doubtful or disconnected that it would have caused unnecessary disturbance to have mentioned them before.

It has been stated, already (sup. p. 48), that in the 3rd century B.C. the abbreviation of the Greek arithmetical symbolism attracted the attention of (*inter alios*) Archimedes and Apollonius. This remark, however, though it has been often made with less reserve, seems to some writers to convey a *suggestio falsi*, inasmuch as abbreviation of the symbolism was not the ostensible object of the works in which Archimedes and Apollonius proposed their improvements in arithmetical nomenclature. It is, therefore, desirable that some fuller account of these works should be given than could be conveniently inserted elsewhere.

In a pamphlet entitled ψαμμίτης[1] (in Latin trans. *arenarius* 'the sand-reckoner') addressed to Gelon, king of Syracuse, Archimedes begins by saying that some people think the sand cannot be counted, while others maintain that, if it can, still no arithmetical expression can be found for the number. "Now I will endeavour" he goes on "to show you, by geometrical proofs which you can follow, that the numbers which have been named by us (? me) and are included in my letter[2] addressed to Zeuxippus, are sufficient to exceed not only the number of a sand-heap as large as the whole earth but of one which is as large as the universe. You understand, of

---

[1] Torelli's Archimedes, pp. 319 sqq. It is printed also, in Heiberg's *Quaestiones Archimedeae*. It is probable that Archytas of Tarentum, whom Horace (*Od.* I. 28. 1) calls 'numero carentis arenae mensorem,' had been busied with the same problem as that of the ψαμμίτης. Archytas was a contempo-

rary of Plato and at least 100 years earlier than Archimedes.

[2] It appears from c. i. sec. 7 that this letter was entitled ἀρχαί. It is clear that it was concerned only with the nomenclature which Archimedes is now about to introduce again (cf. also c. iii. sec. 1) and not with any special problem.

course, that most astronomers mean by 'the universe' the sphere of which the centre is the centre of the earth and the radius is a line drawn from the centre of the earth to the centre of the sun." (But Archimedes himself would be willing to suppose the universe a sphere as large as that of the fixed stars, according to Aristarchus of Samos[1].) Assume the perimeter of the earth to be 3,000,000 stadia[2], and in all the following cases take extreme measurements. The diameter of the earth is larger than that of the moon, and that of the sun is larger than that of the earth. The diameter of the sun is 30 times[3] that of the moon and is larger than the side of a *chiliagon* inscribed in a great circle of the sphere of the universe[4]. (This is proved geometrically.) It follows from these measurements that the diameter of the universe is less than 10,000 times that of the earth[5] and is less than 10,000,000,000 stadia[6].

Now suppose that 10,000 grains of sand *not* < 1 poppy-seed, and the breadth of a poppy-seed *not* < $\frac{1}{40}$th of a finger-breadth. Further, using the ordinary nomenclature, we have numbers up to a myriad myriads (100,000,000). Let these be called the *first order* ($\pi\rho\hat{\omega}\tau\sigma\iota\ \dot{\alpha}\rho\iota\theta\mu\sigma\acute{\iota}$) and let a myriad myriads be the

---

[1] It is at this point that Archimedes mentions the theory of Aristarchus of Samos (advanced in his $\dot{\upsilon}\pi\sigma\theta\acute{\epsilon}\sigma\epsilon\iota\varsigma$) that the earth goes round the sun and that the orbit of the earth is comparatively a mere spot at the centre of the sphere of the fixed stars. Archimedes does not seem to have understood this language and certainly did not adopt this theory. See Heiberg's note, *op. cit.* p. 202. A treatise of Aristarchus *De distantia lunae et solis* is extant. See Wallis' Works Vol. III. and Delambre *Astron. Anc.* I. ch. v. and IX.

[2] 'Though some,' adds Archimedes, 'take it at only 200,000 stadia. I will take it at 10 times the approved size.' He refers to Eratosthenes, who calculated the circumference to be 252,000 stadia. Delambre I. ch. 7. A *stadium* was nearly 200 yards.

[3] Eudoxus, he says, made it 9 times, Pheidias the son of Acupater 12 times and Aristarchus between 18 and 20 times larger than that of the moon.

[4] Aristarchus, he says, made it $\frac{1}{720}$th of the zodiacal circle, but his instruments cannot have been able to make so nice a measurement. Archimedes goes on to describe his own apparatus, by which he found that the diameter of the sun is between $\frac{90^0}{200}$ and $\frac{90^0}{164}$ (i.e. 27' 0" and 32' 56"). See Delambre *Astr. Anc.* I. c. IX.

[5] Following the rule that the diameter of a circle is less than $\frac{1}{3}$ of the perimeter of any inscribed regular polygon, above a hexagon.

[6] Following the rule that the circumference of a circle is more than 3 times its diameter.

unit of the *second order* (δεύτεροι ἀριθμοί) and let us reckon units, tens, etc. of the second order up to a *myriad myriads:* and let a myriad myriads of the second order be the unit of the *third order* (τρίτοι ἀριθμοί) and so on *ad lib.* If numbers be arranged in a geometrical series of which 1 is the first term and 10 is the radix, the first eight terms of such a series $(10^0—10^7)$ will belong to the *first order:* the next eight to the *second order* and so on. Thus, the terms from 1 to 10 millions may be called the first *octad:* 10^8 to 10^{15} may be called the second *octad* and so on[1]. Using these numbers, and following the rule that spheres are to one another in the triplicate ratio of their diameters, Archimedes ultimately finds that the number of grains of sand which the sphere of the universe would hold is less than a thousand myriads or ten millions of the 8th octad. This number would be expressed in our notation by 10^{63} or 1 with 63 ciphers annexed.

40. Now though this work is ostensibly devoted to a fanciful subject and though it is full of references to recondite discoveries in astronomy, geometry and ἀριθμητική, yet it is plain that it contains matter which might have had, and perhaps was intended to have, an important bearing on the

---

[1] At this point Archimedes incidentally adds that it will be convenient (χρήσιμον) to observe the following fact. In any geometrical series beginning with 1, if any two terms be multiplied, the product will be a term as far from the greater of the two multiplied as the lesser was from unity, and as far from unity as the sum of the distance of both the multiplied terms, less 1. E.g. in the geometrical series a, b, c, d, e, f, g, h, i, k, l, where a is unity, $d \times h = l$, and l is as many terms (less 1) from a as d and h together. It will be seen that Archimedes is referring to the fact which we express by saying that $a^m \times a^n = a^{m+n}$. He does not again refer to this fact, and does not otherwise, as some say, anticipate the method of logarithms. I have omitted from the text a further nomenclature which Archimedes suggests. There may be n octads *of the first period,* of which the last number will be 10^{8n-1}. This number, 10^{8n-1}, will then begin the *first octad of the second period* and so on. See Heiberg *op. cit.* p. 59. Nesselmann pp. 122—125.

It may be mentioned here also that the Greeks always began a geometrical series from 1, though they could give no reason for the practice. They did not know that $1 = n^0$. Theon Smyrnaeus (ed. Hiller, p. 24) says explicitly that "1 is not a number, but is the beginning of number," and this was the common Greek notion, though inconsistent with their practice. See Cantor *Vorles.* pp. 134, 368. Aristotle *Metaph.* XIII. 8, etc.

Greek symbolism, which belongs to λογιστική. This matter also had previously been published, without any practical application, in the lost Ἀρχαί, addressed to Zeuxippus. It deals indeed, as we know it, entirely with nomenclature and not a word is said of symbolism. But it is pretty clear (see especially the last note) that Archimedes' procedure was to write down the powers of 10 in order from 1 as far as necessary, then to divide the series into groups of 8 terms each and, when it was necessary to multiply two terms together, merely to add the numbers of their places in the series and so find at a glance their product and the name of the product. It can hardly be doubted that in writing down the powers of 10, in order to bring them within a manageable space, he employed a symbolism[1]. He would have required a symbolism of only 10 signs. Thus if $a, \beta, \gamma, \delta, \epsilon, \zeta, \eta, \theta, \kappa, \lambda$ were his symbols for numbers from 1 to 10, then

his first octad might be, $a, \lambda^a, \lambda^\beta$—$\lambda^\theta$,

„ second „ „ $\lambda^{a'}$—$\lambda^{\theta'}$,

„ third „ „ $\lambda^{a''}$—$\lambda^{\theta''}$.

On this principle, such a number as 1,957,362 would be written $\lambda^\zeta \kappa \lambda^\epsilon \epsilon \lambda^\delta \eta \lambda^\gamma \gamma \lambda^\beta \zeta \lambda^a \beta$. This symbolism, of course, is more cumbrous than ours but it is far shorter than the Herodianic and far more convenient than the common Greek alphabetic signs. If adopted, it would have immensely simplified procedure in the four rules of arithmetic;—would have brought it in fact nearly to the perfection of the Indian method. Yet, whatever symbolism Archimedes himself used (if any), it is quite certain either that he did not publish it or that it never obtained any vogue. No allusion to it occurs in the ψαμμίτης or in any other Greek mathematical work. But a good many reasons may be suggested why a new symbolism would, in Archimedes' time, have been singularly inopportune. The alphabetic numerals

---

[1] To take only the second octad, the last term of this is 10^{15} or a thousand million millions. This, in Greek, is χιλιάκις μυρίαι μυριάδες μυριάδων. The force of language, even Greek, will not go much further. The symbolism suggested in the text is not intrinsically improbable. Compare the Μα, Μβ, etc. of Apollonius to be presently mentioned, and compare the proposition quoted from Iamblichus below § 63.

had in all probability been lately introduced at Alexandria. The professors of ἀριθμητική did not require a new symbolism, since geometrical figures were sufficient for the problems which they dealt with. Thirdly, the efficacy of Archimedes' symbols would at first appear principally in calculations with very high numbers and there was then hardly anybody, save Archimedes himself, who was interested in calculations with such numbers.

41. That it is not at all far-fetched to suppose that Archimedes had in mind, when he invented his new nomenclature, the improvement of customary methods of calculation, will be apparent if we consider the similar work of Apollonius.

At the end of his commentary on the *Circuli Dimensio,* Eutocius says that he had done his best to explain the numbers used by Archimedes, but that Apollonius, using other numbers, had in his ᾿Ωκυτόκιον [1] obtained a closer approximation to the arithmetical value of the ratio $\dfrac{\text{circumference}}{\text{diameter}}$.

He then mentions some other persons who had maliciously criticised Archimedes and adds 'They use multiplications and divisions of myriads, which it is not easy to follow unless one has been through a course of Magnus' *Arithmetic* [2]'; and concludes by recommending Ptolemy's method with sexagesimal fractions. The passage is so vaguely worded that it is impossible to feel sure whether the ᾿Ωκυτόκιον, or 'Aid to Delivery,' of Apollonius has any connexion whatever with the multiplications of myriads mentioned afterwards. The book itself is lost and its name does not occur elsewhere. At any rate, Apollonius did invent a system of multiplication connected with a nomenclature in which myriads played a large part. Some account of both is to be gathered from the fragmentary 2nd Book of Pappus, but unfortunately the first half of the book is lost and with it the name of Apollonius' work and much precise information have doubtless disappeared.

---

[1] The first ed. had ᾿Ωκυτόβοον. The emendation was originally Halley's (*pref.* to his ed. of Apollonius) and was subsequently found to be correct by reference to two Paris MSS. Friedlein, *Zahlz.* p. 78, thinks it was a 'ready-reckoner' or multiplication-table only.

[2] This work is not elsewhere mentioned. It is tantalising to think what it may have contained.

Apollonius, taking, like Archimedes, a geometrical progression of the powers of 10 from 1 to 10^n, divided them into groups of 4 terms, *tetrads* and not octads. The first tetrad (1—1000) he called μονάδες : the 2nd (10,000—ten millions) μυριάδες ἁπλαῖ: the 3rd μυριάδες διπλαῖ etc. The first number of each tetrad is the unit of that tetrad, and the higher tetrads are (at least in Pappus) distinguished by the signs Μα, Μβ, Μγ etc. Thus the number 5,601,052,800,000 or, according to the Greek division, 5,6010,5280,0000 is written by Pappus Μγ . $\overline{\epsilon}$ καὶ Μβ . $\overline{\int\iota}$ καὶ Μα . $\overline{\int\epsilon\sigma\pi}$.

The fragment of Pappus contains examples, selected by that writer, illustrative of Props. XIV.—XXV. in the original work of Apollonius. The examples are of the following kind :

Prop. XIV. Let there be given several numbers, each less than 100 but divisible by 10. It is required to find their product without multiplying them. Let the numbers be 50, 50, 50, 40, 40, 30. The *pythmenes* of these are 5, 5, 5, 4, 4, 3, which, multiplied together, produce 6000. There are also 6 tens, which, divided by 4, give quotient 1 and remainder 2. The product of these tens is therefore 100 of the μυριάδες ἁπλαῖ. This, multiplied by 6000, produces 60 of the μυριάδες διπλαῖ. This is the product of the numbers proposed.

The other examples are all of this sort (the numbers in each case being varied[1]) and all are designed to illustrate a new rule, viz. that in multiplying numbers together, the *coefficients* of the powers of 10 only need be multiplied. These coefficients are called πυθμένες or fundamental numbers. Thus

---

[1] The concluding part of Apollonius' book is given (though not in Ap.'s words) by Pappus (II. 25). The last prop. was "Let two or more numbers be given, each less than 1000 but divisible by 100: and other numbers each less than 100 but divisible by 10: and finally other numbers less than 10. It is required to find their product." After performing this, Apollonius returned to the problem which he had originally set himself, viz. to multiply together

all the numerals contained in the line

Ἀρτέμιδος κλεῖτε κράτος ἔξοχον ἐννέα κοῦραι.

The product is μυριάδες τρισκαιδεκαπλαῖ ρϘϚ´ δωδεκαπλαῖ τξη´ ἑνδεκαπλαῖ δω´, or

$196.10000^{13}+368.10000^{12}+4800.10000^{11}$.

Pappus then tries his own skill on another line :

Μῆνιν ἄειδε θεὰ Δημήτερος ἀγλαοκάρπου

which seems carefully chosen to avoid high numbers.

8 is the $\pi\upsilon\theta\mu\dot{\eta}\nu$ of 80: 5 of 500: 7 of 7000 etc. In multiplying 80 by 600, it is necessary only to multiply 8 by 6. The product of the powers of 10 may be discovered by reference to the geometrical progression of *tetrads*. It is evident that this latter part of the rule is borrowed from the discovery of Archimedes mentioned above. We have here, in fact, the suggestion of the $\psi a\mu\mu\dot{\iota}\tau\eta s$, with only an easy alteration (hardly an improvement) of the nomenclature, put into actual practice. Yet still there is no reference to an abbreviated symbolism. In many places[1] Pappus says that Apollonius used $\gamma\rho a\mu\mu a\dot{\iota}$, which made his solutions far more readily intelligible: but these $\gamma\rho a\mu\mu a\dot{\iota}$ were beyond question *straight lines*. They were used presumably to represent the terms in the progression of tetrads and also to represent other numbers and the $\pi\upsilon\theta\mu\dot{\epsilon}\nu\epsilon s$ of these, but they can hardly have served any higher purpose than merely to prevent mistake. It must have been a certain convenience to distinguish the *pythmenes* in this way: e.g.

$$\iota\text{——}a \qquad \rho\text{——}a$$
$$\kappa\text{——}\beta \qquad \sigma\text{——}\beta \text{ etc.}$$
$$\lambda\text{——}\gamma \qquad \tau\text{——}\gamma$$

where a, β, γ etc. are *pythmenes* not only of tens (ι, κ, λ etc.), but of hundreds (ρ, σ, τ etc.)[2]. In spite, however, of the absence of evidence, it is difficult to believe that Apollonius wrote his book without using some special symbolism and it is still more marvellous if, having written the book, he did not see that it could not become popular without an accompanying symbolism. His symbolism, however, if he had any, was not published or never attracted attention, and thus he, as well as Archimedes, lost the chance of giving to the world once for all its numeral signs. That honour was reserved, by the irony of fate, for a nameless Indian of an unknown time, and we know not whom to thank for an invention which has been as important as any to the general progress of intelligence.

---

[1] E.g. II. 6. 5: 8. 28: 18. 10. (Hultsch's Ed.)

[2] The $\pi\upsilon\theta\mu\dot{\epsilon}\nu\epsilon s$ of tens and hundreds, though absolutely concealed by the symbols, were discernible in the *names* of the numbers. Thus it does not appear that γ' is the $\pi\upsilon\theta\mu\dot{\eta}\nu$ of λ' or τ', but it is easy to see that $\tau\rho\epsilon\hat{\iota}s$ is the $\pi\upsilon\theta\mu\dot{\eta}\nu$ of $\tau\rho\iota\acute{a}\kappa o\nu\tau a$ and $\tau\rho\iota a\kappa\acute{o}\sigma\iota o\iota$.

42. Another arithmetical symbolism is also attributed to the Greeks. It was first mentioned by Noviomagus (*De Numeris,* Cologne, 1539. Lib. I. c. 15) who said it was used by ' Chaldaei et astrologi[1].' It consisted of a curious set of signs (somewhat resembling railway-signals) in which the value of the symbol is determined, as it were, by the position of an arm attached to a post. Thus └ is 1 : ┌ is 10 : ┘ is 100 and ┐ is 1000 : ┴ is 2 : ┬ is 20 : ⊥ is 200 and ┳ is 2000. ∠ is 3 : ⊾ is 4 : ⊒ is 5 : ⊑ is 6 : ⊏ is 7 : ⊐ is 8 and ⊏⊐ is 9. All these signs, when reversed, represent ten times higher values, as with those for 1 and 2 above exhibited. The 'post' was often drawn upright: Γ, ⊦, ⌐ etc. and several 'arms' might be attached to one post. Thus ⅏ = 5543 : ⼍ = 2454 : ⽥ = 3970 etc. It cannot be said that this is a first-rate symbolism: but it is compact in form and it preserves also, to the eye, the analogies which are the greatest aids to calculation. It is impossible to say what is the origin of these signs, or where or at what date they came into use. Friedlein thinks they may be really Chaldaean and have belonged to the mediaeval art of horoscopy[2], which Noviomagus professed.

43. Calculation seems to have been regularly taught in Greek schools as early as there were any schools at all[3]. It became also a favourite subject of the Sophists, among whom the polymath, Hippias of Elis, was its most famous professor[4]. Socrates himself seems to have had a limited liking for it. According to Xenophon[5], he told his pupils to learn λογισμούς, but to beware of the idle pursuit of this as of other branches of learning: so far as was useful (or beneficial, ὠφέλιμον) he was always willing to forward them. As we have just previously

---

[1] Nesselmann (pp. 83—84) took them from Heilbronner's *Historia Matheseos* (pub. 1742). Heilbronner said he got them from Geminus and Hostus, a German antiquary of the 16th cent. Cantor (*Math. Beitr.* pp. 166—167) found the passage in Hostus, who refers to Noviomagus. Friedlein finally unearthed the passage of *Johannes Noviomagus* who says he had the signs from *Joh. Paludanus Noviomagus.* The MS. of Geminus, which Heilbronner saw, remains undiscovered.

[2] Friedlein, *Zahlz.* pp. 12, 13.

[3] See the Excursus on *Education* in Becker's *Charicles.*

[4] Cf. Plato *Protagor.* 318 E. *Hipp. Min.* 367—368.

[5] *Memor.* IV. 7. 8. For geometry, see IV. 7. 2.

been informed, by the same authority, that Socrates thought there was no need for more geometry than would enable a man to measure or parcel out a field, it may be presumed that he preferred the practical art of *logistic* to the theories of ἀριθμη-τική. Plato, however, was not of the same mind. His dislike of the sophists extended to the subjects which they taught and he is, on many occasions, as was seen at the beginning of this chapter, careful to distinguish the vulgar *logistic* from the philosophical *arithmetic*. But calculation cannot be discarded by the philosopher any more than by the merchant: so Plato, in his ideal constitution (*Legg.* 819 B), directs that free boys shall be taught calculation, a "purely childish" art, by pleasant sports, with apples, garlands etc. It makes men "more useful to themselves and wide-awake." Contemptuous language of this sort, used by the most influential of Greek thinkers, set the fashion to too many generations of mathematicians. Euclid is said to have been a Platonist : he certainly never meddled with *logistic*. His successors, with few exceptions, were affected by the same prejudice. The contributions of Archimedes and Apollonius to the art of calculation have been already mentioned. Hipparchus calculated a table of *sines* (so to say) and thus probably introduced the art of reckoning with sexagesimal fractions for astronomical purposes. The brilliant and above all things practical Heron of Alexandria seems, in his Μετρικά, to have offered some improvements in Greek calculation. A long era of Neo-Platonism and Neo-Pythagorism followed, but to this time belonged probably the Apollodorus, whom Diogenes Laertius (VIII. 12) mentions, and Philo of Gadara and the Magnus whom Eutocius praises. Nothing of these writers now survives, and it is very unlikely, judging from the calculations of Theon and Eutocius himself, that they produced any stir in their own day. *Logistic* was practically abandoned as hopeless after Apollonius' time. Ἀριθμητική became the hobby of the more ingenious spirits and to this science belongs the last brilliant achievement of Greek mathematics, the invention of algebra.

CHAPTER IV.

44. THE history of ἀριθμητική, or the scientific study of numbers in the abstract, begins in Greece with **Pythagoras** (*cir.* B.C. 530), whose example determined for many centuries its symbolism, its nomenclature and the limits of its subject-matter. How Pythagoras came to be interested in such inquiries is not at all clear. It cannot be doubted that he lived a considerable time in Egypt[1]: it is said also, though on far inferior authority, that he visited Babylon. In the first country, he would at least have found calculation brought to a very considerable development, far superior to that which he can have known among his own people : he would have also found a rudimentary geometry, such as was entirely unknown to the Western Greeks. At Babylon, if he ever went there, he might have learnt a strange notation (the sexagesimal) in arithmetic and a great number of astronomical observations, recorded with such numerical precision as was possible at that time[2]. But Pythagoras

---

[1] It is asserted by Isocrates, *Laud. Busir.* c. 11. 28, p. 227, and Callimachus ap. Diod. *Excerpt. Vatic.* VII—X. 35. It is implied in Herod. II. 81, 123, Aristotle *Metaph.* I. 1. These are the most ancient authorities. The Egyptian origin of Greek geometry is attested in many more passages, to be cited below. The visit to Babylon is first mentioned Strabo XIV. 1. 16.

[2] Both these statements may be illustrated by one example. One Babylonian document contains a statement of what portions of the moon's face are successively illuminated in the first fifteen days of the month. These are stated as 5, 10, 20, 40, 1. 20, 1. 36, 1. 52 etc. where 1. 20, 1. 36, 1. 52 etc. stand for 80, 96, 112 etc. parts out of 240 into which the moon was divided.

was not the first to be initiated into this foreign learning, for the Asiatic Greeks had certainly, before his time, acquired a good deal of Chaldaean astronomy and had even improved upon Egyptian geometry[1]. Nor was the bent of his mind altogether singular in his time. Among the Greeks everywhere, a new speculative spirit was abroad and they were burning to discover some principle of homogeneity in the universe. Some fundamental unity was surely to be discerned either in the matter or the structure of things. The Ionic philosophers chose the former field: Pythagoras took the latter. But the difficulty is to determine whether mathematical studies led him to a philosophy of structure or *vice versa*. The evidence seems to favour the former view. The geometry which he had learnt in Egypt was merely practical. It dealt mainly with such problems as how to find the area of *given* plane figures, the volume of *given* solids: its highest flight was to find roughly the ratio between the diameter and the circumference of a circle. Its *data* generally, its discoveries always, were numerical expressions. Given the number of a certain straight line, it could find the number of a certain curve: given the numbers of two or three straight lines it could find the numbers of a superficies or a solid. It was natural to nascent philosophy to draw, by false analogies and the use of a brief and deceptive vocabulary[2], enormous conclusions from a very few observed facts: and it is not surprising if Pythagoras, having learnt in Egypt that number was essential to the exact description of forms and of the relations of forms, concluded that number was the cause of form and

See Hincks in *Trans. Royal Irish Acad. Polite Lit.* xxii. 6, p. 406 sqq. Cantor, *Vorles.* pp. 72—76.

[1] Thus Thales had invented some propositions in scientific geometry. He had also predicted an eclipse and is said by many different authorities to have had much astronomical knowledge. Herodotus (ii. 109) says expressly that the knowledge of the *polos* and *gnomon* (on these sundials see below, p. 145 n.) came to the Greeks from Babylon. Pliny (*Hist. Nat.* ii. 76) attributes the introduction of the *gnomon* to Anaximenes: Suidas to Anaximander (*sub voc.*).

[2] Primitive men, on seeing a new thing, look out especially for some resemblance in it to a known thing, so that they may call both by the same name. This developes a habit of pressing small and partial analogies. It also causes many meanings to be attached to the same word. Hasty and confused theories are the inevitable result.

so of every other quality. Number, he inferred, is quantity and quantity is form and form is quality[1].

The genesis of the Pythagorean philosophy here suggested has strong historical warrant. It is certain that the Egyptian geometry was such as I have described it: the empirical knowledge of the land-surveyor, not the generalised deductions of the mathematician. If not certain, it is at least undeniable that Pythagoras lived in Egypt and there learnt such geometry as was known. It is certain that Pythagoras considered number to be the basis of creation[2]: that he looked to arithmetic for his definitions of all abstract terms and his explanation of all natural laws : but that his arithmetical inquiries went hand in hand with geometrical and that he tried always to find arithmetical formulae for geometrical facts and *vice versa*[3].

45.   But the details of his doctrines are now hopelessly lost. For a hundred years they remained the secrets of his school in Italy and when at last a Pythagorean philosophy was published[4], it was far more elaborate than the teaching of its founder. Even the tenets of this later school come to us only by hearsay. Of Pythagoreans we know something from Plato and Aristotle

---

[1] It was Pythagoras who discovered that the 5th and the octave of a note could be produced on the same string by stopping at $\frac{2}{3}$ and $\frac{1}{2}$ of its length respectively. Harmony therefore depends on a numerical proportion. It was this discovery, according to Hankel, which led Pythagoras to his philosophy of number. It is probable at least that the name *harmonical* proportion was due to it, since

$$1 : \tfrac{1}{2} :: (1 - \tfrac{2}{3}) : (\tfrac{2}{3} - \tfrac{1}{2}).$$

Iamblichus says that this proportion was called ὑπεναντία originally and that Archytas and Hippasus first called it *harmonic*. Nicomachus gives another reason for the name: viz. that a cube, being of 3 equal dimensions, was the pattern ἁρμονία : and having 12 edges, 8 corners, 6 faces, it gave its name to harmonic proportion, since

$$12 : 6 :: 12 - 8 : 8 - 6.$$

Vide Cantor, *Vorles.* p. 152. Nesselmann, p. 214 n. Hankel, p. 105 sqq.

[2] Some such vague term must necessarily be chosen. Aristotle (*Metaph.* I. 5) says that the Pythagoreans held that number was the ἀρχὴ καὶ ὡς ὕλη τοῖς οὖσι καὶ ὡς πάθη τε καὶ ἕξεις. It is not possible to extract from these words a definite theory of the functions of number in the cosmogony: it seems to be 'everything by turns.'

[3] See Diog. Laert. VIII. 12 and 14. In the second passage Pythagoras is said, on the authority of Aristoxenus, to have introduced weights and measures into Greece.

[4] By Philolaus. See Diog. Laert. VIII. 15. 85. The silence of Pythagoras was proverbial. On this and the facts stated in the text cf. Ritter and Preller, *Hist. Phil.* §§ 96, 97, 102—128.

and the historians of philosophy, but hardly anything remains which is attributed, by any writer of respectable authority, to Pythagoras himself[1]. He is probably responsible for some of the fantastic metaphysics of his followers. Aristotle expressly says that he referred the virtues to numbers and perhaps he agreed with Philolaus that 5 is the cause of colour, 6 of cold, 7 of mind and health and light, 8 of love and friendship and invention. Plutarch says that he held that the earth was the product of the cube, fire of the pyramid, air of the octahedron, water of the eicosahedron, and the sphere of the universe of the dodecahedron[2]. But doctrines of this kind, though they imply an interest in mathematics, are not themselves contributions to mathematical knowledge and do not require to be discussed in this place. For our present purpose, it is sufficient only to consider what advances in arithmetic are due to Pythagoras or his school, without speculating on the mode or order in which they were obtained or their place in the Pythagorean philosophy.

The following discoveries, at any rate, with the accompanying nomenclature, are as old as Plato's time. All numbers were classified as *odd* or *even* (ἄρτιοι or περισσοί). Of these the odd numbers were *gnomons* (γνώμονες) and the sum of the series of gnomons from 1 to $2n + 1$ was a *square* (τετράγωνος)[3]. The root of a square number was called its *side* (πλευρά). Some compound numbers have no square roots. These latter were *oblongs* (ἑτερομήκεις or προμήκεις)[4]. Products of two numbers were *plane* (ἐπίπεδοι), of three *solid* (στερεοί)[5]. A number multiplied twice into itself was a *cube* (κύβος)[6]. Some more classifications

---

[1] Porphyrius, a Syrian, late in the 3rd century after Christ, and Iamblichus both wrote a 'life of Pythagoras.'

[2] See Ritter and Preller, pp. 72, 79, §§ 116, 117, 127.

[3] Aristotle, *Phys.* III. 4. The *gnomon* is properly a carpenter's instrument, a T square with only one arm. The name was afterwards used in other senses.

[4] See Plato *Theaet.* 147 D—148 B. A *surd* was probably at this early time called *inexpressible* or *irrational* (ἄρρη-

τος or ἄλογος), but this is not certain. Plato calls it a δύναμις.

[5] Cf. Aristotle on Plato in *Pol.* v. 12. 8.

[6] Plato, *Rep.* VIII. p. 246. The same passage invites one or two other little remarks. ἀριθμὸς ἀπὸ in later Greek writers means 'the square of': ἀριθμὸς ὑπὸ means 'the product of.' ἀπόστασις once in Plato (*Timaeus* 43 D) means the 'interval' between the terms of an arithmetical progression. αὔξησις may (like αὔξη, *Rep.* VII. 528 B) mean ' mul-

are given by authorities of less antiquity. Any number of the form $\dfrac{n\,(n+1)}{2}$ was called *triangular* (τρίγωνος). *Perfect* (τέλειοι) numbers are those which are equal to the sum of all their possible factors (e.g. $28 = 1 + 2 + 4 + 7 + 14$): for similar reasons numbers are *excessive* (ὑπερτέλειοι) or *defective* (ἐλλιπεῖς)[1]. *Amicable* (φίλιοι) numbers are those of which each is tne sum of the factors of the other (e.g. $220 = 1 + 2 + 4 + 71 + 142$: $284 = 1 + 2 + 4 + 5 + 10 + 11 + 20 + 22 + 44 + 55 + 110)$[2]. Beside this work in classification of single numbers, numbers were treated in groups comprised either in a series (ἔκθεσις or ἀναλογία συνεχής) or a proportion (ἀναλογία). Each number of such a series or proportion was called a *term* (ὅρος). The mean terms of a proportion were called μεσότητες[3]. Three kinds of proportion, the *arithmetical, geometrical* and *harmonical* were certainly known[4]. To these Iamblichus[5] adds a fourth, the *musical*, which, he says, Pythagoras introduced from Babylon. It is composed between two numbers and their arithmetical and harmonical means, thus $a : \dfrac{a+b}{2} :: \dfrac{2ab}{a+b} : b$ (e.g. $6 : 9 :: 8 : 12$).

Plato knew that there was only one expressible geometrical mean between two square numbers, two between two cubes[6]. It is a familiar fact that the geometrical proposition, Euclid I. 47, is ascribed to Pythagoras. It follows that a right-angled triangle may be always constructed by taking sides which are to one an-

---

tiplication.' For other terms see *Journal of Philology*, XII. p. 92.

[1] Theon Smyrnaeus (ed. Hiller) pp. 31, 45.

[2] Iamblichus *in Nicom. Ar.* (ed. Tennulius) pp. 47, 48.

[3] It will have been observed that much of our modern nomenclature (e.g. 'square,' 'cube,' 'surd,' 'term,' 'mean') is taken from the Latin translation of the Greek expressions.

[4] Philolaus in Nicomachus *Introd. Ar.* (ed. Hoche) p. 135, Archytas in Porphyrius, *ad Ptol. Harm.* cited by Gruppe, *Die Fragm. des Archytas*, etc.

p. 94. This quotation (with one or two more) I take from Cantor, *Vorl.* p. 140 sqq. The statement in the text might be easily confirmed from other sources. See for instance Simplicius on Ar. *de Anima* 409, b. 23. Dr Allman doubts (*Hermathena* v. p. 204) whether these proportions were first applied to number, but see Ar. *An. Post.* I. 5. 74, and Hankel p. 114.

[5] *In Nicom. Ar.* (ed. Tennulius) pp. 141—2, 168.

[6] *Timaeus*, 32 B. Nicomachus, *Introd. Ar.* II. c. 24.

other in the ratios $3 : 4 : 5$, and to these numbers therefore great importance was attached in Pythagorean philosophy[1]. To Pythagoras himself is ascribed a mode of finding other numbers which would serve the same purpose. He took as one side an odd number $(2n + 1)$: half the square of this *minus* 1 is the other side $(2n^2 + 2n)$: this last number *plus* 1 is the hypotenuse $(2n^2 + 2n + 1)$. He began, it will be noticed, with an odd number. Plato[2] invented another mode, beginning with an even number $(2n)$: the square of half this *plus* 1 is the hypotenuse $(n^2 + 1)$: the same square *minus* 1 is the other side $(n^2 - 1)$.

46. A few more details expressly alleged by, or inferred from hints of, later authors might be added to the foregoing but it is impossible to frame with them a continuous history even of the most meagre character. We cannot say precisely what Pythagoras knew or discovered, and what additions to his knowledge were successively made by Philolaus or Archytas or Plato or other inquirers who are known to have been interested in the philosophy of numbers[3].

Proclus says[4] that the Pythagoreans were concerned only with the questions 'how many' ($\tau\grave{o}$ $\pi o\sigma\acute{o}\nu$) and 'how great' ($\tau\grave{o}$ $\pi\eta\lambda\acute{\iota}\kappa o\nu$) that is, with number and magnitude. *Number absolute* was the field of arithmetic : *number applied* of music: *stationary magnitude* of geometry, *magnitude in motion* of

---

[1] This rule was known to the Egyptians, the Chinese and perhaps the Babylonians at a very remote antiquity, v. Cantor, *Vorles.* pp. 56, 92,153—4. The discovery is expressly attributed to Pythagoras (Vitruvius, IX. 2). Cantor (*Vorles.* p. 153 sqq.) is of opinion that Pythagoras knew this empirical rule for constructing right-angled triangles *before* he discovered Eucl. I. 47.

[2] Proclus (ed. Friedlein), p. 428. It will be noticed that both the Pythagorean and Platonic methods apply only to cases in which the hypotenuse differs from one side by 1 or 2. They would not discover such an eligible group of side-numbers as $29 : 21 : 20$. See

Nesselmann, pp. 152—3. These are provided for by the first lemma to Euclid, x. 29. *Infra*, p. 81 n.

[3] Plutarch, *Quaest. Conv.* VIII. 9. 11—13, says that Xenocrates, the pupil of Plato, discovered that the number of possible syllables was 1,002,000,000,000. This looks like a problem in combinations, but the theory of combinations does not appear in any Greek mathematician, and the number seems too round to have been scientifically obtained. (Cantor, *Vorles.* pp. 215, 220.)

[4] Ed. Friedlein, pp. 35, 36. For the distinction between number and magnitude compare Aristotle, *An. Post.* I. 7 and 10, and *Cat.* c. 6.

spheric or astronomy[1]. But they did not so strictly dissociate discrete from continuous quantity. An arithmetical fact had its analogue in geometry and *vice versa;* a musical fact had its analogue in astronomy and *vice versa.* Pythagorean arithmetic and geometry should therefore be treated together, but there is so little known of either, that it seemed unadvisable, for this purpose only, to alter the plan of this book. The history of Greek geometry is so much fuller and more important and proceeds by so much more regular stages than that of arithmetic, that it deserves to be kept distinct.

The facts above stated are sufficient to show that, from the first, Greek ἀριθμητική was closely connected with geometry and that it borrowed, from the latter science, its symbolism and nomenclature. It had not yet wholly discarded the *abacus*[2], but its aim was entirely different from that of the ordinary calculator and it was natural that the philosopher who sought in numbers to find the plan on which the Creator worked, should begin to regard with contempt the merchant who wanted only to know how many sardines, at 10 for an obol, he could buy for a talent.

47. Whensoever and by whomsoever invented, most[3] of the known propositions of ἀριθμητική were collected together, not much later than 300 B. C. by **Euclid** in his *Elements.* Only the seventh, eighth and ninth books are specially devoted to numbers, but it cannot be doubted that the second and the tenth, though they profess to be geometrical and to deal with

---

[1] These four sciences became, through the Pythagorean influence of Alexandria, the *quadrivium* of early mediaevalism. The subjects of this fourfold education are mentioned in the familiar line "*Mus* canit: *Ar* numerat: *Ge* ponderat: *Ast* colit astra." To this, however, another *trivium* Rhetoric, Dialectic and Grammar, were added ("*Gram* loquitur: *Dia* vera docet: *Rhet* verba colorat") and these seven are the goddesses of science and art who attend at the nuptials of Philology and Mercury

celebrated by Martianus Capella (cir. A.D. 400). The same seven branches of education are discussed by Cassiodorus (born about A.D. 468), *De Artibus ac Disciplinis Liberalium Litterarum.*

[2] E.g. Plato, *Legg.* 737 E, 738 A says that 5040 has 59 divisors including all the numbers from 1 to 10. A fact of this sort must have been discovered empirically by means of the *abacus.*

[3] Archimedes uses one or two propositions which are not in the *Elements.*

magnitudes, are intended also to be applicable to numbers. The first 8 propositions of the second book, for instance, are for geometrical purposes proved by inspection. No one can doubt them who looks at the figures. But as arithmetical propositions they are not self-evident if stated with any arithmetical symbolism. In such a form, the first 10 propositions (the 9th and 10th are not treated in the same way as the first 8) are as follows[1]:

(1) $ab + ac + ad + \ldots\ldots = a\,(b + c + d + \ldots\ldots)$.

(2) $(a + b)^2 = (a + b)\,a + (a + b)\,b$.

(3) $(a + b)\,a = ab + a^2$.

(4) $(a + b)^2 = a^2 + b^2 + 2ab$.

(5) $\left(\dfrac{a}{2}\right)^2 = (a - b)\,b + \left(\dfrac{a}{2} - b\right)^2$.

(6) $(a + b)\,b + \left(\dfrac{a}{2}\right)^2 = \left(\dfrac{a}{2} + b\right)^2$.

(7) $(a + b)^2 + a^2 = 2\,(a + b)\,a + b^2$.

(8) $4\,(a + b)\,a + b^2 = (2a + b)^2$.

(9) $(a - b)^2 + b^2 = 2\left(\dfrac{a}{2}\right)^2 + 2\left(\dfrac{a}{2} - b\right)^2$.

(10) $b^2 + (a + b)^2 = 2\left(\dfrac{a}{2}\right)^2 + 2\left(\dfrac{a}{2} + b\right)^2$.

The eleventh proposition[2] is the geometrical way of solving the quadratic equation $a\,(a - b) = b^2$ and the fourteenth solves the quadratic $a^2 = bc$. From this statement, in algebraical form, of the chief contents of the 2nd Book, it will at once be seen what an advantage Greek mathematicians found in a geometrical symbolism. These propositions are all true for *incommensurable*, as well as commensurable, magnitudes, irrational as well as rational, numbers. But in numbers the Greeks had no symbolism at all for *surds*. They knew that surds existed,

---

[1] It will be observed that Theon's method of finding a square root, cited above, is founded on Eucl. II. 4. So also Diophantus (*infra*, p. 104) uses Euclid II. as an arithmetical book.

[2] This is the famous problem of 'the golden section,' which is used again in Euclid IV. 10 for the purpose of constructing a regular pentagon. Euclid's solution of the quadratic would be in algebraical form,

$$b = \sqrt{a^2 + \left(\dfrac{a}{2}\right)^2} - \dfrac{a}{2}.$$

(Cantor, *Vorles.* pp. 226, 227.)

that there was no exact numerical equivalent, for instance, for the root of 2 : but they knew also that the diagonal of a square : side :: $\sqrt{2}$: 1 [1]. Hence lines, which were merely convenient symbols for other numbers [2], became the *indispensable* symbols for surds. Thus, Euclid's 10th book, which deals with incommensurables, is in form purely geometrical, though its contents are of purely arithmetical utility : and every arithmetical proposition, in the proof or application of which a surd might possibly occur, was necessarily exhibited in a geometrical form. It is not, therefore, surprising that a linear symbolism became habitual to the Greek mathematicians and that their attention was wholly diverted from the customary arithmetical signs of the unlearned.

48. It is in the 7th book of the Elements [3] that Euclid first turns to the consideration of numbers only.

It begins with 21 definitions which serve for the 7th 8th and 9th books. The most important of these are the following :

(1) Unity ($\mu o\nu \acute{a}s$) is that by virtue of which everything is called ' one ' ($\grave{\epsilon}\nu\ \lambda\acute{\epsilon}\gamma\epsilon\tau a\iota$) [4].

(3) and (4) A less number, which is a measure of a

---

[1] This fact, according to an old scholiast (said to be Proclus) on the 10th book of Euclid, remained for a long time the profoundest secret of the Pythagorean school. The man who divulged it was drowned. See Cantor, *Vorles.* pp. 155, 156, quoting Knoche, *Untersuch. über die Schol. des Proklus* etc., Herford, 1865, pp. 17—28, esp. p. 23.

[2] The use of lines of course avoided the necessity of calculation. A rectangle represented a product: its side a quotient. Thus, for instance, Euclid (x. 21), wishing to show that a rational number divided by a rational gave a rational quotient, states that 'if a rational rectangle be constructed on a rational line, its side is also rational.'

[3] In the 7th 8th and 9th books, no geometrical figures are given, as indeed

none are necessary. In the 7th book according to our MSS. numbers are generally represented by *dots* (in Peyrard's edition by *lines*), in the 8th book particular numbers are given by way of illustration: in the 9th book both dots and particular numbers occur. Euclid probably used lines only, except where a number was to be represented as odd or even, in which case perhaps he used dots. At any rate, he does not, any more than in the geometrical books, use *division*, and his treatment of the propositions is purely synthetic, as elsewhere.

The arithmetical books of Euclid are included in Williamson's translation, Oxford 1781—1788.

[4] In the 2nd definition $\mu o\nu \acute{a}s$ means 'the unit.' 'Number' is there defined as '$\tau\grave{o}\ \grave{\epsilon}\kappa\ \mu o\nu \acute{a}\delta\omega\nu\ \sigma\nu\gamma\kappa\epsilon\acute{\iota}\mu\epsilon\nu o\nu\ \pi\lambda\hat{\eta}\theta os.$'

greater, is a μέρος (*part*) of it : but if not a measure, it is μέρη (*parts*) of the other[1].

(6) and (7) 'Odd' and 'even' numbers (περισσοί and ἄρτιοι).

(11) 'Prime' numbers (πρῶτος ὁ μονάδι μόνον μετρούμενος)·

(12) Numbers 'prime to one another' (πρῶτος πρὸς ἀλλήλους).

(13) Composite numbers (σύνθετοι).

(16) Products of two numbers are 'plane' (ἐπίπεδοι) and each factor is a 'side' (πλευρά).

(17) Products of three numbers (πλευραί) are 'solid' (στερεοί).

(18) 'Square' numbers (τετράγωνος ὁ ἰσάκις ἴσος).

(19) 'Cubes' (κύβος ὁ ἰσάκις ἴσος ἰσάκις).

(20) Numbers are 'proportional' (ἀνάλογον εἰσί) when the 1st is the same multiple, part or parts of the 2nd as the 3rd of the 4th.

(21) Plane and solid numbers are 'similar' when their sides are proportional.

(22) A 'perfect' (τέλειος) number is that which is the sum of all its factors (μέρη).

It will be seen that this nomenclature is purely Pythagorean. The class of 'prime' (πρῶτοι) numbers is not indeed mentioned by any earlier writer now known, but it can hardly be doubted that they were defined by the Pythagoreans, as a sub-class of odd numbers. The book deals with the following matters :

Prop. I. If of two given unequal numbers the less be subtracted from the greater as often as possible and the remainder from the less and the next remainder from the preceding remainder and so on, and no remainder is a measure of the preceding remainder until 1 is reached, the two given numbers are prime to one another. This (which is proved by *reductio ad absurdum*) leads to

Propp. II., III. To find the greatest common measure of two or more numbers. (The procedure is identical with ours.)

Propp. IV.—XXII. These deal with submultiples and fractions

---

[1] μέρος ἐστὶν ἀριθμὸς ἀριθμοῦ, ὁ ἐλάσσων τοῦ μείζονος, ὅταν καταμετρῇ τοῦ μείζονα. μέρη δὲ, ὅταν μὴ καταμετρῇ.

This word μέρη is the plural of μέρος, and is a very inconvenient expression.

and apply to numbers the doctrines of proportion which had been previously proved for magnitudes in the 5th book [1].

Propp. XXIII.—XXX.   Of numbers prime to one another. E. g. XXIX.   If two numbers are prime to one another, all their powers are prime to one another.

Propp. XXXI.—XXXIV.   Of prime numbers in composition. E. g. XXXIV.   Every number is prime or is divisible by a prime.

Propp. XXXV.—XLI.   *Miscellanea*: e. g. XXXV.   To find the lowest numbers which are in the same ratio with any given numbers. XXXVI.   To find the L. C. M. of two, and XXXVIII. of three, numbers. XLI.   To find the lowest number which is divisible into given parts.

49.   The 8th book deals, in the first half, chiefly with numbers in continued proportion (ἀριθμοὶ ἑξῆς ἀνάλογον) e. g. III.   If any numbers are in a continued proportion and are the least which have the same ratio to one another [2], the extreme terms will be prime to one another. VII.   If the 1st term is a divisor of the last, so is it of the 2nd. But a few other propositions are inserted, e. g. V.   Plane numbers are to one another in the ratio which is compounded of their sides. XI.   There is one mean proportional between two squares and XII. two between two cubes. The last half of the book (Propp. XIV. to XXVII.) is entirely devoted to the mutual relations of squares, cubes and plane numbers, e. g. XXII.   If three numbers are in continued proportion and the first is a square, so is the third. XXIII.   If four numbers are in continued proportion, and the first is a cube so is the fourth.

50.   The 9th book continues the same subject for a few propositions: e.g. III.   If a cube be multiplied by itself the product is a cube. Then follow (VIII.—XV.) some more propositions on numbers in continued proportion, or geometrical series: e. g. IX.   If in a series, commencing from unity, the 2nd term is a square, so are the following terms. And if the 2nd

---

[1] E.g. IV.   Every number is either a μέρος or μέρη of every higher number. V. VI. If A is the same μέρος (or μέρη) of B as C of D, $A + C$ is the same of $B + D$. XIX.   If $A : B :: C : D$, then $AD = BC$ and conversely.

[2] i.e. are the least which can form a continued proportion of the same number of terms, bearing the same ratio to one another, as in the given case.

term is a cube, the following terms are cubes. A few propositions on prime numbers (XVI.—XX.) are then given of which the most important is XX. The number of primes is greater than any given number. The discussion of odd and even numbers is then introduced (XXI.—XXXIV.), the propositions being of such a character as XXIV. If an even number be subtracted from an even number, the remainder is even. Then suddenly, appears the following proposition, XXXV. "If any numbers be in continued proportion, and the first term be deducted from the 2nd and also from the last, the remainder of the 2nd will be to the 1st as the remainder of the last to the sum of all the preceding terms[1]." Stated in another form, this proposition is : If a, ar, ar^2, ar^3... ar^n be a geometrical series, then

$$ar - a : a :: (ar^n - a) : a + ar + ar^2 ... + ar^{n-1}.$$

It is an easy step further to conclude that

$$a + ar + ar^2 ... + ar^{n-1} = \frac{a(ar^n - a)}{ar - a}$$

and thus to sum the series, but Euclid does not take this step. The proposition, as it stands, is apparently introduced solely for the purpose of proving the next (XXXVI.), the last in the book. This is, in effect, that in a geometrical series of the powers of 2 from 1 onwards, the sum of the first n terms (*if a prime number*) multiplied by the nth term is a *perfect* number[2]. In the proof[3], which is too long to be here inserted, the sum of n terms is assumed to be known by simple addition.

---

[1] Euclid takes only four numbers. His proof, put shortly, is as follows:

Let $a : \beta\gamma :: \beta\gamma : \delta :: \delta : \epsilon\zeta$. Take $\gamma\eta = \zeta\theta = a$, $\zeta\kappa = \beta\gamma$, $\zeta\lambda = \delta$. Then $\zeta\theta : \zeta\kappa :: \zeta\kappa : \zeta\lambda :: \zeta\lambda : \zeta\epsilon$. *Dividendo* $\zeta\theta : \theta\kappa :: \zeta\kappa : \kappa\lambda :: \zeta\lambda : \lambda\epsilon$ and *componendo* $\zeta\theta : \theta\kappa :: \zeta\theta + \zeta\kappa + \zeta\lambda :: \theta\kappa + \kappa\lambda + \lambda\epsilon$. By substitution (taking the terms backwards) $\epsilon\theta : a + \beta\gamma + \delta :: \beta\eta : a$. i.e. $\epsilon\zeta - a : a + \beta\gamma + \delta :: \beta\gamma - a : a$. Q. E. D.

[2] Ἐὰν ἀπὸ μονάδος ὁποσοιοῦν ἀριθμοὶ ἑξῆς ἐκτεθῶσιν ἐν τῇ διπλασίονι ἀναλογίᾳ κ.τ.λ.

[3] A short proof is easy :

$$1 + 2 + 4 + 8 + ... + 2^n = 2^{n+1} - 1 = p.$$

$2^n p$ is a perfect number if

$$2^n p = 1 + 2 + 4 + ... + 2^n + p(1 + 2 + 4 + ... + 2^{n-1})$$

which is obviously the case.

From this also it is evident that the proposition is untrue unless p is a prime

Such was the ἀριθμητική of rational numbers known in
Euclid's time. Not all of it was of Euclid's invention, but
it contains much the importance of which the later Greek
arithmeticians did not perceive and which, neglected by them,
was only in modern times resumed into consideration and
made the elementary foundation of a scientific theory of
numbers.

51. The 10th book treats of *irrational* magnitudes and
treats them geometrically through a symbolism of irrational lines.

Definitions occur at intervals throughout the book. It
starts with the following:

1, 2. Magnitudes are commensurable (σύμμετρα) when they
are measurable by one and the same measure : *contra,* incommensurable (ἀσύμμετρα).

3, 4. Straight lines are commensurable in square (δυνάμει
σύμμετροι) when their squares may be measured by the same
unit of space (χωρίον): *contra,* δυνάμει ἀσύμμετροι[1].

5. Hence, to any given straight line, there are an infinite

number. Nesselmann (*Alg. Gr.* p. 164 n.), after remarking that it is not
very easy to know whether a high number is prime or not, quotes from Fermat
(*Varia Opp. Math.* Toulouse 1679, p. 177) the following rule. Write down the
powers of 2 *minus* 1 each and above them the corresponding exponents of the
powers: thus

1 2 3 4 5 6 7 etc.
1 3 7 15 31 63 127 etc.

If the exponent is not prime, neither is the
power *minus* 1. If the exponent n is prime, the
power *minus* 1 is divisible *only* by numbers of the
form $2mn+1$. These can easily be tried. Fermat gives no proofs for his rule,
and his accuracy is not above suspicion. (Jevons, *Elem. Logic,* p. 222.)

[1] As Euclid does not define the word
δύναμις (whence *potentia,* 'power') it
may be desirable here to give some
account of it. The word δύνασθαι means
to be the square root of (Plato, *Theaet.*
148 A is probably the earliest instance):
hence δύναμις, as a rule, means *the
square;* but sometimes (Plato *loc. cit.*)
means a *square root* or rather a *surd,*
i.e. a square root which cannot be
otherwise described. Δύναμις is a more
general term than τετράγωνος, which
is used only when a square figure is
contemplated. There is no evidence

to show how δύναμις acquired this
mathematical sense. If the passage
Eudemi Fragm. (ed. Spengel) pp. 128—
129 is really *quoted* from Hippocrates
of Chios, this is the earliest which contains the technical δύναμις. Alexander
Aphrodisiensis (ed. Bonitz, 1847, p.
56) says that δυναμένη was the hypotenuse, δυναστευόμεναι the sides of a right-
angled triangle. These words, in this
connexion, probably mean 'equalling'
and 'equalled.' If these names are
ancient, perhaps the technical use of
δύναμις grew out of them.

number of straight lines commensurable or incommensurable, some both in length and in their squares, some in square only. Let this given straight line be called ῥητή, 'rational.' Then

6, 7, ῥηταὶ, rational straight lines, are commensurable with it in length and square or in square only: lines incommensurable with it in length and square or in square, are called ἄλογοι[1].

8, 9. The square of the ῥητή is also 'rational' and so is every square which is commensurable with it.

10, 11. Squares incommensurable with that of the ῥητή are ἄλογοι: so also are the sides of such squares. If a rectilineal figure be irrational, the sides of the square which is of equal area with it are also irrational.

The book begins with 21 propositions on incommensurables generally. Of these the most important are:

I. If two unequal magnitudes be given and from the greater more than half be subtracted and from the remainder more than half and so on with successive remainders, the final remainder will be less than the less of the two given magnitudes. So also, if only halves be deducted. This proposition, that a magnitude less than any given magnitude can be found, is the basis of the method of *Exhaustion* of which so much and so brilliant use is made in Greek geometry [2].

II. If of two given unequal magnitudes the less be deducted from the greater as often as possible and the remainder from the less and the next remainder from the preceding remainder and so on, and no remainder is a measure of the preceding, the two magnitudes are incommensurable. (Compare VII. 1.)

III. IV. To find the G. C. M. of commensurable magnitudes.

---

[1] It will be observed that Euclid's nomenclature differs from the modern. We call *irrational* all that he calls *incommensurable:* but with him $a : \sqrt{b}$ is *rational*, because $a^2 : b$ is rational. On the other hand, $a\sqrt{b}$ or any other multiple of an incommensurable, is with Euclid *irrational*, because $a\sqrt{b}$ and the rest are rectangles already and cannot be squared. The ῥητή of Euclid serves the same purpose as what we call 'the standard unit' of length or space.

[2] It is curious that Euclid does not add the further proposition that 'if two given magnitudes are incommensurable, there can be found a third, commensurable with one of the given and differing as little as we please from the other.' See Cantor, *Vorles.* p. 230.

v.—ix. Commensurable magnitudes are to one another as numbers to numbers: and their squares as square numbers. *Contra*, of incommensurables and conversely.

xi. If four magnitudes[1] are in proportion and the 1st is commensurable (or incommensurable) with the 2nd so is the 3rd with the 4th. Here follow various propositions on commensurables and incommensurables in proportion, on the sum of two commensurables or incommensurables etc.

xxi. If a rational rectangle be constructed on a rational line, the side of the rectangle is also rational (i.e. a rational number divided by a rational gives a rational quotient). To this is appended a *Lemma*, proving that the line or number, whose square is irrational, is also irrational (a fact which was provided for in the definitions). This lemma, which introduces, so to say, the consideration of the expression $\sqrt{\sqrt{ab}}$, leads to the discussion of the *medial line* (μέση) in the 2nd part of the book.

The definition of μέση[2] is given in Prop. xxii. viz. The rectangle contained by *rational lines commensurable only in square* (i.e. $a\sqrt{b}$ or $\sqrt{a}\sqrt{b}$) is irrational and the side of the square which is equal to this rectangle is also irrational and may be called μέση (i.e. $\sqrt{a\sqrt{b}}$, or else $\sqrt{\sqrt{ab}}$, where, if numbers be contemplated either a or b must not be square). The following propositions xxiii.—xxxv. deal with μέσαι or medials only. They are of the following kind:

xxiv. Medials may be commensurable with one another in length and square or in square only.

xxv. Given two medials commensurable in length (e.g. $m\sqrt[4]{ab}$ and $n\sqrt[4]{ab}$), the rectangle contained by them ($mn\sqrt{ab}$) is medial.

xxvi. Given two medials, commensurable only in square (e.g. $\sqrt{a\sqrt{b}}$ and $\sqrt{c\sqrt{b}}$), the rectangle contained by them ($\sqrt{ac b}$)

---

[1] This proposition is numbered x. in Gregory's edition (Oxon. 1703) the 10th and 11th exchanging places.

[2] The reason for the name is given in the same place. If AB, ΒΓ are the sides of an irrational rectangle, the side of the square which is equal to this rectangle, is a *mean* proportional (μέση ἀνάλογον) to AB, ΒΓ. The name 'medial' is used in the text, as more convenient.

is either rational or medial (i.e. according as *acb* is square or not).

Upon these two propositions follow several problems[1], to find medials, commensurable in line or in square, whose rectangle or square is of a given character: e.g. XXXIII. To find two medials commensurable only in square, such that their rectangle is medial and that the square of the greater exceeds the square of the less, by the square of a line either (*a*) commensurable or (*b*) incommensurable with the former. Two similar problems on lines incommensurable in square conclude the second part. All these lines are intended ultimately to form part of binomial expressions (cf. *infra*, p. 83, n. 2).

At the XXXVIIth proposition, some editor has introduced a new heading, viz. Ἀρχὴ τῶν κατὰ σύνθεσιν ἑξάδων and again at Prop. LXXIV. Ἀρχὴ τῶν κατ᾽ ἀφαίρεσιν ἑξάδων. These *hexads* are six groups, of six propositions each, on irrational binomials. There is thus a set of 36 propositions (XXXVII.—LXXII.) on binomials "formed by addition" and another of 36 exactly corresponding propositions (LXXIV.—CIX.) on those "formed by subtraction." The enunciation of Prop. XXXVII. runs: 'If two rational lines, commensurable only in square, are added together, the sum is irrational and may be called a *biterminal*' (ἐκ δύο ὀνομάτων). The difference of two such lines is, in Prop. LXXIV., called *apotomé*. The *biterminals* are $\sqrt{a} + \sqrt{b}$, and $a + \sqrt{b}$: the *apotomae* are $\sqrt{a} - \sqrt{b}$, $a - \sqrt{b}$ and $\sqrt{a} - b$: but altogether, twelve kinds of irrational binomials are distinguished. Of these twelve, six are formed by addition and are described and named in the first hexad: the other six are the corresponding binomials formed by subtraction and are described and named in the seventh hexad. The third hexad describes six

---

[1] To Prop. XXIX. two lemmas are appended, the first of which is 'To find two square numbers, such that their sum is a square number.' This is solved with the help of Eucl. II. 6. By that proposition, in the line

$$\overline{\underset{\text{A}}{\quad} \underset{\text{D}}{\quad} \underset{\text{C}}{\quad} \underset{\text{B}}{\quad}}$$

$AB \cdot BC + CD^2 = BD^2$. $AB \cdot BC$ will represent a square number, if AB and BC are both square or similar rectangular numbers. AC is assumed to be *even*.

The numbers which follow are those of Gregory's edition. Nesselmann, who used the Basle edition, (pub. 1537, 1546, 1558) cites XXXII. for XXXIII. etc., the 30th proposition in that

kinds of *biterminals*, the ninth six kinds of *apotomae*, and these are shown, in the following hexads, to be the squares of the binomials of addition and subtraction first defined.

These few remarks being premised, to show the structure and style of the remainder of the book, the effect of the whole may best be given in the words of a most competent critic, as follows. "Euclid investigates," says Prof. De Morgan[1], "every possible variety of lines which can be represented by $\sqrt{(\sqrt{a} \pm \sqrt{b})}$, a and b representing two commensurable lines. He divides lines which can be represented by this formula into 25 species and he succeeds in detecting every possible species. He shows that every individual of every species is incommensurable with all the individuals of every other species[2]; and also that no line of any species can belong to that species in two different ways or for two different sets of values of a and b[3]. He shows how to form other classes of incommensurables in number how many soever, no one of which can contain an individual line which is commensurable with an individual of any other class[4], and (?) he demonstrates the incommensurability of a square and its diagonal[5]. This book has a completeness which none of the others (not even the fifth) can boast of: and we could almost suspect that Euclid, having arranged his materials in his own mind, and having completely elaborated the 10th book, wrote the preceding books after it, and did not live to revise them thoroughly."

---

edition being divided into two parts, which are treated by Gregory as two separate propositions.

[1] Article *Eucleides* in Smith's *Dict. of Gr. and Rom. Biography.*

[2] This sentence gives the effect of the *sixth* hexads (Props. 67—71 and 104—108) which, however, contain only 5 propositions each. They are devoted to proving, by separate cases, that "every line, commensurable in length with a binomial irrational line, is an irrational line of the same species." Nesselmann, p. 179.

[3] This sentence gives the effect of

the *second* hexads (Props. 43—48 and 80—85). They are devoted to proving, by separate cases, that "every binomial irrational line can be divided into its terms *only in one point:*" i.e. that $\sqrt{a} + \sqrt{b}$ cannot $= \sqrt{x} + \sqrt{y}$, unless $a = x$ and $b = y$. Nesselmann, p. 177.

[4] This sentence gives the effect of Prop. 116. Nesselmann, p. 182.

[5] This refers to Prop. 117, which is clearly not Euclid's, as we have it. The enunciation, for instance, begins "Let it be proposed to prove" etc. and two proofs are given.

52. "The preceding enumeration," says the same writer in another place[1], "points to one of the most remarkable pages in the history of geometry. The question immediately arises, had Euclid any substitute for algebra? If not, how did he contrive to pick out, from among an infinite number of orders of incommensurable lines, the whole, and no more than the whole, of those which were necessary to a complete discussion of all lines represented by $\sqrt{} \ (\sqrt{a} \pm \sqrt{b})$, without one omission or one redundancy? He had the power of selection, for he himself has shown how to construct an infinite variety of other species, and an algebraist could easily point out many more ways of adding to the subject, which could not have been beyond Euclid. If it be said that a particular class of geometrical questions, involving the preceding formula and that one only, pointed out the various cases, it may be answered that no such completeness appears in the 13th book, in which Euclid applies his theory of incommensurables. It is there proved that each of the segments of a line divided in extreme and mean ratio is an *apotomé*—that the side of an equilateral pentagon inscribed in a circle is, relatively to the radius, the irrational line called a *lesser line*[2], as is also the side of an icosahedron inscribed in a sphere—and that the side of a dodecahedron is an *apotomé*. The *apotomé* then and the *lesser*

---

[1] In the English (also in the Penny) Cyclopedia, Art. "Irrational Quantity." A most complete summary of the contents of Euclid's 10th book is here given, followed by the remarks quoted in the text. The book was evidently a favourite with De Morgan. Nesselmann, p. 184, after remarking on the unsuggestiveness of the linear symbolism, says "Abstract thought alone has extracted from these lines their hidden secrets, which our formulae, almost unasked, declare. Indeed I think it is not too much to say that this book, almost useless in its geometrical form and therefore little esteemed, is the very one which shows

us the old mathematician in his highest glory."

[2] See Prop. LXXVII. compared with XL. If two lines, incommensurable in square and such that the sum of their squares is rational but their rectangle is medial, are combined, their sum and their difference are both irrational. The former is called ἡ μείζων, the latter ἡ ἐλάττων. Two such lines are found earlier in Prop. XXXIV. They are represented algebraically by

$$\sqrt{\frac{a^2 + a\sqrt{b}}{2}} \text{ and } \sqrt{\frac{a^2 - a\sqrt{b}}{2}} : \text{ or}$$

$$\sqrt{\frac{a + \sqrt{ab}}{2}} \text{ and } \sqrt{\frac{a - \sqrt{ab}}{2}}.$$

line are the only ones applied......The most conspicuous propositions of elementary geometry which are applied in the 10th book are the 27th, 28th and 29th of the 6th book, of which it may be useful to give the algebraical significance [1]. The first of these (the 27th) amounts to showing that $2x - x^2$ has its greatest value when $x = 1$, and contains a limitation necessary to the conditions of the two which follow.   The 28th is a solution of the equation $ax - x^2 = b$, upon a condition derived from the preceding proposition, namely, that $\frac{1}{4} a^2$ shall exceed b.   It might appear more correct to say that the solution of this equation is one particular case of the proposition, namely, where the given parallelogram is a square: but nevertheless the assertion applies equally to all cases.   Euclid however did not detect the *two* solutions of the question: though if the diagonal of a parallelogram in his construction be produced to meet the production of a line which it does not cut, the second solution may be readily obtained.   This is a strong presumption against his having anything like algebra; since it is almost impossible to imagine that the propositions of the 10th book, deduced from any algebra, however imperfect, could have been put together without the discovery of the second root.   The re-

---

[1] Cantor, *Vorles.* p. 228, gives practically the same algebraical equivalents, which, he says, first appeared in Matthiesen, *Grundzüge der antik. u. mod. Algebra* etc. 1878.   He does not seem to have heard of De Morgan. As these propositions are not usually printed, the enunciations may be here subjoined:

xxvii.   Of all parallelograms applied to the same straight line and defective by parallelograms similar and similarly situate to that which is described on half the line, the greatest is that which is applied to half the line, and is similar to its defect.

xxviii.   To a given straight line to apply a parallelogram equal to a given rectilineal figure and having its defect similar to a given parallelogram: provided that the given rectilineal figure

be not greater than that which can be applied to half the line, so that the defects of the given rectilineal figure and of that which is applied to half the line be similar.

xxix.   To a given straight line to apply a parallelogram equal to a given rectilineal figure, and excessive by a parallelogram which is similar to a given parallelogram.

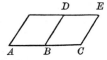

In the figure AD if applied to the line AC is *defective* (ἐλλείπει). AE if applied to the line AB is *excessive* (ὑπερβάλλει).   See also Simson's note on these Propositions.

maining proposition (the 29th) is equivalent to a solution of $ax + x^2 = b$: but the case of $x^2 - ax = b$ is wanting, which is another argument against Euclid having known any algebraical reasoning."

53. It must be added, before quitting this book, that Euclid nowhere alludes to any familiar example of an incommensurable. Some editor (whose language and style of proof differ noticeably from Euclid's) has added, at the end of the book, Prop. CXVII. proving that 'the diagonal of a square is incommensurable with the side[1].' Prof. De Morgan in one place[2] suggests that Euclid's interest in incommensurables was perhaps due to a suspicion that the circumference of a circle was incommensurable with the diameter. In another place[3], he suggests that Euclid had discussed some known examples of incommensurables in his lost work on Fallacies (περὶ ψευδαρίων), which, he thinks, was intended to be prefatory to the Elements. Both suggestions, of course, are purely conjectural. Hardly anything is known of the Greek theory of incommensurables before Euclid's time. Their discovery is expressly attributed to Pythagoras[4], but for a long time, the sole known fact was that the diagonal of a square : the side :: $\sqrt{2} : 1$. To this, according to Plato[5], Theodorus of Cyrene added the fact that sides of squares represented by $\sqrt{3}$, $\sqrt{5}$ etc., up to $\sqrt{17}$ were irrational. Theaetetus, a pupil of Theodorus, made the generalization that the side of any square, represented by a surd, was incommensurable with the linear unit. At a later date, perhaps, he improved this into the form of Euclid x. 9: Two magnitudes, whose squares are (*or* are not) to one another

---

[1] The proof is as follows. 'Suppose the diagonal : side :: $p : q$, p and q being whole numbers prime to one another. Then $p^2 = 2q^2$. p^2 and p are, therefore, *even* numbers. It follows that q, which is prime to p, must be *odd*. But p, being even, $= 2r$. Therefore $(2r)^2 = 2q^2$ and q must be *even*. Which is absurd.' This proof is twice referred to by Aristotle (*An. Prior.* I. c. 23. 41, a 26, and c. 44. 55, a 37). It may be very old, yet the method of *reductio ad absurdum*

is attributed to Plato.

[2] Art. *Eucleides* in Smith's *Dic. of Biogr.*

[3] Art. 'Irrational Quantity' in Penny Cyclop.

[4] See *supra* p. 74 n.

[5] *Theaetetus*, pp. 147 D—148 B. In this passage, the young Theaetetus says he made the same generalization for *cube* roots as for square roots. Cube roots are not mentioned anywhere in Euclid.

as a square number to a square number, are commensurable (*or*
incommensurable) and conversely[1]. Democritus is said[2] to have
written a treatise περὶ γραμμῶν ἀλόγων καὶ ναστῶν, but no
trace of it remains nor does any clue exist to the meaning of
ναστῶν. It cannot be doubted that Euclid's work contains at
least all that was known of the theory of incommensurables
before his time, and as Euclid left it, so it remained, untouched,
down to the 15th century, when Lucas Pacioli de Burgo
resumed the study[3].

54.   After the death of Euclid, the astonishing successes of
geometry in the hands of Archimedes and Apollonius and the
growing interest of astronomy seem to have attracted all attention to those sciences[4], and, so far as we know, no substantive
work on the theory of numbers was produced for nearly four
centuries. Some small additions, however, were made *en
passant* to the theory of rational numbers by various mathematicians. Thus, some theory of combinations was perhaps invented. The problem, attributed to Xenocrates by Plutarch,
has been mentioned above (p. 71 n.), and Plutarch in the same
passage[5] states, without more, that Chrysippus (B. C. 282—209)
found that the number of possible combinations of 10 axioms
was over a million: but that Hipparchus showed that the
axioms, if affirmed, admitted of 101,049, and, if denied, of

---

[1] The scholiast to Euclid x., said to
be Proclus (ed. Knoche, *sup. cit.* p. 74 n.),
expressly attributes Euclid x. 9 and 10
to Theaetetus. See Hankel, pp. 100—
103.

[2] Diog. Laertius IX. 47.

[3] Nesselmann, p. 183.

[4] Nesselmann, p. 187, gives many
instances of the changes of fashion in
mathematics. From the time of Nicomachus (A.D. 100) the theory of numbers became the Greek fashion. When
Leonardo Bonacci (A.D. 1292) brought
the Arabian algebra into Europe, this
also became the fashion for 400 years.
When Diophantus became known (Xylander 1571, Bachet 1621) indeterminate equations became the favourite

study e.g. of Bachet, Fermat, Pell,
Frenicle. The differential calculus
followed and occupied all attention till
Euler brought back the Diophantic
analysis, which was in fashion with
Lagrange, Legendre, Gauss, Jacobi and
their contemporaries.

[5] *Quaest. Conv.* VIII. 9, 11—13. Also
De Stoicorum Repugn. XXIX. 3 and 5,
(Reiske's ed. Vol. x. p. 330). Cantor
Vorles. pp. 215, 220. (The first number attributed to Hipparchus is quoted
as 103,049 in *De Stoic. Rep.* where also
he is said to be 'one of the arithmeticians'.) The *Quaestiones Convivales*
are also known as *Symposiacon* and
Disputationes Convivales.

310,952, combinations. Such results, however, may have been obtained empirically, and certainly no theory of combinations appears in any extant mathematical writer.

55. Eratosthenes the famous librarian of Alexandria (B.C. 275—194) invented a mode of distinguishing prime numbers, which was called, after him, 'the sieve' (κόσκινον, *cribrum*) of Eratosthenes. All composite numbers are 'sifted' out in the following manner[1]. The odd numbers are set out in order from 3 to as high a number as possible. Then every 3rd number from 3 is a multiple of 3 and may be rejected : every 5th number from 5 is a multiple of 5 and may be rejected : every 7th number from 7 is a multiple of 7 and may be rejected, and so on. The numbers ultimately retained are prime. **Hypsicles** (*circa* B.C. 180), the author of the 14th and 15th books added to Euclid's elements, made some contributions to the theory of *arithmetical progression*, which Euclid entirely neglects. The first three propositions of his ἀναφορικός (a little work on the 'risings of the stars,' ἀναφοραί)[2] are to the following effect. (1) In an arithmetical series of $2n$ terms, the sum of the last n terms exceeds the sum of the first n by a multiple of n^2 : (2) in such a series of $2n + 1$ terms, the sum of the series is the number of terms multiplied by the middle term : (3) in such a series of $2n$ terms, the sum is half the number of terms multiplied by the two middle terms. Some more general formula for the summation of arithmetical series perhaps led to the following definitions, most of which are entirely unknown to, or neglected by, Euclid[3]. "If as many numbers as you please be set out at equal intervals from 1, and the interval is 1, their sum is a *triangular* number : if the interval is 2, a *square* : if 3, a *pentagonal* : and generally the number of angles is greater by 2 than the interval." This statement is quoted from 'Hypsicles ἐν ὅρῳ' by Diophantus[4]; but whether

---

[1] Nicomachus, *Introd. Ar.* ed. Hoche, p. 29 sqq., and Iamblichus' Commentary (ed. Tennulius) pp. 41, 42.

[2] Described by Delambre *Astron. Anc.* I. and Cantor *Vorles.* p. 312.

[3] Philippus Opuntius, a pupil of Socrates and Plato, and earlier than Eu-

clid, is said to have written a work on polygonal numbers. See Cantor *Vorles.* p. 143, quoting Westermann's Βιόγραφοι, p. 446.

[4] Prop. 8 of the treatise on *Polygonal Numbers.* Nesselmann, p. 466. Cantor *Vorl.* p. 312.

Ὅρος (' the term ') was the name of a book, or ὅρος here means only the 'definition' of polygonal numbers, cannot now be ascertained. This extension of polygonal numbers (however originated) became a very favourite subject of later arithmeticians, to be presently mentioned.

Hipparchus (*cir.* B.C. 150) is said, by Arabian authorities, to have written on the solution of Quadratic Equations[1]. Heron, the ingenious mechanician and land-surveyor (B.C. 100), evidently knew some algebraical processes which were strange to Euclid, but he was not an *arithmetician* proper[2], and the more particular account of his work may be left for the history of geometry. From this time ἀριθμητική may be said to disappear at least from history for two centuries.

56.   It was revived by **Nicomachus**, a native of Gerasa, probably a town in Judaea. The date at which he lived may be determined roughly by the two facts, that he himself quotes one Thrasyllus[3], who seems to have been the astrologer, friend of the Emperor Tiberius, and that his work was translated into Latin by Apuleius of Madaura, in the time of the Antonines. He may be taken, therefore, to have flourished about 100 A.D. He is said to have been a Pythagorean and to have written a work on arithmetical theosophy, but the curious *farrago*, entitled θεολογούμενα τῆς ἀριθμητικῆς, is not his, for here Anatolius is cited, the Bishop of Laodicea (A.D. 270) who wrote a commentary on Nicomachus[4]. Two treatises of Nicomachus are extant, the *Enchiridion Harmonices* in two books, and the *Introductio Arithmetica* (εἰσαγωγὴ ἀριθμητική) also in two

---

[1] Cantor *Vorles.* p. 313, quoting Woepcke's ed. of '*L'Algèbre d'Omar Alkhayyâmî.*' Paris 1851, Pref. XI and *Journal Asiat.* v. (5th Series) pp. 251—253.

[2] For a specimen of his skill, see below, p. 106.

[3] In the *Enchiridion Harmonices* I. p. 24 (ed. Meibomius, 1652). In the same work, II. p. 36, Ptolemy is cited, but this is clearly an interpolation, for it would be inconsistent with the translation by Apuleius, which is attested by

Cassiodorus *de Arithmet.* p. 555.

[4] All the facts about Nicomachus are collected, and the errors corrected, by Nesselmann (pp. 188—191) who alleges also that none of the mathematical historians (including Montucla) can have read Nicomachus at all. Nesselmann seems to have used an edition of the *Introductio* published by Ast in 1817. Quite recently an edition has been published in the Teubner series (ed. Hoche) with a good preface.

books. It is with the latter only that we are here concerned. It was an extremely famous book in its day, and earned for its author a distinction similar to that so long enjoyed in England by Mr Cocker. Thus Lucian, wishing to compliment a calculator, says "You reckon like Nicomachus of Gerasa[1]." The number of commentaries on the *Introductio* also sufficiently attests its importance. Beside the translation of Apuleius and the notes of Anatolius, mentioned above, we know of a commentary by Iamblichus[2], another (not extant) by Heronas[3], a translation (extant) by Boethius, commentaries (extant in MS.) by Asclepius Trallianus and Johannes Philoponus and another (not extant) by Proclus: extracts in Arabic by Thabit-ibn-Corra (A.D. 836—901) and a commentary by Camerarius of the 16th century[4]. Nicomachus in fact inaugurates the final era of Greek mathematics. From his time onwards, ἀριθμητική is the favourite study, and geometry is neglected in its turn.

57. After a philosophical introduction, the first book of Nicomachus proceeds (c. 8—10) to the classification of numbers, as *even* and *odd* (ἄρτιοι and περιττοί)[5]. *Even* numbers are ἀρτιάκις ἄρτιοι (2^n), ἀρτιοπέριττοι $\{2(2m+1)\}$, and περισσάρτιοι $\{2^{n+1}(2m+1)\}$ i.e. they are either powers of 2 or 2 multiplied by an odd number or 2 multiplied by an even number, which is itself a multiple of an odd number[6]. *Odd* numbers (c. 11—13) as either 'prime and uncompounded' (πρῶτοι καὶ ἀσύνθετοι), 'compounded' (δεύτεροι καὶ σύνθετοι) or 'compounded but prime to one another.' The habit of dividing numbers into

<hr/>

[1] *Philopatris*, 12.

[2] Ed. Tennulius (very badly as Nesselmann shows) at Arnheim, 1667. The commentary of Iamblichus forms the 4th part of his treatise on the Pythagorean philosophy.

[3] Mentioned by Eutocius (ad Archimed. *de Sphaera et Cyl.* II.).

[4] See Nesselmann, pp. 220—223.

[5] Nicomachus begins by saying that every number is the half of the sum of the preceding and succeeding numbers. 1, however, has no predecessor and is half of 2 only. From this it is evident

how far the Greeks still were from the conception of 0 as a number.

[6] This is an improvement on Euclid's definitions to Book VII. There ἀρτιάκις ἄρτιος is the product of two even numbers, ἀρτιάκις περισσός of an even multiplied by an odd number. Hence in IX. 34 he has to confess that 'numbers which are not powers of 2 and which, when divided by 2, give an even quotient, are *both* ἀρτιάκις ἄρτιοι and ἀρτιάκις περισσοί.' Nicomachus, as usual, gives a table for finding his three species.

3 groups ($\epsilon\emph{ἴδη}$) has here led Nicomachus into great confusion of thought. His second class contains all the third: his second and third classes might very well contain even numbers, and lastly his third class defines numbers by their relation to others, whereas in c. 17 he says he has hitherto been considering numbers in themselves. Chapters 14—16 contain the definitions of perfect, excessive and defective numbers ($\tau\acute{\epsilon}\lambda\epsilon\iota o\iota$, $\acute{v}\pi\epsilon\rho\tau\acute{\epsilon}\lambda\epsilon\iota o\iota$, $\grave{\epsilon}\lambda\lambda\iota\pi\epsilon\hat{\iota}s$). Nicomachus then proceeds to the classification of the relations in which numbers stand to other numbers. Of *inequality* between two numbers, 5 kinds may be distinguished (c. 18—23). These are

1. When the greater divided by the less, gives a whole number as quotient. The greater is then called $\pi o\lambda\lambda a\pi\lambda\acute{a}\sigma\iota os$, a 'multiple,' the less $\acute{v}\pi o\pi o\lambda\lambda a\pi\lambda\acute{a}\sigma\iota os$, 'a submultiple.'

2. When the greater : the less :: $m + 1 : m$. The greater is $\grave{\epsilon}\pi\iota\mu\acute{o}\rho\iota os$ (*superparticularis*), the less $\acute{v}\pi\epsilon\pi\iota\mu\acute{o}\rho\iota os$ (*subsuperparticularis*). Thus $\frac{4}{3}$ is $\grave{\epsilon}\pi\acute{\iota}\tau\rho\iota\tau os$, $\frac{5}{4}$ $\grave{\epsilon}\pi\iota\tau\acute{\epsilon}\tau a\rho\tau os$ etc. $\frac{3}{4}$ is $\acute{v}\pi\epsilon\pi\acute{\iota}\tau\rho\iota\tau os$, $\frac{4}{5}$ $\acute{v}\pi\epsilon\pi\iota\tau\acute{\epsilon}\tau a\rho\tau os$ etc. But $\frac{3}{2}$ is specially named $\acute{\eta}\mu\iota\acute{o}\lambda\iota os$[1].

3. Greater : less :: $2m + n : m + n$. The greater is $\grave{\epsilon}\pi\iota\mu\epsilon\rho\acute{\eta}s$ (*superpartiens*), the less $\acute{v}\pi\epsilon\pi\iota\mu\epsilon\rho\acute{\eta}s$ (*subsuperpartiens*). As a general rule, the fractions here contemplated are of the form $\dfrac{m}{m+1}$, and in the nomenclature the denominator is not mentioned. Thus

$1 + \frac{2}{3}$ is $\grave{\epsilon}\pi\iota\delta\iota\mu\epsilon\rho\acute{\eta}s$, *superbipartiens*,
$1 + \frac{3}{4}$ is $\grave{\epsilon}\pi\iota\tau\rho\iota\mu\epsilon\rho\acute{\eta}s$, *supertripartiens*, etc.

But Nicomachus himself does not always use this nomenclature and was evidently equal to finding names for a fraction of the form $\dfrac{m}{m+n}$. Thus $1 + \frac{2}{3}$ is $\grave{\epsilon}\pi\iota\delta\acute{\iota}\tau\rho\iota\tau os$, $1 + \frac{3}{4}$ is $\grave{\epsilon}\pi\iota\tau\rho\iota$-

[1] Here Nicomachus refers to a table of 10 rows, divided into 10 columns. The first horizontal row contains the numbers 1 to 10: the second these numbers multiplied by 2: the third, multiplied by 3, etc. up to 10 times. It is, in fact, the earliest known multiplication table. Every number in a lower row is $\pi o\lambda\lambda a\pi\lambda\acute{a}\sigma\iota os$ of the corresponding number in the 1st row: the numbers in successive rows (except the 1st and 2nd) are related, so that the lower rows are $\grave{\epsilon}\pi\iota\mu\acute{o}\rho\iota o\iota$ of the next higher, the higher are $\acute{v}\pi\epsilon\pi\iota\mu\acute{o}\rho\iota o\iota$ of the next lower. This table is referred to also in the succeeding chapters.

τέταρτος and in the same way, $1 + \frac{3}{5}$ etc. *might* have been called ἐπιτρίπεμπτος, etc.

4. Greater : less :: $mn + 1 : n$. The greater is πολλαπλασιεπιμόριος, the less ὑποπολλαπλασιεπιμόριος.

$2\frac{1}{2}$ is διπλασιεφήμισυς, *duplex sesquialter.*

$2\frac{1}{3}$ is διπλασιεπίτριτος, *duplex sesquitertius.*

$3\frac{1}{4}$ is τριπλασιεπιτέταρτος, *triplex sesquiquartus,* etc.

5. Greater : less :: $p(m+1) + m : m+1$, where p is more than 1. The nomenclature, so far as regards the whole number, is the same as in class 4, and as regards the fraction, the same as in class 3 (e.g. διπλασιεπιδιμερής $= 2\frac{2}{3}$ etc.). Another table is here appended, showing how to find numbers which shall be to one another in the foregoing ratios. There is, in fact, little of mathematical value in the 1st book of Nicomachus, but it is of some historical interest to observe how complicated the Greek treatment of fractions still remained. It should be remembered also that the nomenclature of Nicomachus was translated into Latin, and became habitual in Western Europe down to the introduction of the Arabian arithmetic.

58. The 2nd book begins with another table, showing how to find series of ἐπιμόριοι, and various comments on this table[1]. In c. 6, Nicomachus turns to the theory of *polygonal* numbers. These he describes (c. 8—11) in precisely the same manner as that which is attributed to Hypsicles by Diophantus (*supra* p. 87), save that in Nicomachus the terms of the arithmetical

---

[1] The table is thus constructed. Write out a geometrical series beginning from 1. Take the sum of each pair of successive terms and set these sums in a row below the 2nd and succeeding terms of the 1st, and continue this process *ad lib.* E.g.

1	3	9	27	81	etc.
	4	12	36	108	etc.
		16	48	144	etc.
			64	192	etc.

Here each column is a geometrical series of which the radix is ἐπίτριτος.

The first numbers of the rows form a geometrical series of which the radix is 4 (chaps. 3 and 4). Legendre, in the preface to his *Théorie des Nombres,* says that this science becomes a sort of passion with those who take it up: whereupon De Morgan remarks that this is probably because the curious character of the conclusions is not lessened by the demonstration. The explanation is peculiarly appropriate to Nicomachean ἀριθμητική, with its unsuggestive symbolism.

series by which the polygonal numbers are found are all called
gnomons. This word therefore, which in Euclid means the
difference between one *square* and the next, means in Nicomachus the difference between *any* polygonal number and the
next of the same order. Then follow (c. 12) some analogies
between arithmetical and geometrical facts: e.g. as every
square can be divided by a diagonal into two triangles, so every
square number is the sum of two triangular: every square
number *plus* a triangular makes a pentagonal etc. Then, as
usual, a table is given of the polygonal numbers of each order,
with remarks thereon. Chaps. 13—17 deal with *solid* numbers.
The sum of a series of polygonal numbers from 1 upwards is a
pyramid, triangular or square etc. according to the order of the
polygonal numbers. The highest of such polygonal numbers is
the *base,* 1 is the *apex,* of each such pyramid. If 1 be omitted,
the pyramid is truncated (κόλουρος): if 1 and the next polygonal number be omitted, the pyramid is δικόλουρος and so on.
Besides pyramids, there are *cubes, beams* (δοκίδες), *tiles* (πλιν-
θίδες), *wedges* (σφηνίσκοι), *spheres* and *parallelepipeds.* *Wedges*
are numbers of the form $(m \times n \times p)$, where all 3 dimensions are
different[1]: *tiles* are $m^2 (m - n)$: *beams* (or *columns,* στηλίδες in
Iamblichus) are $m^2 (m + n)$. A number of the form $m (m + 1)$
is ἑτερομήκης: $m (m + n)$ is oblong (προμήκης) if $n > 1$: a
parallelepiped is of the form $m^2 (m + 1)$. The powers of 1, 5 and
6 always end in 1, 5 and 6: the squares of these numbers may
therefore be called *circular,* their cubes *spherical.* In c. 18—20
square numbers and ἑτερομήκεις (2, 6, 12 etc.) are set out in
parallel rows and attention is drawn to a number of curious
coincidences, thus exhibited: e.g. the differences between successive squares form the series of *odd,* those between successive
ἑτερομήκεις the series of *even,* numbers: in the series of odd
numbers from 1, the first term is the first cube, the sum of the
2nd and 3rd terms is the 2nd cube, the sum of the 4th 5th and
6th terms is the 3rd cube and so on.

59. At this point (c. 21) Nicomachus turns to the dis-

---

[1] Also called σφηκίσκοι 'stakes' or
βωμίσκοι 'altars.' All solid numbers
of 3 unequal dimensions are *scalene*
(c. 16). On the origin of this classification of numbers, see Dean Peacock in
Ency. Metrop. I. pp. 422, 423.

cussion of proportion ($\dot{\alpha}\nu\alpha\lambda o\gamma\acute{\iota}\alpha\iota$, $\mu\epsilon\sigma\acute{o}\tau\eta\tau\epsilon\varsigma$)[1], which, he says, is very necessary for "natural science, music, spherical trigonometry and planimetry and particularly for the study of the ancient mathematicians." He begins with a slovenly definition: "ratio ($\lambda\acute{o}\gamma o\varsigma$) is the relation between two terms: proportion is the composition of ratios[2]." When the same term is "on both sides," consequent ($\acute{\upsilon}\pi\acute{o}\lambda o\gamma o\varsigma$, *comes*) to the highest number, antecedent ($\pi\rho\acute{o}\lambda o\gamma o\varsigma$, *dux*) to the least, the proportion is called 'continued' ($\sigma\upsilon\nu\eta\mu\mu\acute{\epsilon}\nu\eta$). When the middle terms are different from one another, the proportion is 'disjunct' ($\delta\iota\epsilon\zeta\epsilon\upsilon\gamma\mu\acute{\epsilon}\nu\eta$)[3]. He goes on to say that Pythagoras, Plato and Aristotle knew only six kinds of proportion, viz. the arithmetical, geometrical and harmonical, and "their three sub-contraries[4], which have no names." Later writers added four more. He then describes (c. 23—25) the first three kinds, with a few remarks on each. In a continued arithmetical proportion ($a - b = b - c$), he has discovered a "most splendid rule, which has escaped most mathematicians," viz. that $b^2 - ac = (a - b)^2 = (b - c)^2$. In a continued geometrical proportion ($a : b :: b : c$), he notices[5] that $a - b : b - c :: a : b$, and that between 2 square numbers there is one, between 2 cubes two geometrical means. In a harmonic proportion ($a : c :: a - b : b - c$) he observes, among other things, that $(a + c) b = 2ac$. In c. 27, he states that between any two numbers, even or odd, three mean terms may always be found, one arithmetical, one geometrical and one

---

[1] Properly (i.e. originally) $\dot{\alpha}\nu\alpha\lambda o\gamma\acute{\iota}\alpha$ means *geometrical* proportion: $\mu\epsilon\sigma\acute{o}\tau\eta\varsigma$ any other kind. But this distinction was practically lost by Nicomachus' time. See an excellent note in Nesselmann, pp. 210—212.

[2] Euclid v. Deff. 3 and 8 is more precise. Iamblichus himself corrects Nicomachus on this point.

[3] Euclid's name for a continued proportion is $\dot{\alpha}\rho\iota\theta\mu o\grave{\iota}$ $\dot{\epsilon}\xi\hat{\eta}\varsigma$ $\dot{\alpha}\nu\acute{\alpha}\lambda o\gamma o\nu$: Theon's is $\sigma\upsilon\nu\epsilon\chi\grave{\eta}\varsigma$ $\dot{\alpha}\nu\alpha\lambda o\gamma\acute{\iota}\alpha$. Theon's name for an ordinary proportion of 4 terms is $\delta\iota\eta\rho\eta\mu\acute{\epsilon}\nu\eta$. It is also called $\delta\iota\epsilon\chi\acute{\eta}\varsigma$.

[4] Iamblichus (*In Nicom.* pp. 141—2)

says that the first three only were known to Pythagoras, the second three were invented by Eudoxus. The remaining four he attributes (p. 163) to the Pythagoreans, Temnonides and Euphranor. All ten are treated in the Euclidean manner by Pappus, *Math. Coll.* III. (ed. Hultsch) pp. 85 sqq.

[5] A similar rule is true of any geometrical proportion, not necessarily continued. Euclid v. 17 and 19. The next proposition mentioned in the text is the Platonic theorem proved in Euclid VIII. 11, 12.

harmonical[1]. He next (c. 28) states the remaining seven kinds of proportion viz.

(1) $a : c :: b - c : a - b$ (e.g. 6, 5, 3).

(2) $b : c :: b - c : a - b$ (e.g. 5, 4, 2).

(3) $a : b :: b - c : a - b$ (e.g. 6, 4, 1).

(4) $a : c :: a - c : b - c$ (e.g. 9, 8, 6).

(5) $a : c :: a - c : a - b$ (e.g. 9, 7, 6).

(6) $b : c :: a - c : b - c$ (e.g. 7, 6, 4).

(7) $b : c :: a - c : a - b$ (e.g. 8, 5, 3).

We have previously been told (c. 22) that the number of proportions was expressly raised to 10, because that was held by the Pythagoreans to be the most perfect number. It is rather surprising, therefore, to find that Nicomachus has yet another in reserve, the *musical*, which he calls τελειοτάτη, the " most perfect, comprehending 3 dimensions and embracing all the other proportions." This, as was above stated (p. 70), is of the form $a : \dfrac{a + b}{2} :: \dfrac{2ab}{a + b} : b$, the 2nd term being the arithmetical, and the 3rd the harmonical, mean between the two extremes.

60. The foregoing summary is sufficient to show that, in the interval of 400 years or so between Euclid and Nicomachus, something had been done, though we know not by whom, for the theory of numbers. In plane numbers, Euclid knows, or at least uses, only the square and the gnomon: in solids, only the cube: in proportions, only the geometrical. Almost the whole learning of polygonal numbers and solids and proportions was elaborated after his time, and before that of Nicomachus, for it is evident that the *Introductio* contains little that is original. In the meanwhile, again, mathematics had passed from the study of the philosopher to the lecture-room of the under-graduate. We have no more the grave and orderly proposition, with its deductive proof. Nicomachus writes a continuous

---

[1] He omits to mention that the two given numbers multiplied must produce a square, else the geometrical mean will be irrational. He also fails to notice that the three means will be in geometrical proportion, viz.

$$\frac{a + b}{2} : \sqrt{ab} : \frac{2ab}{a + b}.$$

Nesselmann, p. 215.

narrative, with some attempt at rhetoric, with many inter-
spersed allusions to philosophy and history. But more important than any other change is this, that the ἀριθμητική of
Nicomachus is *inductive*, not deductive. It retains from the old
geometrical style only its nomenclature. Its sole business is
classification, and all its classes are derived from, and are
exhibited in, actual numbers. But since arithmetical inductions
are necessarily incomplete, a general proposition, though *prima
facie* true, cannot be strictly proved save by means of an
universal symbolism. Now though geometry was competent to
provide this to a certain extent, yet it was useless for precisely
those propositions in which Nicomachus takes most interest.
The Euclidean symbolism would not show, for instance, that all
the powers of 5 end in 5 or that the square numbers are the
sums of the series of odd numbers. What was wanted, was a
symbolism similar to the ordinary numerical kind, and thus
inductive ἀριθμητική led the way to algebra.

61. Contemporary with, or not much later than Nicomachus, was **Theon** of Smyrna, author of a treatise on "the
mathematical rules necessary for the study of Plato[1]." The date
of this author may be roughly determined by the fact that,
in citing all the writers on music since Pythagoras, he stops at
Thrasyllus (the friend of Tiberius) and does not quote the
ἀρμονική of Ptolemy. Ptolemy himself also quotes from a
certain Theon four observations of Mercury and Venus taken in
the years A.D. 139—142. There seems no reason to doubt that
this was Theon Smyrnaeus, whose *Expositio* is largely devoted
to astronomy. The book itself[2] contains almost exactly the same
matter as Nicomachus (without the chapters on proportion),
but is very ill-arranged, so that rules are anticipated, one class
of numbers is treated in two or three widely separate chapters,

---

[1] Cited as '*Expositio rerum mathe-
maticarum ad legendum Platonem uti-
lium.*' Ed. Hiller, Leipzig, 1878.

[2] The *Expositio*, as we have it, was
formerly thought to be only a fragment.
We have it in two books, one on arithmetic, the other on astronomy. It was
supposed that three more were missing,

on geometry, stereometry and the
music of the spheres. Cantor and the
most recent editor, Hiller, are of opinion, however, that we have the entire
work. See Cantor *Vorles.* p. 367.
Nesselmann p. 231 (quoting Bouil-
laud).

the same facts are many times repeated[1]. It contains, however, two novelties, which may be thus stated. (1) Every square[2], or every square *minus* 1, is divisible by 3 or 4 or both: if the square is divisible only by 3, then *minus* 1, it is divisible by 4: and if it is divisible only by 4, then *minus* 1, it is divisible by 3: if it is not divisible by 3 or 4, then *minus* 1 it is divisible by both 3 and 4. (2) Theon introduces a new kind of numbers, called *diameters* ($\delta\iota\alpha\mu\epsilon\tau\rho\iota\kappa\omicron\iota$ c. 31). These are numbers whose squares are of the form $2n^2 \pm 1$. They are obtained in the following way. If 1 and 1 be the side and diameter of a square, then $1 + 1$ is the side of the next, and 3 or $2 + 1$ is its diameter: $2 + 3$ is the next side, $4 + 3$ is its diameter: $5 + 7$ is the next side, $10 + 7$ is its diameter etc., each successive side being the sum of the last side + last diameter, and each successive diameter being twice the last side + last diameter. Each diameter is the whole number nearest to the root of twice the square of the corresponding side[3]. It is curious that the ratios between these diameters and the corresponding sides are represented by the successive convergents of the continued fraction

$$1 + \frac{1}{2+} \frac{1}{2+} \frac{1}{2+} \frac{1}{2+} \text{etc.}$$

which represents the approximate value of $\sqrt{2}$. Theon, however, says nothing either of $\sqrt{2}$ or of continued fractions[4].

62. At some unknown date, certainly before Iamblichus, (i.e. before A.D. 300) lived one **Thymaridas**, the inventor of a certain proposition, known as his $\epsilon\pi\acute{\alpha}\nu\theta\eta\mu\alpha$ or 'after-blossom.' A brief and obscure account of this is preserved by Iamblichus[5],

---

[1] See Nesselmann, pp. 226—227.

[2] C. 20. The same rule is given by Iamblichus, *In Nicom.* p. 126.

[3] These diameters are the $\dot{\rho}\eta\tau\alpha\iota$ $\delta\iota\acute{\alpha}\mu\epsilon\tau\rho o\iota$, '*rational* diameters' to which Plato seems to allude in the famous passage about the 'nuptial number'. *Rep.* VIII. 246.

[4] See Cantor *Vorles.* pp. 229, 272—274, 369—370. Nesselmann, pp. 229, 230 observes that Theon has here stated a mode of finding all the solutions, in

rational numbers, to two indeterminate quadratic equations, viz. $2t^2+1=u^2$ and $2x^2-1=y^2$. He does not, indeed, suppose that Theon knew this, but the fact is interesting as bearing on the work of Diophantus.

[5] *In Nicom.* p. 88. Cantor, in his *Math. Beiträge*, pp. 97 and 380, identified this Thymaridas with him of Tarentum, who is said by Iamblichus to have been a pupil of Pythagoras. In the *Vorles.* (Pref. VII.) he abandons

and has been brilliantly explained by Nesselmann (pp. 232—236). The proposition, which is curiously worded, is as follows: "When any *defined or undefined* ($\dot{\omega}\rho\iota\sigma\mu\acute{\epsilon}\nu o\iota\ \mathring{\eta}\ \dot{a}\acute{o}\rho\iota\sigma\tau o\iota$) quantities amount to a given sum and the sum of one of them *plus* every other (in pairs) is given, the sum of these pairs *minus* the first-given sum is (if there be 3 quantities) equal to the quantity which was added to all the rest (in the pairs): or (if there be 4 quantities) to $\frac{1}{2}$ of it: (if 5) to $\frac{1}{3}$: (if 6) to $\frac{1}{4}$" etc. That is, if $x_1 + x_2 + x_3 = S$, be given, and $x_1 + x_2 = s_1$ and $x_1 + x_3 = s_2$, then $x_1 = s_1 + s_2 - S$. If four quantities $x_1 + x_2 + x_3 + x_4 = S$ be given, and $x_1 + x_2 = s_1$, $x_1 + x_3 = s_2$, $x_1 + x_4 = s_3$ be given, then

$$x_1 = \frac{s_1 + s_2 + s_3 - S}{2}.$$ And generally, if $x_1 + x_2 + x_3 + \ldots + x_n = S$,

and $x_1 + x_2 = s_1$, $x_1 + x_3 = s_2 \ldots x_1 + x_n = s_{n-1}$, then

$$x_1 = \frac{s_1 + s_2 + \ldots\ldots + s_{n-1} - S}{n - 2}.$$

What is chiefly of importance in this proposition[1] is the use of the word $\dot{a}\acute{o}\rho\iota\sigma\tau o\varsigma$ for an "unknown quantity." It does not, indeed, appear whether Thymaridas had or had not a corresponding symbol, but at least he has here stated an algebraical theorem and used an algebraical expression. He has gone beyond Nicomachus and nearly approached Diophantus.

63. The $\sigma\upsilon\nu\alpha\gamma\omega\gamma\acute{\eta}$, or Mathematical Collections, of **Pappus** the Alexandrian must have been written about A.D. 300. Probably the first two books were arithmetical, since a fragment of the 2nd Book contains an account of the tetrads of Apollonius already described (*supra*, p. 62) and the remaining eight deal almost entirely with geometry and mechanics. **Iamblichus**, who has been so often quoted in these pages, is a little later. He was born at Chalcis in Coele-Syria and may have been alive

this supposition. Two other facts about Thymaridas are mentioned by Iamblichus, (1) that he called unity the 'terminating quantity' ($\pi\epsilon\rho\alpha\acute{\iota}\nu o\upsilon\sigma\alpha$ $\pi o\sigma\acute{o}\tau\eta\varsigma$), and (2) that he called prime numbers $\epsilon\dot{\upsilon}\theta\upsilon\gamma\rho\alpha\mu\mu\iota\kappa o\acute{\iota}$, because they cannot form plane figures.

[1] A very similar proposition appears in *Strophe* 29 of the Algebra of Aryabhatta (ed. L. Rodet, pp. 14, 15, 38, 39 in *Journal Asiatique* for 1879). Cantor (*Vorles.* pp. 529—530) maintains that the Indian (who was born A.D. 476) has purposely disguised the *epanthem*, in order to conceal his plagiarism.

as late as A.D. 360[1]. It has been already stated (*supra*, p. 89) that his commentary on Nicomachus forms the 4th Book of his treatise on Pythagorean philosophy, the greater part of which is still extant. In this commentary Iamblichus includes some new matter, most of which is unimportant and need not be here quoted[2]. One very singular statement, however, should not be omitted. Iamblichus says that the Pythagoreans called 10 'the unit of the second course,' 100 'the unit of the third course,' 1000 of the fourth and so on[3]. Upon this he founds the following proposition: "If the units of any three consecutive numbers, whereof the highest is divisible by 3, be added together and the units (i.e. *digits*) of their sum be added together again and so on, the final sum will be 6." E.g. $7 + 8 + 9 = 24$ and $2 + 4 = 6$: $997 + 998 + 999 = 2994$: $2 + 9 + 9 + 4 = 24$, $2 + 4 = 6$. It will at once be seen that this was, for a Greek, a very difficult and remarkable discovery, and it tends very much to confirm the suspicion that the *octads* and *tetrads* of Archimedes and Apollonius were in fact accompanied by a symbolism which, if applied to tens, hundreds etc., would closely have resembled the Arabic numeral system.

64. The extracts given, in previous pages, from Nicomachus and Thymaridas will have led the reader to expect that algebra is not far distant. This expectation becomes the more lively, when we find that about this time problems leading to equations were a common form of puzzle. Between 50 and 60 riddles of this kind are preserved in the Palatine Codex of Greek epigrams (usually called the Palatine Anthology) and elsewhere. At least 30 of these are attributed to one Metrodorus, of the time of the Emperor Constantine (A.D. 306—337)[4].

---

[1] The Emperor Julian (A.D. 361—363) is supposed to have corresponded with Iamblichus, but the extant letters are of doubtful authenticity.

[2] See Nesselmann, pp. 237-242. Can-

tor *Vorles.* 390—392.

[3] μονὰς δευτερωδουμένη, τριωδουμένη, &c. The name was suggested by a singular fancy of arranging numbers in a kind of race-course; thus:

(ὕσπληξ, ' start ') 1 . 2 . 3 . 4 . 5 . 6 . 7 . 8 . 9
(νύσσα, 'goal')   1 . 2 . 3 . 4 . 5 . 6 . 7 . 8 . 9 . 10 (καμπτήρ.)

The next course begins at 10 and goes on to 100 and back and so on. Iamblichus makes great use of this figure.

[4] Jacobs, *Comm. in Gr. Anthol.* Pt. XIII. p. 917.

A few of them are older, though perhaps their metrical form is of this date: many, no doubt, are much later, for the anthology was not collected till the 10th century[1]. One of them, attributed to Euclid[2], is to this effect. A mule and a donkey were walking along laden with corn. The mule says to the donkey, "If you gave me one measure I should carry twice as much as you : if I gave you one, we should both carry equal burdens. Tell me their burdens, O most learned master of geometry." It will be allowed that this problem, if authentic, was not beyond Euclid, and the appeal to geometry smacks of antiquity. Another and a far more difficult puzzle is the famous 'cattle-problem' ($\pi\rho\delta\beta\lambda\eta\mu a$ $\beta o\epsilon\iota\kappa\delta\nu$) which Archimedes is said to have propounded to the Alexandrian mathematicians[3]. It is to the following effect. The sun had a herd of bulls and cows, of different colours. (1) *Of Bulls,* the white (W) were, in number, $(\frac{1}{2} + \frac{1}{3})$ of the blue (B) and yellow (Y): the B were $(\frac{1}{4} + \frac{1}{5})$ of the Y and piebald (P): the P were $(\frac{1}{6} + \frac{1}{7})$ of W and Y. (2) *Of Cows,* which had the same colours, (w, b, y, p), $w = (\frac{1}{3} + \frac{1}{4})$ $(B + b)$: $b = (\frac{1}{4} + \frac{1}{5})$ $(P + p)$: $p = (\frac{1}{5} + \frac{1}{6})$ $(Y + y)$: $y = (\frac{1}{6} + \frac{1}{7})$ $(W + w)$. Find the number of bulls and cows[4]. This is a very difficult problem, leading to excessively high numbers, and may very well have been invented by Archimedes. The problems of Metrodorus are shorter. One of them is of a kind still very familiar to schoolboys. It runs (Jacobs, XIV. no. 130) : "Of four pipes, one fills the cistern in one day, the next in two days, the third in three days, the fourth in four days: if all run together, how soon will they fill the cistern ?" There are several more of the same pattern. Another (Jacobs

---

[1] See art. *Planudes* in Smith's *Dic. of Biogr.* Most of the algebraical epigrams are in Pt. XIV. of Jacobs' Anthology, but a few more are in the Appendix (e. g. Nos. 19, 25, 26). Those attributed to Metrodorus are in XIV. 116—146. See Nesselmann, pp. 477 sqq. Cantor *Vorles.* pp. 393—4.

[2] Jacobs' Appendix, No. 26.

[3] Discovered and printed by Lessing, *Zur Gesch. der Lit.* I. pp. 421—446.

Nesselmann, who gives a translation and discusses it exhaustively (pp. 481—491), stoutly denies its authenticity. Heiberg (*Quaest. Archim.* p. 26) is inclined to admit it. It is not in Jacobs.

[4] Solution in Nesselmann, pp. 484—485. Some later hand has added some further difficulties : $W + B$ is a square-number, $P + Y$ is a triangular. On this, see also Nesselmann's comments.

XIV. 127) is : "Demochares has lived $\frac{1}{4}$th of his life as a boy : $\frac{1}{5}$th as a young man ; $\frac{1}{3}$rd as a man, and 13 years as an old man. How old is he ?" and there are more of this sort[1]. Another, not by Metrodorus (Jacobs, XIV. 49) is : 'Make me a crown of gold and copper and tin and iron, weighing 60 minae. Copper and gold shall be $\frac{2}{3}$rds of it : gold and tin $\frac{3}{4}$ths : gold and iron $\frac{3}{5}$ths. How much gold, copper, tin and iron are in the 60 minae ?' This is a problem on the *epanthem* of Thymaridas. None of these problems, of course, lead to more than simple equations, in which a line would be as good a symbol for the unknown quantity as any other. But they are all arithmetical problems requiring analytical treatment, and they all involve the consideration of an unknown quantity, for which some quasi-arithmetical symbol would be most convenient. They became especially popular just about the time of Diophantus[2], and they are therefore, as will be seen presently, of some historical importance.

65. Contemporary with Iamblichus, or perhaps rather earlier, lived **Diophantus** of Alexandria, the last and one of the most fruitful of the great Greek mathematicians. His date indeed can hardly be determined exactly. An arithmetical epigram on his age is attributed to Metrodorus. From this, it would appear that he died at the age of 84 years, some time

---

[1] One of this kind is on the life of Diophantus (Jacobs XIV. 126.) On these problems, Dean Peacock (art. *Arithmetic* in *Ency. Metrop. Pure Sci.* I. §§ 244—248) remarks that many of them may have been solved (as similar problems were by the Indians, Arabians and early Italians) by the rule of '*falsa positio*' or '*regula duorum falsorum*' : which dispensed with any algebraical symbol. The simple '*falsa positio*' was the assigning of an assumed value to the unknown quantity: which value, if wrong, could be corrected, in effect, by a 'rule of three' sum (as in the modern rules for Interest, Discount or Present Worth). This was used by the Egyptian Ahmes and by the Indian Bhâskara (born

A.D. 1114, author of the *Lîlâvatî*, forming part of the larger work *Siddhânta-çiromani*). By the '*regula duorum falsorum*' (Arabic *el Cataym*, i.e. 'the two errors') two false assumptions were taken and x was found from their difference. The simple '*falsa positio*' is called in the Lîlâvatî '*Ishta carman*', or 'operation with an assumed number'. Both are continually used by Luca Pacioli and Tartaglia. See Dean Peacock's art. *supra cit.*: Hankel. p. 259, Cantor *Vorles.* pp. 524, 628—9. Also below p. 116 n.

[2] An epigram is actually included in Dioph. *Arithm.* (v. 33) and the problem solved. The epigram is printed in Jacobs, *App.* no. 19.

before A.D. 330 or thereabouts. But he is not quoted by any writer before the younger Theon[1], who was working A.D. 365— 372 and later. Theon's daughter, the famous Hypatia (died A.D. 415) is said by Suidas (*s. v.*) to have written a commentary on Diophantus. Abulpharagius, a Syrian historian of the 13th century, says positively that Diophantus was a contemporary of Julian the Apostate, who was emperor A.D. 361—363. If the date of Metrodorus were certain, and the epigrams ascribed to him were undoubtedly authentic, the epigram above cited would be conclusive. But it is not so, and Abulpharagius may be right[2]. It would suit either testimony if we assign Diophantus to the first half of the 4th century, a time at which algebra certainly ought to have appeared, but he may have been much earlier. Doubts were at one time felt whether his name might not be Diophantes, for the passages in which he is mentioned, generally have the genitive Διοφάντου, which would suit either nominative[3], but Theon and Abulpharagius both call him Diophantos, and this may be taken to be his real name.

Only one work by Diophantus is cited, viz. the Ἀριθμητικά. Two, however, are extant, viz. an Ἀριθμητικά and a pamphlet on polygonal numbers, both mutilated. Diophantus himself, in the opening words of his Ἀριθμητικά, announces it as a work in 13 books: yet all the existing copies (save one) have it in 6 books, and the one exception (Vatican MSS. no. 200) has it in 7[4]. Yet it is evident that these 6 or 7 books are

---

[1] *Comm. on the Almagest.* Ed. Halma. I. 111. The 6th definition of Diophantus is there quoted *verbatim.*

[2] On the question of Diophantus' date, see an exhaustive discussion in Nesselmann, pp. 243—256. Here, however, Hypsicles, whom Diophantus quotes, is assigned to far too late a date, and Diophantus is (probably wrongly) identified with him who, according to Suidas, was teacher of Libanius the sophist (*cir.* A.D. 314—400). Cantor *Vorles.* Pref. p. VII cites Tannery, in *Bulletin de Sci. Math. et Astron.*, but I cannot find the article.

[3] One MS. of Suidas, (*s.v.* Hypatia) has Διοφάντην, but others have Διόφαντον. Nesselmann, indeed (pp. 244, 247—249), thinks all are wrong and that the true reading of the passage (Küster's) is ὑπόμνημα εἰς Διοφάντου ἀστρονομικὸν κανόνα. This would be the name of a commentary by Hypatia on the astronomical tables of *some other* Diophantus.

[4] The known MSS. of Diophantus are enumerated by Nesselmann, p. 256 *n.* They are three Vatican (nos. 191, 200 and 304) one at Paris, one in the Palatine library (at Heidelberg). The

not a reasonable, and therefore probably not the original, division. Some propositions contained in the 2nd Book (1—5 and 18, 19) clearly belong to the 1st, others, as clearly ought to belong to the 3rd (especially the last two, 35 and 36). Similar suspicions are aroused in other books. Evidence of mutilation is afforded also by many propositions (e.g. II. 19 and several of the 5th Book) which are not proved at all[1]. Two subjects, which Diophantus must have treated, are entirely omitted, viz. the solutions of determinate quadratic and of indeterminate simple equations. On the other hand, the last books of our copies are pretty clearly the limit of Diophantus' learning. For all these reasons, Nesselmann[2] comes to the conclusion that the 6 or 7 books of the Ἀριθμητικά, as we have them, do substantially represent the original, *minus* the two omitted subjects: that the omitted subjects were treated between the 1st and 2nd Books of our editions, and that the mutilation took place before the date of the earliest MS. (*i.e.* before the 13th century). A further question arises, whether the fragment on polygonal numbers ever formed part of the Ἀριθμητικά or not. Nesselmann thinks it did: Hankel and Cantor hold that it did not. It can hardly be doubted that the latter are right. The Ἀριθμητικά is purely algebraical and analytic (with the single exception of v. 13): the fragment on polygonal numbers is purely geometrical and synthetic. A similar question arises as to a work called Πορίσματα (? 'Corollaries') which Diophantus quotes in at least three places (v. 3, 5 and 19) : but as this is lost, it is not worth while in the present place to consider what it may

extant works of Diophantus were published with a commentary by Bachet de Meziriac (Paris, 1621): further notes were afterwards added by Fermat (Toulouse, 1670). A German translation was published by O. Schulz (Berlin, 1822). A Latin paraphrase by Xylander (Holzmann) Basle, 1571, first brought Diophantus into general notice, though many scholars knew of his existence before.

[1] A very striking instance occurs in v. 22, where the solution has nothing to

do with the problem to which it is attached. Nesselmann (pp. 410—413) suggests that here *the problem* properly follows on v. 21 : its solution is lost: the two next problems with their solutions are lost: the next problem is lost, but *its solution* remains.

[2] See esp. pp. 265 sqq. Cantor *Vorles.* pp. 397—398. Hankel, pp. 157—158. Nesselmann quotes, in support of his views (p. 272) Colebrooke's *Algebra of the Hindus*, note M, p. LXI.

have contained, and how it would have fitted into the extant works[1].

66. Of these latter, the fragment of Polygonal numbers may be first dismissed, both because it is very short and because, also, it is in the antique geometrical style. The only difference, here, between Diophantus and his predecessors is that he treats of polygonal numbers generally, without specially handling the different classes of them. The book, as we have it, is divided into 10 propositions, which is an excessive number for the actual matter contained[2]. It begins by stating (as well-known) that all numbers above 3 are polygons, containing as many angles as units: and that the side of each such polygon is 2. Then follows a statement of the purpose of the work to this effect: " As a square number is known to be the product of a number multiplied by itself, so every polygonal number, multiplied by one number and added to another, both of which depend upon the number of its angles, produces a square number. I shall prove this, and shall show also how from a given side to find its polygon and conversely. Some auxiliary propositions must first be proved." Then follow some propositions on arithmetical progression, proved geometrically. Their result may be stated algebraically thus: Prop. II[3]. If a, $a + b$, $a + 2b$ be three terms of an A. P. then $8(a + 2b)(a + b) + a^2 = [(a + 2b) + 2(a + b)]^2$. Prop. III. If $a, a + b, a + 2b$ etc. be an A. P. the difference between the 1st term and the nth is $(n - 1)b$. Props. IV. V. Summation of an A. P. of n terms, proved first where n is even, secondly where n is odd. The following propositions introduce the more familiar progressions, in which the first term is 1. Thus VI.[4] if S be the sum of n terms of the series 1, $1 + b$, $1 + 2b$ etc. then $8bS + (b - 2)^2 = [b(2n - 1) + 2]^2$. Prop. VII. contains the geometrical proof that $b^2(2n - 1)^2 = [b(2n - 1)]^2$. The most important is VIII.: in the series 1, $1 + b$, $1 + 2b$ etc. the sum of n terms is a

---

[1] The curious may consult Nesselmann, pp. 269—270, and his 10th Chapter, pp. 437 sqq.

[2] See a very full abstract in Nesselmann, ch. XI. pp. 462—476. A sum-mary of Props. VIII. and IX. in Cantor *Vorles.* p. 414.

[3] Proof given in full by Nesselmann, pp. 471—472.

[4] Proof in Nesselmann, pp. 473—4.

polygon, containing $b + 2$ angles, and its side is the sum of the preceding $(n - 1)$ terms (compare Hypsicles, whom Diophantus here quotes)[1]. Thence follows a "definition" of a polygonal number to this effect: "Every polygonal number of n angles, multiplied by 8 $(n - 2)$ and added to $(n - 4)^2$, is a square number"[2]. This with the IXth Prop., in fact, completes the promise of the introduction, but a Xth proposition is added "To find in how many orders a given number is polygonal[3]." Only a fragment of this remains, from which it is impossible to discern how Diophantus intended to complete his proof. All these propositions are given in the Euclidean manner, with an enunciation, a linear symbolism, and a synthetic proof ending ὅπερ ἔδει δεῖξαι (Q. E. D.). But it is to be remembered that lines, with Diophantus, are symbols for numbers only (as in Euclid VII—IX), and not for magnitudes (as in Euclid II. or X.) Nevertheless, he adopts for arithmetical purposes propositions proved for geometrical by Euclid (e.g. II. 3, 4 and 8) and from this it is evident that the arithmetical uses of Euclid were known to the later Greek mathematicians[4].

---

[1] The proof is given in full by Nesselmann, pp. 474—476.

[2] If P be the polygonal number, $8(n-2)P + (n-4)^2 =$ a square number. This proposition was known of *triangular* numbers as early as Plutarch (*Plat. Quaest.* v. 2, 4), and is so repeated by Iamblichus. The general proof is probably Diophantus' own. Bachet remarks that the converse is not necessarily true. If $8(n-2)P + (n-4)^2$ is a square, it does not follow that P is a polygon of the nth order, unless n is 3 or 4. E.g. If $n=5$, $24P + 1$ is a square, if $P=2$: but 2 is not a pentagonal number. See Nesselmann, p. 467.

[3] E.g. 36 has 3, 4, 13 or 36 angles: 225 has 4, 8, 24, 76 or 225 angles, etc. Nesselmann, pp. 468—470.

[4] The proof of the 1st proposition is short enough to be inserted here. It runs thus:

$$E \quad\quad A \quad\quad\quad B \quad\quad D \quad G$$

"Three numbers, AB, BG, BD have a constant difference $(=GD)$. It is to be proved that $8AB \cdot BG + BD^2 =$ a square of which the side is $AB + 2BG$. Since $AB = BG + GD$, $8AB \cdot BG = 8BG^2 + 8BG \cdot GD$, and $4AB \cdot BG = 4BG^2 + 4BG \cdot GD$. But (Eucl. II. 8) $4BG \cdot GD + BD^2 = AB^2$. The inquiry therefore is how $AB^2 + 4AB \cdot BG + 4BG^2$ give a sum which is a square number. Take $AE = BG$. Then (Euclid II. 3) $4BA \cdot AE + 4AE^2 = 4BE \cdot EA$ and $4BE \cdot EA + BA^2 = (BE + EA)^2$. But $(BE + EA) = AB + 2AE = AB + 2BG$, Q. E. D." It will be seen that this proof is wholly Euclidean, but omits many steps which Euclid would certainly have inserted, and that it uses Euclid II. 3 and 8 for arithmetical purposes.

67. The *Arithmetica* is a work of infinitely greater importance. It is a treatise on *algebra,* and, if not the first that ever was written, is by far the earliest now extant. It is devoted to the solution of equations, which Diophantus expresses with *algebraical symbols* and treats always *analytically.* But as Diophantus does not claim for himself the credit of inventing either the symbols which he uses or his method of proof, a short recapitulation should be here inserted to account, as far as possible, for both inventions.

The ancient geometers knew two modes of proof, which they called *synthetic* and *analytic*[1]. With the former, a proposition is proved directly by steps advancing from the known to the unknown. With the latter (of which the *reductio ad absurdum* is a particular kind), a proposition to be proved is *assumed* to be true or untrue, and the assumption is shown to be consistent or inconsistent with some simpler facts already known, or is shown to be so upon certain conditions. Algebraic proof is of this latter, the *analytical,* kind. The invention of this kind of proof is expressly attributed to Plato[2]. We have already seen (*supra,* p. 18), that calculation with an unknown quantity (called *Hau* or ' heap') was practised by the Egyptians in very remote antiquity, and that some conventional signs at least for addition and subtraction were then used. So useful an art can hardly have disappeared entirely from the later Egyptian civilization. Aristotle first, so far as we know, employed letters of the alphabet to indicate unknown magnitudes, though not for purposes of calculation[3]. But this

---

[1] Euclid XIII. 1, (Schol.) Pappus, Bk. VII. Preface. Ed. Hultsch, p. 634. Proclus, ed. Friedlein, pp. 211—212. Todhunter's Euclid, Notes, pp. 309 sqq.

[2] Proclus, *loc. cit.* and Diog. *Laert.* III. 24.

[3] See, for example, *Physics,* VII. VIII. *passim:* but esp. VII. 5 (pp. 249—250 of the Berlin ed.), where it is stated: "If A be the mover, B the moved thing, Γ the distance and Δ the time

of the motion, then A will move $\frac{B}{2}$ twice the distance Γ in the time Δ or the whole distance Γ in half the time Δ," etc. Poggendorf, *Gesch. der Physik.* p. 242, sees in this the germ of the principle of virtual velocity. It is evident also that Aristotle understood the advantage of these alphabetic symbols, for he explains (*Anal. Post.* I. 5, p. 74 a 17) how much time and trouble is saved by a general symbolism.

suggestion could hardly be followed up because the alphabet, soon after Aristotle's time, came to be used for ordinary arithmetical purposes. Euclid uses lines as symbols for magnitudes, including numbers, and though he solves quadratic equations and performs other operations of universal arithmetic, he uses always the synthetic mode of proof, and confines himself strictly to geometrical conditions. He will not add a line to a square, or divide a line by another line or name a particular number. The limitations imposed upon universal arithmetic by the linear symbolism were too great. Algebra could come only from the practical calculator who was not hampered by such difficulties. The first step seems to have been taken, not by a Greek, but by the Egyptian Heron. Thus, in a proposition now included in the *Geometria* (p. 101 in Hultsch's edition) but originally part of "another book" unknown, Heron does not scruple to add an area to a circumference[1]. In modern symbols, the proposition runs: 'If S be the sum of the area (A), the circumference (C), and the diameter (D) of a circle, find the diameter. The answer which he gives is $d = \dfrac{\sqrt{154S + 841} - 29}{11}$.

The proof, which he does not give, is obviously as follows:

A is $\dfrac{\pi}{4} d^2$: C is πd : π is $\frac{22}{7}$. Then $S = \dfrac{\pi}{4} d^2 + (\pi + 1) d = \frac{11}{14} d^2 + \frac{29}{7} d$. Multiply each term by 154 $(= 11 \times 14)$. Then $121 d^2 + 638 d + 841 = 154S + 841$ or $(11 d + 29)^2 = 154 S + 841$, from which Heron's answer immediately follows. It cannot be doubted that Heron could solve an impure quadratic equation in a way which, but for the want of a symbolism, would be simply algebraical[3]. Two centuries or more afterwards, we find,

---

[1] Compare Diophantus, who, in his 5th Book, means by a "right-angled triangle" three numbers such that $a^2 + b^2 = c^2$, adds its area $\dfrac{ab}{2}$ to its side, treats its side upon occasion as a cube, etc. Hankel, p. 159. Of the great Greek mathematicians, Archimedes alone (in his *Circuli Dimensio*) ventures to introduce actual numbers into a geometrical discussion, and to divide a line by another line. He finds the value of π and some other similar ratios but does not himself pursue such investigations further and is not followed by any other writer. Trigonometry was used only for astronomical purposes and did not form part of geometry at all.

[3] Cantor *Vorles.* pp. 341—2.

in the Semite Nicomachus, that the practical calculator has taken to proving, by induction from numbers themselves, the theories which hitherto had been proved deductively by the geometrician. In the 2nd Book of Pappus, the Aristotelian use of the alphabet appears again, but whether this was due to Apollonius or his predecessor Archimedes cannot be discovered. Pappus, at any rate, uses A for 20, B for 3, Γ for 4, Z for 2 and T for an unknown number[1]. Thymaridas calls an unknown number ἀόριστος, and proves an algebraical theorem. Problems leading to simple equations become a common form of amusement; and finally, in Diophantus, the method of forming and handling equations appears almost complete, accompanied by an algebraical symbolism which was probably not new but of which no trace has been found in previous writers. The foregoing statement is sufficient to raise a very strong suspicion that there are large gaps in the history of Greek ἀριθμητική, and that the later Greek mathematicians were not by any means so futile as they are sometimes represented to be[2]. Nevertheless hardly any writers are quoted save those of whose works large portions are still extant. It is therefore not an improbable supposition that there were in Alexandria and Pergamum and elsewhere, as in the English universities at the present day, many mathematicians of great ability and inventiveness, who did not write books at all but were content to allow their knowledge to ooze out in lectures and private communications[3]. What little evidence there is, and the absence of more, alike suggest that these mathematicians were of Semitic or Egyptian origin. On the other hand, it is still possible that

[1] See Hultsch's Ed. pp. 8 and 18. Cantor *Vorles.* pp. 298, 387.

[2] Thus Hankel (p. 157) says, "Of the performances of the Greeks in arithmetic our judgment may be stated shortly thus: they are, in form and contents, unimportant, childish even: and yet they are not the first steps which science takes, as yet ignorant of her aim, tottering upon shaky ground: they are the work of a people which had once produced an Euclid, an Archimedes, an Apollonius. It is dotage without a future which wearies us in these writings. In the midst of this dreary waste appears suddenly a man with youthful energy, Diophantus."

[3] What, for instance, was the *Logistic* of Magnus, and how did it assist Porus and Philo of Gadara in those researches which Eutocius (*ad Arch. Cir. Dim.*) mentions, but which he could not understand?

Diophantus actually invented the symbolism, and the rules which first appear in his book but for which he claims no credit. In any case, Diophantus must always be esteemed one of the most brilliant of Greek mathematicians.

68. The *Arithmetica* begins with a prefatory letter to one Dionysius, to whom Diophantus says, "Knowing that you are anxious to learn the solution of arithmetical problems, I have tried to systematise (or 'state in a handy form', ὀργανῶσαι) the method, beginning from the foundations of the matter. You will think it hard before you get thoroughly acquainted with it" etc. He then proceeds to definitions (11 altogether), in which he does not (as an inventor would) use the imperatives, ἔστω, καλείσθω, 'let it be', 'let it be called', etc. but the indicatives, ἐστὶ, καλεῖται, 'it is', 'it is called', etc. With such expressions he states the symbolism and the rules of algebraical multiplication. He gives as a fact, without explanation, that 'a negative term multiplied by a negative produces a positive'[1], and (after recommending continual practice in the use of the previous rules) he states shortly how to reduce an equation to its simplest form. This is the evidence from which it is concluded that Diophantus was not the inventor of the method which he employs. The method itself may be shortly described as follows.

Diophantus uses *only one unknown*, which he calls ὁ ἀριθμός or ὁ ἀόριστος ἀριθμός[2]. Its symbol is ς′ or ς^{ο′} in the plural ͞ςς or ςς^{οὶ}, and, as in this last case, all inflexions may be appended to the symbol as ς^{οῦ}, ςς^{οὺς} etc[3]. The *square of the unknown* x^2

---

[1] Def. 9, λεῖψις ἐπὶ λεῖψιν πολλαπλα- σιασθεῖσα ποιεῖ ὕπαρξιν· λεῖψις δὲ ἐπὶ ὕπαρ- ξιν ποιεῖ λεῖψιν. This should properly be translated, "A difference multiplied by a difference makes an addition" etc. For it is to be remembered that *Diophantus has no notion of a negative term standing by itself or of subtracting a greater term from a less.* $8x - 20$ is to him an absurdity unless $x = 2\frac{1}{2}$ at the least.

[2] In the definition described as πλῆθος μονάδων ἄλογον.

[3] This symbol ς, as it appears in our MSS. is always assumed to be the final *sigma*, adopted here because it was the final letter of ἀριθμός, and because also it was the only Greek letter which had not a numerical value. It must, however, be remembered (1) that it is only *cursive* Greek which has a final *sigma* and that the cursive form did not come into use till the 8th or 9th century: (2) that inflexions are *appended* to Diophantus' symbol ς′ (e.g. ς^{οῦ}, ςς^{οί} etc.) and that his other symbols (except

is called $\delta \acute{v} \nu a \mu \iota s$, its symbol is $\delta^{\hat{v}}$: x^3 is called $\kappa \acute{v} \beta o s$ (symbol $\kappa^{\hat{v}}$): x^4 is called $\delta \nu \nu a \mu o \delta \acute{v} \nu a \mu \iota s$, (symbol $\delta \delta^{\hat{v}}$): x^5 is called $\delta \nu \nu a \mu \acute{o} \kappa \nu \beta o s$ (symbol $\delta \kappa^{\hat{v}}$): x^6 is called $\kappa \nu \beta \acute{o} \kappa \nu \beta o s$ (symbol $\kappa \kappa^{\hat{v}}$): but beyond this *sixth power of the unknown* Diophantus does not go. These terms and symbols are not applied to the powers of any number except the unknown. All known numbers are called $\mu o \nu \acute{a} \delta \epsilon s$, (symbol $\mu^{\hat{o}}$) and unity itself is always written $\mu^{\hat{o}}$ $\acute{a}$ or $\mu^{\hat{o}}$ $\mu \acute{\iota} a$. The coefficients are written after the symbols (e.g. $\varsigma \varsigma^{o\iota}$ $\kappa' = 20x$: $\mu^{\hat{o}}$ $\kappa' = 20$). The sign of subtraction is the word $\lambda \epsilon \acute{\iota} \psi \epsilon \iota$ (*minus*), its symbol is ⋔, a truncated and inverted ψ (Def. 9). The symbol of equality is ι, the initial of $\emph{\'{\iota}\sigma o s}$, $\emph{\'{\iota}\sigma o \iota}$[1]. In a composite expression, the negative terms are placed after all the positive, but there is no sign of addition save mere juxtaposition. Thus $\delta \delta^{\hat{v}} \theta$ $\delta^{\hat{v}} \bar{\varsigma}$ $\mu^{\hat{o}} \bar{a}$ ⋔ $\kappa^{\hat{v}} \bar{\delta} \varsigma \varsigma \iota \bar{\beta}$ means

$$9x^4 + 6x^2 + 1 - 4x^3 - 12x \quad \text{(IV. 29)}.$$

⋔) are *initial* letters or syllables. The objection (1) might be disposed of by the fact that the Greeks had two *uncial* sigmas C and Σ, one of which might have been used by Diophantus, but I do not see my way to dismissing objection (2). It would be of great historical importance if we could discover what symbol Diophantus used, and of what word the inflexions appended to the symbol were supposed to form part. Both word and symbol may be Egyptian or Indian or Babylonian, and may reveal an entirely unknown chapter in the history of mathematics. Since, however, the only distinct anticipations of Diophantus' art are found in Egypt in Ahmes and Heron (who also is believed to have been an Egyptian) I am inclined to look for the origin of Diophantus' symbols in some hieratic characters. The Greek sign ς' is in form practically identical with two hieratic signs (1) for a *papyrus-roll*, $s'\bar{a}$, a determinative of unknown force, which, as it happens, is the last character of the four with which Ahmes wrote his *hau* (Eisenlohr I. p. 60, II. pl. xi):

(2) for 'sum-total', *tmt*. The hieratic signs differ slightly in form, and are said to be derived from different hieroglyphic pictures (see Levi, *Raccolta dei Segni Ieratici*, 1875, Plates 37 and 52): but Dr Birch tells me that he thinks the sign for a 'sum-total' is identical with the papyrus-roll. So also I should expect to find ⋔ in some hieratic character. If I could prove these points, I would recast this chapter. See p. 317.

[1] Luca Pacioli (1491) uses p and m for *plus* and *minus*: Tartaglia (1556) uses ϕ for *plus:* Vieta has $+$ and $-$, also $=$ (later ∞) for the sign of *difference* ($A \infty B$): Oughtred first has $\times$: Harriot (1634) writes factors consecutively without any sign of multiplication. Descartes uses ∞ for *equal*. Wallis turned this into $=$. See Nesselmann, p. 305. Hallam, however, (*Europ. Lit.* Pt. I. Ch. ix. s. 6) rightly ascribes $+$ and $-$ to Stifel (1544): and says also that Xylander in his Diophantus used $\parallel$ for $=$. As to $+$ and $-$ see De Morgan's *Arithmetical Books—A Bibliography*—1847, pp. 19—20, and his art. in *Trans. Camb. Philos. Soc.* Vol. IX.

Fractions, of which the denominator is some power of the unknown, e.g. $\frac{1}{x}$, $\frac{1}{x^2}$, $\frac{1}{x^3}$ etc. are described as $\dot{\alpha}\rho\iota\theta\mu o\sigma\tau\grave{o}\nu$, $\delta\upsilon\nu\alpha\mu o\sigma\tau\acute{o}\nu$, $\kappa\upsilon\beta o\sigma\tau\acute{o}\nu$ etc.: in the symbolism generally the denominator is written above and after the numerator, $\overline{\delta}^{\varsigma o\hat{\upsilon}\,a}$ means $\frac{4}{x}$, $\overline{\gamma}^{\delta\hat{\upsilon}\,\acute{a}}$ means $\frac{3}{x^2}$: but if the numerator itself contains a fraction, then the whole word $\dot{\alpha}\rho\iota\theta\mu o\sigma\tau\grave{o}\nu$ etc. is written before the numerator (as $\dot{\alpha}\rho\iota\theta\mu o\sigma\tau\grave{o}\nu\ \ \overline{a}\ \ \overline{a^\beta}=\frac{1\frac{1}{2}}{x}$ etc.). If, however, the numerator and denominator are composite expressions (also if they are very high numbers) Diophantus writes the numerator first, then $\dot{\epsilon}\nu\ \mu o\rho\acute{\iota}\omega$ or $\mu o\rho\acute{\iota}o\upsilon$, then the denominator : e.g. $\mu^\delta\ \overline{\eta}\ \mu o\rho\acute{\iota}o\upsilon\ \delta^{\hat{\upsilon}}\ \overline{a}\ .\ \varsigma^{o\hat{\upsilon}}\ \overline{a}$ means $\frac{8}{x^2+x}$, $\delta\delta^{\hat{\upsilon}}\ \overline{\beta}$, $\overline{\epsilon\chi}\ \dot{\epsilon}\nu$ $\mu o\rho\acute{\iota}\omega\ \overline{\rho\kappa\beta}$, $\overline{,\alpha\kappa\epsilon}$ means $\frac{25600x^4}{1221025}$. Some further details might be added but they are not necessary for the present purpose. Suffice it to say that Diophantus often writes a name in full where a symbol would have served, that his symbols are only abbreviations of the words (except $\pitchfork$), that inflexions are appended to symbols (not to $\delta^{\hat{\upsilon}}$, $\kappa^{\hat{\upsilon}}\ \mu^{\delta}$) as if they were words, and that he states, in grammatical sentences, the nature and the result of each step in an operation[1]. The following brief

---

[1] Nesselmann (p. 302) divides algebraical *styles* into 3 classes:

(*a*) the *Rhetorical*, where no symbols are used and every term and operation is described in full. This is the style of Thymaridas, Iamblichus, all the Arabian and Persian algebraists, and the early Italians (e.g. Leon. Bonacci, of the 13th, Regiomontanus and Luca Pacioli of the 15th century.)

(*b*) the *Syncopated*, where abbreviations are used for the most common words and operations, but in other respects syntactical rules are observed. This is the style of Diophantus and of the later Europeans down to the middle of the 17th century.

(*c*) the *Symbolical*, the modern style, where no words are used at all. Vieta (1540—1603), who in time belongs rather to the early Italians, uses a style which is very nearly symbolical and which was not generally adopted till more than a century later. Before his time, the Italians used R (*res* or *radix*), Z (*zensus*), C (*cubus*), etc. for x, x^2, x^3, etc.: and Bachet and Fermat long afterwards have N (*numerus*), Q (*quadratus*), etc. in the style of Diophantus. Vieta, however, wrote *A*, *Aq*, *Ac*, *Aqq* etc.

examples will illustrate all these points. The Prop. V. 3 concludes καὶ γίνεται ὁ τετράγωνος δ̅ῦ̅ δ μ̅ο̅ λ̅ϛ̅ λείψει ϛ̅ϛ̅ κ̅δ̅ ἴσος δυνάμεσι δ̅ ϛϛοῖς κ̅η̅ . μ̅ο̅ λ̅δ̅ καὶ γίνεται ὁ μ̅ο̅ α̅ϰ̅ϛ̅ etc. i.e. The square $4x^2 + 36 - 24x$ is equal to $4x^2 + 28x + 34$ and x is $\dfrac{1}{26}$.

In IV. 42 we read λοιπὸν δὲ τὰ ὑπὸ τοῦ πρώτου καὶ τρίτου συναμφοτέροις ἔσται πεντάκις. ἀλλ' ὁ ὑπὸ τοῦ πρώτου καὶ τρίτου ἐστὶ δ̅ῦ̅ ι̅β̅ ἐν μορίῳ δ̅ῦ̅ α̅ μ̅ο̅ι̅β̅ λείψει ϛ̅ϛ̅ ζ̅' : i.e. It remains that the product of the first and third shall be 5 times their sum. But their product is $\dfrac{12x^2}{x^2 + 12 - 7x}$, etc.

69. It might have been expected that Diophantus, in introducing a new method of inquiry, which consists mainly in applying to a number, *pro tem.* unknown, the ordinary rules of calculation, would have called his work λογιστικά. But it has been already pointed out that the distinction between ἀριθμητική and λογιστική, though originally perhaps only one of method, soon became one of purpose. Logistic seeks only to find an answer to a question about some particular numbers, while ἀριθμητική endeavours to define classes of numbers or to find rules which are applicable to all numbers. Ostensibly, the problems which Diophantus sets himself are generally of this latter kind: e.g. II. 33. To find three numbers such that the square of each plus the next number is a square: III. 7. To find three numbers such that their sum, and also the sum of any two of them, shall be a square: IV. 22. 'To find three

---

thus admitting of more than one unknown (as *Bq*, *Cqq* etc.) and he also introduced general coefficients as (*mA* etc.) Harriot (1631) and Wallis (1685) used to write *aaa*, etc. for a^3, etc. Descartes is sometimes said to have introduced the numeral exponents (which Wallis also uses) but Hallam (*loc. cit.*) ascribes this to Michael Stifel (1544). See Nesselmann, pp. 58, 296, 302 sqq. See also preceding note and the preface to Wallis's *Algebra*, 1685. The word *zensus*, from which the symbol *Z* was derived, is a mis-spelling of *census*, which is a bad Latin translation of *mâl* (i.e. 'wealth', or 'possession'), the Arabic name for the square of the unknown. The Arabs called the unknown *shai*, 'thing', translated in Latin *res*, in Italian *cosa*, whence algebra used to be called the *Cossic* art. See Colebrooke *Algebra of the Hindus*, p. xiii. The same writer (p. x. n) says that Robt. Recorde (A.D. 1540) first used the sign =. The history of such signs seems to require investigation. See p. 317.

numbers in continued proportion, such that the difference between any two of them is a square :' v. 17. 'To divide a given
number into four parts, such that the sum of any three parts is
a square:' VI. 14. 'To find a right-angled triangle such that
its area *minus* either of the sides is a square.' Problems of
this sort should be capable of general solutions: they are intended to discover classes of numbers having a common property, and are therefore rightly ascribed to ἀριθμητική. But
Diophantus does not, in fact, treat them generally. He is
satisfied with a solution which gives only one case or a few
cases. Usually he arrives at an equation to which he finds
only one particular solution. Even where the problem leads to
a quadratic equation, which may be solved for two positive
roots, *he never gives more than one*[1]. With a symbolism which
admitted of only one unknown quantity, he could not have
been expected to find a perfectly general solution, but he might
have done much more than he does[2]. It must be added also
that he will not accept a result which is either a negative or an
irrational quantity[3]. Equations which lead to such are 'impossible' or 'absurd' (ἀδύνατον IV. 28, ἄτοπον, v. 2). On the
other hand, he does not by any means object to a fractional
result, and he is the first of the Greeks to whom a fraction was
a number and not a ratio.

   70. Of the 6 Books of the Ἀριθμητικά now extant, the
first, as has been said already, is mainly devoted to determinate
equations of the first degree, the remainder to indeterminate
equations of the second. The problems, however, which Diophantus sets before the reader, do not as a rule lead immediately

---

[1] In such a case, says Nesselmann,
(p. 320) the Arabs and the earliest
Italians always gave *both* roots.

[2] Hankel (p. 162) suggests that Diophantus' habit of only giving one solution, was a relic of the old geometrical
practice. It seems to me more probable
that algebra was originally the invention of practical men, who only wanted
one solution.

[3] Hence, for instance to v. 30, 'To

find two numbers such that their sum
and product shall = given numbers'
he adds (as a προσδιορισμός or 'determination') 'If the square of their
sum be subtracted from twice the sum
of their squares, the remainder must
be a square' {i.e. $2x^2 + 2y^2 - (x^2 + 2xy + y^2)$ must = a square number with a
rational root.} Similarly v. 33. Nes-
selmann, p. 326.

and easily to equations with only one unknown. His art therefore distinguishes itself in two separate departments, the construction of equations and their solution. The second of these may be treated here first.

In Def. XI. Diophantus gives a rule for the solution of *pure* equations in the following manner: " If a problem leads to an equation containing the same powers of the unknown ($\epsilon i\delta\eta$ $\tau\grave{a}$ $a\dot{v}\tau\acute{a}$) on both sides but not with the same coefficients ($\mu\grave{\eta}$ $\dot{o}\mu o\pi\lambda\acute{\eta}\theta\eta$)[1], you must deduct like from like till only two equal terms remain. But when on one side or both some terms are negative ($\dot{\epsilon}\nu\epsilon\lambda\lambda\epsilon\acute{\iota}\pi\epsilon\iota$), you must add the negative terms to both sides till all the terms are positive ($\dot{\epsilon}\nu\nu\pi\acute{a}\rho\chi\epsilon\iota$) and then deduct as before stated[2]." He then promises to give the method of solving mixed or *adfected quadratic equations*[3], but this rule does not appear in our texts, and unfortunately Diophantus, though he often arrives at such equations, never goes through the process of solving them. He merely states a root or says that the equation is soluble (e.g. VI. 6 "$84x^2 + 7x = 7$ *whence x is found* $=\frac{1}{4}$": or VI. 8 "$630x^2+73x=6$, *whence the root is rational*[4]"). But it is evident that he did not solve them empirically, for where a root is irrational, he sometimes gives approximations to it (e.g. V. 33). His method of solution seems to have differed from ours only in this, that in an equation $mx^2 + px = q$, he first *multiplied* the terms by m instead of dividing them[5]. Three forms of adfected quadratics occur in Diophantus viz. (1)

---

[1] $\pi\lambda\hat{\eta}\theta os$ is the ordinary Diophantic expression for ' coefficient '.

[2] The addition of the negative terms was called by the Arabs *al-jebr* (or ' restitution '): the deduction was called *al-mukâbalah* ('comparison'). These two names were used together for Algebra until the end of the 16th century, when the second was discarded. Nesselmann, pp. 47 sqq. and 315. Comp. the 1st chap. of Wallis's *Algebra*.

[3] $\ddot{v}\sigma\tau\epsilon\rho o\nu$ $\delta\acute{\epsilon}$ $\sigma o\iota$ $\delta\epsilon\acute{\iota}\xi o\mu\epsilon\nu$ $\kappa a\grave{\iota}$ $\pi\hat{\omega}s$, $\delta\acute{v}o$ $\epsilon\dot{\iota}\delta\hat{\omega}\nu$ $\ddot{\iota}\sigma\omega\nu$ $\dot{\epsilon}\nu\grave{\iota}$ $\kappa a\tau a\lambda\epsilon\iota\phi\theta\acute{\epsilon}\nu\tau\omega\nu$, $\tau\grave{o}$ $\tau o\iota o\hat{v}\tau o\nu$ $\lambda\acute{v}\epsilon\tau a\iota$.

[4] Cf. Nesselmann, p. 318.

[5] Nesselmann, p. 319. On p. 324 sqq. Nesselmann discusses from what source Diophantus obtained his method of solution. The ancients, from Euclid's time or earlier, could solve the equations $x^2 \pm px = q$ and $px - x^2 = q$, *geometrically considered*. Thus $x(x+p) = q$ would be in geometrical language: To produce a given straight line p to a length $p + x$, so that the rectangle between the whole line so produced and the part produced i.e. $x(p+x)$ shall be equal to a given figure q. The other cases are equally easy to put geometrically. All three are solved in Euclid

$mx^2 + px = q$ (e.g. VI. 6), (2) $mx^2 = px + q$ (e.g. IV. 45) and
(3) $mx^2 + q = px$ (e.g. VI. 24). One cubic equation (reducible
at once to $x^3 + x = 4x^2 + 4$) occurs, VI. 19. No example of
indeterminate simple equations occurs in the present text of
Diophantus. Some problems, leading to such, are contained in
the 1st Book (Nos. 14. 25—28), but Diophantus takes a short
way with these by assuming one of the required numbers and
so converts the equations into a determinate form. *Indeterminate quadratics* are confined to the case " that one or two
(never more) functions of the unknown, of the form $Ax^2 + Bx + C$,
must be a rational square (ἴσον τετραγώνῳ). Hence we have to
do only with the equation $Ax^2 + Bx + C = y^2$ or with two equations of the same form." Let the single equation be considered
first. It assumes many forms according as one or another term
is wanting or is eliminated. These need not here be considered,
but it should be mentioned that the complete expression
$Ax^2 + Bx + C = y^2$ is deemed by Diophantus to be soluble[1]
only (1) when A is a positive square number: in which case
$a^2 x^2 + Bx + C = y^2$: he then takes $y = ax + m$: (2) when C is a
positive square number: in which case he takes $y = mx + \sqrt{C}$:

VI. 28, 29 stated above (p. 84 n.). In
the figures

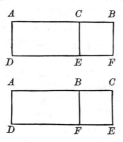

To the line AB, a rectangle AE is to be
applied so that $AE = q$ and BE is
similar to mc^2 (or $BC : CE = m : 1$). If
$AD = x$, then $BC = mx$. $AC = p \pm mx$:
and $AE = x(p \pm mx)$, so that the proof
of the geometrical proposition involves
the solution of the quadratic equations. This fact was first pointed out
by Montucla *Hist. Math.* I. p. 413.

Nesselmann quotes other suggestions
by Cossali and Bachet, but does not
decide for any. It should be stated
that Diophantus nowhere appeals to a
geometrical figure, whereas modern
algebraists (acc. to Hankel, p. 162)
down to the end of the 17th century
always added one as an illustration to
the solution of a quadratic equation.
I do not, however, find this in Harriot
or Wallis.

[1] It must be remembered that Diophantus does not avoid *fractional solutions* for indeterminate equations, hence
the problems which in modern text-
books are called Diophantic (viz. to
find a solution in positive integers for
$ax + by = c$) are wrongly named, since
Diophantus does not treat such equations nor does he solve for integers
those which he does treat. (Hankel,
p. 163.)

(3) when $\dfrac{B^2}{4} - AC$ is a positive square number, a condition which he uses only covertly. In such a case (e.g. IV. 33) he takes $y = mx$. If beside $Ax^2 + Bx + C = y^2$, another function of x $A_1 x^2 + B_1 x + C_1$ is to be made equal to another square number y_1^2 Diophantus calls the problem a "double equation" ($\delta\iota\pi\lambda o\acute{\iota}$-$\sigma\acute{o}\tau\eta\varsigma$, $\delta\iota\pi\lambda\tilde{\eta}$ $\iota\sigma\acute{o}\tau\eta\varsigma$, $\delta\iota\pi\lambda\tilde{\eta}$ $\acute{\iota}\sigma\omega\sigma\iota\varsigma$). He seems unable to solve these simultaneous equations unless A and A_1 are the same square number, but if x^2 is wanting in both expressions, he can solve them either if B and B_1 are to one another as two squares or C and C_1 are both squares. Several examples of indeterminate equations of degrees higher than the second also occur. The opinion of Nesselmann on the methods of Diophantus is shortly as follows: (1) Indeterminate equations of the 2nd degree are treated completely only when the quadratic or the absolute term is wanting: his solution of the equations $Ax^2 + C = y^2$ and $Ax^2 + Bx + C = y^2$ is in many respects cramped. (2) For the 'double equation' of the 2nd degree he has a definite rule only when the quadratic term is wanting in both expressions: even then his solution is not general. More complicated expressions occur only under specially favourable circumstances. (3) The solution of the higher indeterminates depends almost entirely on very favourable numerical conditions and his methods are defective[1].

71. But the extraordinary ability of Diophantus appears rather in the other department of his art, namely the ingenuity with which he reduces every problem to an equation which he is competent to solve. To exhibit completely his cleverness in this respect would be, as Nesselmann says : "to transcribe his book[2]." The same critic, however, has selected a number of

---

[1] The following remarks by an accomplished critic will sufficiently excuse me for saying so little on the Diophantic equations and their solutions. "In 130 indeterminate equations, which Diophantus treats, there are more than 50 different classes...Almost more various than the problems are their solutions... Each calls for a quite distinct method, which is often useless for the most closely-related problems. It is therefore difficult for a modern, after studying 100 Diophantic equations, to solve the 101st." Hankel, pp. 164—165.

[2] Nesselmann, ch. 9, pp. 355 sqq.

typical specimens, exhibiting the most striking characteristics of Diophantus' style. Some of these may be here given. (1) Diophantus shows great *Adroitness in selecting the unknown,* especially with a view to avoiding an adfected quadratic. Thus IV. 38 is a problem 'to find 3 numbers, so that the product of any two + the sum of the same two shall be given numbers[1].' Here $ab + a + b = 8 : bc + b + c = 15 : ac + a + c = 24$. Here he takes $b + 1 = x$, whence $b = x - 1$. Then from the first equation $a = \dfrac{9}{x} - 1$: from the second $c = \dfrac{16}{x} - 1$: from the third $x = \frac{12}{5}$. In I. 16 'To find 3 numbers such that the sum of each pair is a given number,' the three given sums being a, b, c he takes the sum of all three numbers together $= x$. The numbers therefore are $x - a$, $x - b$, $x - c$. Whence $3x - (a + b + c) = x$: and $x = \dfrac{a + b + c}{2}$. (2) The most common and characteristic of Diophantus' methods is his use of *tentative assumptions*[2] which is applied in nearly every problem of the later books. It consists in assigning to the unknown a preliminary value which satisfies one or two only of the necessary conditions, in order that, from its failure to satisfy the remaining conditions, the operator may perceive what exactly is required for that purpose.

---

[1] Cf. also I. 16, 18, 23, II. 33, III. 5, 6, 7, 16, IV. 14, 16, 38 etc. Diophantus, of course, does not, in the selected specimen or elsewhere, use a, b, c, or other symbols. He says 'the first, second, third numbers, the product of the first and second' etc. describing in full every expression which does not contain the unknown s. I have occasionally, for shortness, also altered the wording of a problem, by introducing the *given number* or *given ratio* etc. into the *enunciation.*

[2] Nesselmann quotes too many specimens to be here cited. He calls this procedure '*Falscher Ansatz*', but says that it is to be distinguished from the later "so berühmt gewordene *regula falsi* oder *falsa positio*" (mentioned above p. 100 n.), with which it has nothing in common. Both processes seem to me to go *pari passu* up to a certain point. Here is an Italian specimen of the simple 'falsa positio' given by Dean Peacock. 'I buy a jewel and sell it for 50 *lire* (1 *lira* = 100 *soldi*): I make 3½ *soldi* on each *lira* of the original price. What did I give for the jewel?' The operator says: 'Assume that I gave 30 *lire*: then I should have sold it for 31. But, in reality, I sold it for 50. Therefore the original price was $\dfrac{30 \times 50}{31}$.' In Diophantus, however, the original assumption is completely dismissed, when its falsity, and the reason of this, are discovered, and no further use is made of it.

A good example is IV. 9 : 'To find a cube and its root such
that if the same number be added to each, the sums shall also
be a cube and its root.' Here let x be the number added, $2x$
the root, $8x^3$ the cube. Then $8x^3 + x = (3x)^3 = 27x^3$, whence
$19x^2 = 1$. As x is to have a rational value, $19x^2 = 1$ will not suit.
'Now this 19 arises' says Diophantus in effect 'from the difference of $27x^3$ and $8x^3$, or the cubes of $3x$ and $2x$. There is a
difference of 1 between these last coefficients. Let me now
find two numbers x and $x + 1$ such that the difference between
their cubes is a square number. That difference will be
$3x^2 + 3x + 1$. If I assume this to be $= (2x - 1)^2$, I shall find
$x = 7$, and my two numbers are 7 and 8. Now I return to my
original problem. Let x again be the number to be added, $7x$
the root and $343x^3$ the cube. Then $343x^3 + x = (7x + x)^2 = 512x^3$,
whence $169x^2 = 1$ and $x = \frac{1}{13}$.' This example will serve also to
illustrate a third characteristic of Diophantus, viz. (3) *the
use of the symbol for the unknown in different senses*[1]. The
following is a more complicated instance of both methods. In
IV. 17 the problem is 'to find 3 numbers, such that their sum
is a square and that the square of any one of them + the
following number is a square.' The 3 numbers are first taken
as $x - 1$, $4x$ and $8x + 1$, where $(x - 1)^2 + 4x$ and $(4x)^2 + 8x + 1$
are both square numbers. Two conditions are thus satisfied. But
the sum of all 3 numbers, viz. $13x$, must be a square. 'Take
$13x$ equal to x^2 with some square coefficient, e.g. $169x^2$. Then
$x = 13x^2$.' A new use of x is thus introduced and $13x^2$ is substituted for the original x, the numbers now being $13x^2 - 1$,
$52x^2$ and $104x^2 + 1$. A fourth condition remains, viz. that
$(104x^2 + 1)^2 + (13x^2 - 1)$ shall be a square number. Diophantus,
then, takes this expression equal to $x^2 (104x + 1)^2$, finds $x = \frac{55}{52}$,
and substitutes this value in the expression. The use of
'tentative assumptions' leads, again, to another device which
may be called (4) the *method of limits*[2]. This may best be
illustrated by a particular example. If Diophantus wishes to
find a square lying between 10 and 11, he multiplies these

[1] If in any particular case, confusion
is likely, Diophantus alludes to the
first symbol as ὁ ἀόριστος, ἡ δύναμις etc.
by name. Nesselmann cites I. 22, III.
18, IV. 17, 18, VI. 13, 14, 15 etc.

[2] Compare IV. 45, V. 33, VI. 2, 23 etc.

numbers by successive squares till a square lies between the products. Thus between 40 and 44, 90 and 99 no square lies, but between 160 and 176 there lies the square 169. Hence $x^2 = \frac{169}{16}$ will lie between the proposed limits. The method is very neatly used in the following instance. In IV. 34 the problem is 'to divide 1 into two parts, such that if 3 be added to the one part and 5 to the other, the product of the two sums shall be a square.' If one part be $x - 3$, the other is $4 - x$. Then $x(9 - x)$ must be a square. Suppose it $= 4x^2$: then $x = \frac{9}{5}$. But this will not suit the original assumption, since x must be greater than 3 (and less than 4). Now 5 is $4 + 1$: hence what is wanted is to find a number $y^2 + 1$ such that $\dfrac{9}{y^2 + 1}$ is > 3 and < 4. For such a purpose y^2 must be < 2 and $> 1\frac{1}{4}$. "I resolve these expressions into square fractions" says Diophantus and selects $\frac{128}{64}$ and $\frac{80}{64}$ between which lies the square $\frac{100}{64}$ or $\frac{25}{16}$. He then takes $x(9 - x) = \dfrac{25x^2}{16}$ instead of $4x^2$. Sometimes, indeed, Diophantus solves a problem wholly or in part by (5) *synthesis*[1]. Thus IV. 31 is 'To find 4 squares, such that their sum added to the sum of their roots is a given number.' The solution is as follows. "Let the given number be 12. Since a square + its root $+ \frac{1}{4}$ is a square, the root of which *minus* $\frac{1}{2}$ is the root of the first-mentioned square, and since the four numbers added together $= 12$, which *plus* the four quarters $(12 + \frac{4}{4})$ is 13, it follows that the problem is to divide 13 into four squares. The roots of these *minus* $\frac{1}{2}$ each will be the roots of the four squares sought for. Now 13 is composed of two squares 4 and 9: each of which is composed of two squares, viz. $\frac{64}{25}$, $\frac{36}{25}$, $\frac{144}{25}$ and $\frac{81}{25}$. The roots of these, viz. $\frac{8}{5}$, $\frac{6}{5}$, $\frac{12}{5}$ and $\frac{9}{5}$, *minus* $\frac{1}{2}$ each, are the roots of the four squares sought for, viz. $\frac{11}{10}$, $\frac{7}{10}$, $\frac{19}{10}$, $\frac{13}{10}$: and the four squares themselves are $\frac{121}{100}$, $\frac{49}{100}$, $\frac{361}{100}$ and $\frac{169}{100}$." Although it has been said above, and has been sufficiently shown by the foregoing examples, that Diophantus does not treat his problems generally and is usually content with finding any particular numbers which happen to satisfy the conditions of his problems,

---

[1] Compare also III. 16, IV. 32, v. 17, 23 etc.

yet it should be added that he does occasionally attempt (6) such *general solutions*[1] as were possible to him. But these solutions are not often exhaustive because he had no symbol for a general coefficient. Thus in v. 21 'to find 3 numbers, such that each of them shall be a square *minus* 1 and their sum shall be a biquadrate ($\delta\nu\nu\alpha\mu\omicron\delta\acute{\nu}\nu\alpha\mu\iota\varsigma$)' he finds the 3 numbers in the form $x^4 - 2x^2$, $x^2 + 2x$ and $x^2 - 2x$, and adds 'the problem has been solved in general ($\dot{\alpha}\omicron\rho\acute{\iota}\sigma\tau\omicron\iota\varsigma$) terms,' and at the end of IV. 37 (*comp.* also IV. 20) where a similar solution is given he remarks "A solution in general terms is such that the unknown in the expressions for the numbers sought may have any value you please." The problems IV. 20, 37, 39 and 41 are expressly problems for finding general expressions. He solves them by a 'tentative assumption.' For instance IV. 39 is 'To find two general expressions for numbers such that their product *minus* their sum is a given number.' The solution runs as follows : 'The given number is 8. The first number may be taken as x, the second as 3. Then $2x - 3 = 8$, and $x = 5\frac{1}{2}$. Now $5\frac{1}{2}$ is $\frac{11}{2}$, 11 is the given number *plus* the second : and 2 is the second *minus* 1. Hence at whatever value the second number be taken, if I add it to the given number and divide the sum by the second number *minus* 1, I get the first number. Suppose the second number to be $x + 1$: then $\dfrac{x+9}{x}$ is the first.' These general solutions for *two* numbers are immediately afterwards (IV. 21, 38, 40, 42) used in problems of a similar character for *three* numbers, of which two are first found in general terms and then the third by a determination of x in the usual manner. Sometimes, however (e.g. IV. 26 and frequently in the 6th book[1]), a problem after being solved by particular numbers (as 40, 27, 25) is solved generally (by $40x$, $27x$, $25x$ in IV. 26). But though the defects in Diophantus' proofs are in general due to the limitation of his symbolism, it is not so always. Very frequently indeed Diophantus introduces into a solution (7) *arbitrary conditions and determinations* which are not in the

---

[1] VI. 3, 4, 6, 7, 8, 9, 10, 11, 13, 15, 17. See Nesselmann, pp. 418—421. The problems of the VIth Book deal almost entirely with 'right-angled triangles', i.e. with sets of three numbers, such that $x^2 + y^2 = z^2$.

problem. Of such "fudged" solutions, as a schoolboy would call them, two particular kinds are very frequent. Sometimes an unknown is assumed at a determinate value[1]: as in I. 14 'To find two numbers whose product is three times their sum,' where Diophantus, without a word of apology, takes the first number as x, the second as 12. Sometimes a new condition is introduced, as in VI. 19, where, two numbers being sought such that the cube of one is greater by 2 than the square of the other, Diophantus takes the numbers as $x-1$ and $x+1$, thus introducing a condition that the difference between the two numbers shall be 2. A very remarkable case of the latter kind occurs in IV. 7 where the problem would be, in our symbolism, to find three numbers, a^3, b^2, c^2, so that $a^3 + c^2$ shall be a square, $b^2 + c^2$ a cube. Diophantus begins his solution by taking $b^2 + c^2 = a^3$. Arbitrariness of this kind is of course different from the cases in which Diophantus merely takes a particular number, where any other would evidently do as well. In the latter, he is urged by the defects of his symbolism: in the former he is urged only by the want of a solution to a particular problem: the difference is one of kind and not of degree.

72. From the very brief survey of the *Arithmetica*, it will be obvious to the reader that it is a work of the utmost ingenuity but that it is deficient, sometimes pardonably, sometimes without excuse, in generalization. The book of *Porismata*, to which Diophantus sometimes refers, seems on the other hand to have been entirely devoted to the discussion of general properties of numbers. It is three times expressly quoted in the *Arithmetica*. These quotations, when expressed in modern symbols, are to the following effect. In V. 3 the *porism*[2] is cited : 'If $x + a = m^2$, $y + a = n^2$, and $xy + a = p^2$, then $m = n + 1$': in V. 5: 'If three numbers x^2, $(x+1)^2$, $4x^2 + 4x + 4$, be taken, the

---

product of any two + their sum, *or* + the remaining number, is a square: in v. 19 'the difference between two cubes may be resolved into the sum of two cubes.' Of all these propositions he says ἔχομεν ἐν τοῖς πορίσμασιν, 'we find it in the Porisms'; but he cites also a great many similar propositions without expressly referring to the *Porisms*. These latter citations fall into two classes, the first of which contains mere *identities*, such as the algebraical equivalents of the theorems in Euclid II. For instance in Diophantus II. 31, 32, and IV. 17 it is stated, in effect, that $x^2 + y^2 \pm 2xy$ is always a square (Eucl. II. 4): in II. 35, 36, III. 12, 14 and many more places it is stated that $\left(\dfrac{a-b}{2}\right)^2 + ab$ is always a square (Eucl. II. 5) etc[1]. The other class contains general propositions concerning the resolution of numbers into the sum of two, three or four[2] squares. For instance, in II. 8, 9 it is stated 'Every square number' (in II. 10 'every number which is the sum of two squares') 'may be resolved into the sum of two squares in an infinite number of ways': in v. 12 'A number of the form $(4n + 3)$ can never be resolved into two squares,' but 'every prime number of the form $(4n + 1)$ may be resolved into two squares': in v. 14 'A number of the form $(8n + 7)$ can never be resolved into three squares.' It will be seen that all these propositions are of the general form which ought to have been but is not adopted in the *Arithmetica*. We are therefore led to the conclusion that the Porismata, like the pamphlet on Polygonal Numbers, was a synthetic and not an analytic treatise. It is open, however, to anyone to maintain the contrary, since no proof of any *porism* is now extant.

With Diophantus the history of Greek arithmetic comes to an end. No original work, that we know of, was done afterwards. A few scholiasts appear, such as *Eutocius* of

---

[1] Nesselmann, pp. 446—450, cites 10 such identities, most of which are used more than once by Diophantus.

[2] In IV. 31, 32, v. 17 Fermat thought that Diophantus was using a proposition 'Every number whatever can be resolved into four squares,' but Nesselmann (p. 460—1) inclines to the opinion that Diophantus did not know this proposition generally but was relying on the known properties of certain determinate numbers.

Askalon (*cir.* A.D. 550) who wrote on Archimedes, *Asclepius* of Tralles and his pupil *John Philoponus* (*cir.* A.D. 650) who wrote on Nicomachus, and the unknown commentators who have added *lemmas* to the arithmetical books of Euclid; but though there is evidence that the old mathematicians were still studied in Athens and Alexandria and elsewhere, no writer of genius appears and the history of arithmetic and algebra is continued henceforth by the Indians and Arabs.

PART III. GEOMETRY.

CHAPTER V.

PRE-HISTORIC AND EGYPTIAN GEOMETRY.

73. THE earliest history of Geometry cannot be treated in the same way as that of Arithmetic. There is not for the former, as there is for the latter, a nomenclature common to many nations and languages; and the analysis of a geometrical name in any one language leads only to the discovery of a root-syllable which is common to many very different words and to which only the vaguest possible meaning may be assigned. Nor is any assistance, so far as I know, furnished by travellers among savage and primitive races. Arithmetical operations are matters of such daily necessity that every general arithmetical proposition, of which a man is capable, is pretty certain to be applied in his practice and to attract attention : but a man may well know a hundred geometrical propositions which he never once has occasion to use, and which therefore escape notice. I have sought, in vain, through many books which purport to describe the habits and psychology of the lower races, for some allusion to their geometrical knowledge or for an account of some operations which seem to imply geometrical notions. One would be glad, for instance, to learn whether savages anywhere distinguish a right angle from an acute. Have they any mode of ascertaining whether a line is exactly straight or exactly circular? Do they by name distinguish a square from any other rectilineal figure? Do they attach any mysterious properties to perpendicularity, angular symmetry, etc.? We

have, at present, no answer to these and similar questions and
there is consequently a gap in the history of geometry which
no writer, since Herodotus, has attempted to fill up. Where
this gap occurs will be seen from the following remarks[1].

74. Geometry is the science of space and investigates the
relations existing between parts of space, whether linear, super-
ficial or solid. Some of these relations are obviously capable
of arithmetical expression, so soon as units of length, area
and solid contents are selected. For the first of these, some
measurement of the human body has universally served: the
finger-breadth, palm, span, foot, ell, cubit, fathom have been
and are, all the world over, the units of length. Distances
too great to be exactly ascertained have also generally been
measured by some reference to human capacity, such as 'a
stone's throw,' 'within shouting distance' (ὅσσον τε γέγωνε
βοήσας as Homer has it) 'a day's journey' etc. But the human
body does not furnish any convenient unit of area or solid
contents. Large areas and volumes, like long distances, seem in
primitive times to have been described roughly by reference to
labour; a field, for instance, is a *'morning's* work' (Ger. *morgen*)
or a day's work for a yoke of oxen (Lat. *jugerum*): a barn
contains so many loads: but we do not know how small areas
and volumes were described[2]. Now the oldest exact geometry,
of which we know anything, is concerned almost entirely with
the measurement of various areas or solids by reference to a
square or a *cubical* standard unit. The selection of these particular shapes, out of several which *prima facie* would serve

---

[1] The modern writers on the history
of Greek geometry, whom I have
chiefly consulted, are the following:
Bretschneider, *Die Geometrie und die
Geometer vor Eukleides* (Leipzig,
1870): Hankel, *Zur Geschichte der
Mathematik* (Leipzig, 1875): Dr G.
J. Allman, *Greek Geometry from Thales
to Euclid* in *Hermathena* (Dublin) Nos.
v. and vii. (Vols. iii. and iv. 1877 and
1881), Cantor, *Vorlesungen über Geschichte der Mathematik* (Leipzig 1880),
Prof. M. Chasles, *Aperçu Historique*

sur l'origine etc. de Géométrie (Paris,
1837 and 1875. Both editions
are identical). Bretschneider and the
rest convict Montucla (*Hist. des
Mathém.* 1758) of so many mistakes
in his history of Greek mathematics,
that I have seldom referred to him.
All these authors will in future be
cited generally by name only.

[2] Small volumes were perhaps described by *weight,* as conversely Gr.
δραχμή, properly a 'handful,' came to
be a standard of weight.

just as well, implies a long period of observation and consideration. How did this observation begin ? It must be assumed,
of course, that mankind, like birds and bees, were from the first
familiar with, and able to distinguish, the many symmetrical
figures which occur in nature and that they knew generally that
suspended strings all hang alike and that all posts, to be stable,
must be stuck in the ground in a particular manner[1]. But the
question is, how they were induced to examine the properties
of these figures, to investigate the peculiarities of this particular angle. Herodotus says (II. 109) that Sesostris (Ramses
II. abt. 1400 B. C.) divided the land of Egypt into equal
rectangular (or square) plots for the purpose of more convenient
taxation; that the annual floods, caused by the rising of the
Nile, often swept away portions of a plot, and that surveyors
were in such cases appointed to assess the necessary reduction
in the tax. 'Hence in my opinion' (δοκέει δέ μοι) he goes on
'arose geometry, and so came into Greece.' The same account
is elsewhere[2] repeated as legendary, without reference to
Herodotus, and it is not unlikely to be an Egyptian tradition
which Herodotus appropriated. This history of geometry is
generally scouted[3], but I think it perhaps contains a germ of
truth. Suppose that lands were originally measured roughly by
their produce or by the labour which they demanded. Then,
I imagine, the first attempt at exact *numerical* calculation of
areas was merely the measurement of the *periphery*, a method
which was useful enough so long as the areas were of approximately the same shape. But in process of time areas of one

---

[1] It may be supposed that attention
would be called to the right angle because it is, as Aristotle calls it, the
'angle of stability.' But men might
well recognise a right angle in the
vertical plane without recognising it in
the horizontal. Compare the remarks
of Œnopides, an early Greek geometer,
quoted below p. 147.

[2] Heron Alex. *Rell.* ed. Hultsch, p.
138. Diodorus Sic. I. 69, and 81.
Strabo, XVII. c. 3 (Meineke's ed. p.

1098). The quotations are printed
in full in Bretschneider, *Geometrie etc.
vor Eukleides*, pp. 7—9.

[3] Prof. de Morgan quotes (Art.
Geometry in *Penny Cyclop.*) from "an
obsolete course of mathematics" the
following lines :
'To teach weak mortals property to
  scan
Down came geometry and formed a
  plan.'

shape were exchanged for areas of another shape, and it was then for the first time discovered that figures of equal periphery are not necessarily of the same area[1]. A man who had had a square field, for instance, exchanged it for a rhombus of equal periphery, but found that he got less produce than before. A discovery of this kind would at once call attention to angles and suggest the propriety of establishing a unit of area. The utility of the square unit might have been established by long experience or have been suggested by the aspect of stone or brick-buildings subsequent to the Cyclopean era of architecture.

75. But it is needless to dwell longer on a theory which must, at present, remain purely conjectural. Whatever opinion be ultimately adopted concerning the first steps in geometry, it will always remain true that the word 'geometry' ($\gamma\epsilon\omega\mu\epsilon\tau\rho\acute{\iota}\alpha$) means 'land-measurement[2],' that the Egyptians gave this science to the world and that among the Egyptians, from first to last, it answered to its name and was confined almost entirely to the practical requirements of the surveyor.

The work of Ahmes, which was so frequently cited in the earlier pages of this book, contains, beside sums in arithmetic, a great many geometrical examples which deserve to be cited[3].

Immediately after the examples of *Tunnu-* or difference-calculation cited above (p. 19), Ahmes proceeds to calculate the contents of barns and other similar receptacles, of which unfortunately we do not know the shape, so that the necessary

---

[1] The erroneous assumption that figures of equal periphery are of the same area appears in classical authors. Thucydides (VI. 1) estimates the area of Sicily by the time spent in circumnavigating it. Polybius (IX. 21) mentions that there are some people who cannot understand that camps of the same periphery may not be the same size. Quintilian (I. 10, 39 sqq.) points out the fallacy as one that easily deceives the vulgar. So also Proclus (ed. Friedlein, p. 237). See Cantor, pp. 146—7.

[2] So in Egyptian *hunu* = 'land measurer,' 'geometer,' v. Brugsch's *Hierogl. Demot. Wörterbuch*, p. 967.

[3] It is curious that all the geometrical matter occurs in the middle of the arithmetical and that the calculation of solid contents precedes the calculation of areas. From this it may perhaps be inferred that the geometrical propositions known to Ahmes were empirically obtained and that he was really interested only in the arithmetical problems which they suggested.

clue to the interpretation of the examples is wanting[1]. For the examples in plane geometry (Nos. 49—55), however, the figures given by Ahmes are sufficient, save in a few cases[2] where solutions and figures are given which have no connection whatever with the problems to which they are appended. The rectilineal figures of which Ahmes calculates the areas are the square, oblong, isosceles triangle and isosceles parallel-trapezium (regarded as part of an isosceles triangle cut by a line parallel to the base). As to the last two, the areas which he finds are incorrect. Thus in Ex. 51 he draws an isosceles triangle of which the sides measure 10 *ruths*, the base 4 *ruths*. He multiplies the side by half the base and finds the area at 20 square *ruths*. The real area is 19·6. Similarly in no. 52 the area of an isosceles parallel-trapezium is taken to be 100 square *ruths*, instead of 99·875[3]. The errors in these cases are small but are not on that account the less suggestive. The area of a circle is found (in no. 50) by deducting from the diameter $\frac{1}{9}$th of its length and squaring the remainder. Here π is taken $= (\frac{16}{9})^2 = 3\cdot1604\ldots\ldots$, a very fair approximation.

76. Lastly, the papyrus contains (nos. 56 to 60)[4] some examples which seem to imply a rudimentary trigonometry. In these (except the last) the problem is to find the *uchatebt*,

---

[1] Eisenlohr pp. 93—117, Nos. 41—48. The contents of all the barns are obtained in this way. Of three given linear measurements two are multiplied together and the product is multiplied by *one-and-a-half* of the third. But it does not appear whether the first product is the area of the top or the bottom or the side of the barn or of what line the third given number is the measure.

[2] E.g. nos. 53, 54. Eisenlohr pp. 118—133.

[3] Eisenlohr, pp. 125, 127—129. If in an isosceles triangle the equal sides be a, a, the base b, the area is

$$\frac{b}{2} \sqrt{a^2 - \frac{b^2}{4}}.$$

If in an isosceles parallel trapezium

the equal sides be a, a, the parallel sides b_1, b_2, the area is

$$\frac{b_1 + b_2}{2} \times \sqrt{a^2 - \left(\frac{b_2 - b_1}{2}\right)^2}.$$

Ahmes makes the areas

$$\frac{ab}{2} \text{ and } \frac{a(b_1 + b_2)}{2}$$

respectively, neglecting the difficult square roots.

[4] Eisenlohr, pp. 134—149. On the use of these *seqt* calculations, see below p. 142. In Ahmes, of course, they are only exercises in arithmetic.

piremus or *seqt* of a pyramid or obelisk. *Uchatebt* apparently means 'search for the base' and is clearly a line which has something to do with the base: *piremus* apparently means 'result (or issue) of the saw' and is a line which can be obtained only by section of the pyramid: *seqt* apparently means 'relation' or 'like-making,' and is a number. For the purposes of these problems, the *uchatebt* is always halved. By means of these clues, Eisenlohr and Cantor have very ingeniously explained the purport of Ahmes' examples. In the pyramid figured the *uchatebt* may be either $2DE$ (i.e. DL) or $2BE$ (i.e. BH): the *piremus* may be either AD or AB, according as the pyramid is cut parallel with the base-line or along the diagonal of the base-square[1]. The problems which Ahmes proposes are always of the form 'Given any two of the

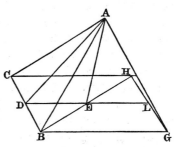

uchatebt, *piremus* and *seqt*, to find the third,' and the solution is always obtained from the fact that the *seqt* is half the *uchatebt* divided by the *piremus*. In the figure above given, therefore, the *seqt* is $\dfrac{DE}{AD}$ or $\dfrac{BE}{AB}$, i.e. *cos ADE* or *cos ABE*. The actual *seqt* given by Ahmes is, in one case, 0·72, in three more 0·75. These are the cosines of the angles 43° 56′ 44″ and 41° 24′ 34″ respectively. The angle *ABE* in most existing pyramids is nearly of these measurements. Further, these cosines of *ABE* correspond to angles of 53° 44′ 7″ and 51° 16′ 40″ respectively at *ADE* and these again are nearly the slopes of most existing pyramids[2]. This explanation being premised, the problems

---

[1] It cannot be that the *uchatebt* is the *visible* base-line, the *piremus* the sloping edge; for it is a property of pyramids upon square bases, such as Ahmes seems to be considering, that half the square of the base-line can never be greater than the square of the sloping edge. But in Ahmes' first example the *uchatebt* is 360 ells, the

piremus 250. Eisenlohr, p. 135.

[2] According to Piazzi Smith the slopes of the largest pyramid at Gizeh are between 51° 49′ and 51° 51′. If the face of a pyramid on a square base were equal to the square of the height, the slope would be 51° 50′. If the base were equal to a circle of which the height is the radius, the

themselves may here be given. In no. 56 it is required to find the *seqt* (S) of a pyramid, whereof the *uchatebt* (U) is 360 ells, the *piremus* (P) is 250[1]. The answer is $\frac{180}{250}$. Reducing ells to palms, (1 *ell* = 7 *palms*) S is $5\frac{1}{25}$ palms, that is, there are $5\frac{1}{25}$ palms in $\frac{U}{2}$ to every ell in P. In no. 57 U is 140 ells: *seqt* is $5\frac{1}{4}$ palms. Find P. The answer is $93\frac{1}{3}$ ells. In no. 58 the dimensions of U and P are as in no. 57. Find S. In no. 59 new dimensions of U and P are given, but S is again found at $5\frac{1}{4}$ palms. No. 60 does not relate to a pyramid at all. It applies to an obelisk of which the *height* (*qāi*) is 30 ells: the base-line (*senti*) 15. The *seqt* here is determined at 4, which is the *tangent* of the angle included between the side and the base-line of a triangular face. The figures appended are very ill-drawn to scale and are all furnished with a pedestal: e.g. the figure to no. 58 is like

77. One or two glimpses of Egyptian geometry are obtained also at a far later time. The most interesting is furnished by the etymology of a Greek word. The philosopher Democritus (*cir.* B.C. 460—370) is quoted by Clement of Alexandria[2] as saying, "In the construction of plane figures (lit. composition of lines) with proof no one has yet surpassed me, not even the so-called *Harpedonaptae* of Egypt." It was evident, of course, that these *Harpedonaptae*[3] were famous geometers, but Prof. Cantor has first pointed out that their name is compounded of two Greek words and means simply

---

slope would $= 51^0$ 51'. Mr. Petrie's measurements (*Pyramids and Temples of Gizeh*, 1883, pp. 42, 97, 112) do not differ substantially from Piazzi Smith's.

The Egyptian ell, according to Lepsius, is O^m· 525. Eisenlohr p. 94.

[2] *Strom.* I. p. 357 (Potter's ed.) γραμμέων συνθεσιος μετὰ ἀποδείξιος οὐδείς κώ με παρήλλαξεν, οὐδ' οἱ Αἰγυπτίων καλεόμενοι Ἀρπεδονάπται.

[3] See also p. 318.

'rope-fasteners' or 'rope-stretchers.' He explains their functions in the following way[1]. There is no doubt that the Egyptians were very careful about the exact orientation of their temples and other public buildings. But inscriptions seem to shew that only the N. and S. line was drawn by actual observation of the stars. The E. and W. line, therefore, was drawn at right-angles to the other. Now it appears, from the practice of Heron of Alexandria and of the ancient Indian and probably also the Chinese geometers, that a common method of securing a right angle between two very long lines was to stretch round three pegs a rope measured into three portions, which were to one another as 3 : 4 : 5. The triangle thus formed is, of course, right-angled. Further, the operation of 'rope-stretching' is mentioned in Egypt, without explanation, at an extremely early time (Amenemhat I.) If this be the correct explanation of it, then the Egyptians were acquainted, 2000 years B.C., with the geometrical propositions familiar to us as Euclid I. 47, 48[2], or with one particular case of them.

78. It will readily be supposed that the Egyptians, who had so early invented so many rules of practical geometry, could not fail in process of time to make many more discoveries of the same kind, and thus be led to geometrical science. But it appears that in Egypt land-surveying, along with writing, medicine and other useful arts, was in the monopoly of the priestly caste[3]; that the priests were the slaves of tradition, and that, in their obstinate conservatism, they were afraid to alter the rules or extend the knowledge of their craft. Of their medicine, Diodorus (I. 82) expressly relates that, even in his day, the Egyptian doctors used only the recipes contained in the ancient sacred books, lest they should be accused of manslaughter in case the patient died. Geometry seems to have been treated with similar timidity. The temple of Horus at Edfu,

---

[1] Vorles. I. pp. 55—57 (Egyptian Temple inscriptions etc.) pp. 324—5 (Heron): pp. 540—542 (the Çulva-sûtras): pp. 580—581 (Chinese 'Figur des Seiles'). Compare also Hankel, p. 83.

[2] Compare Plutarch, *De Is. et Osir.* c. 56.

[3] Isocrates, *Busiris*, c. 9. Aristotle, *Metaph.* I. 1. Diodor. Sic. I. cc. 69, 81, 82. With c. 82 comp. Arist. *Pol.* III. 15.

in Upper Egypt, bears some inscriptions describing the lands
which formed the endowment of the priestly college attached
to the temple. These lands were given by King Ptolemy XI.
(Alexander I.) who reigned B.C. 107—88, but the geometrical
description of them, made 200 years since Euclid died, is un-
worthy of Ahmes himself. It will be remembered (supra, p. 127
n. 3) that Ahmes uses the incorrect formulae $\dfrac{ab}{2}$ and $\dfrac{a\,(b_1 + b_2)}{2}$
for the areas of an isosceles triangle and an isosceles parallel-
trapezium. The Edfu inscriptions retain both these, but they
apply the second for finding the areas of trapezia of every
kind, no matter how irregular[1]. The dulness, or laziness, of
this proceeding is monumental in more senses than one. It
is obvious that the Greek mathematicians had by this time no
more to learn from the native Egyptians, and we may therefore
leave Egyptian geometry with a quiet conscience.

79. It remains only to cite the universal testimony of
Greek writers, that Greek geometry was, in the first instance,
derived from Egypt, and that the latter country remained for
many years afterwards the chief source of mathematical teaching.
The statement of Herodotus on this subject has already been
cited. So also in Plato's *Phaedrus* Socrates is made to say
that the Egyptian god Theuth first invented arithmetic and
geometry and astronomy. Aristotle also (*Metaph.* I. 1) admits
that geometry was originally invented in Egypt, and Eudemus
(see *post* pp. 134, 135) expressly declares that Thales studied
there. Much later Diodorus (B.C. 70) reports an Egyptian
tradition that geometry and astronomy were the inventions
of Egypt, and says that the Egyptian priests claimed Solon,
Pythagoras, Plato, Democritus, Œnopides of Chios and Eudoxus
as their pupils[2]. Strabo gives further details about the
visits of Plato and Eudoxus. He relates that they came to
Egypt together, studied there thirteen years, and that the

----

[1] Hankel, pp. 86, 87. Cantor, pp.
60, 61. In the case of a trapezium
with 4 *unequal* sides (a, b, c, d) the
formula is $\dfrac{a+b}{2}$, $\dfrac{c+d}{2}$. The Edfu inscriptions use this also for *triangles*,
the dimensions here being given e.g.
as "nothing by 5, and 17 by 17."

[2] Diodorus I. cc. 69, 96.

houses where they lived were still shown in Heliopolis[1]. Later
writers, of course, have the same tale, and it is needless to
collect further evidence. Beyond question, Egyptian geometry,
such as it was, was eagerly studied by the early Greek philosophers, and was the germ from which in their hands grew that
magnificent science to which every Englishman is indebted for
his first lessons in right seeing and thinking[2].

80. A word or two should be added also in this place concerning Babylonian mathematics[3]. The Chaldees, at a time
almost contemporaneous with Ahmes, but whether independently
or not cannot now be ascertained, had made advances, similar to
the Egyptian, in arithmetic and geometry, and were especially
busy with astronomical observations. It seems that they had
divided the circle into 360 degrees, and that they had obtained
a fairly correct determination of the ratio of the circumference
of a circle to its diameter. They used, also, in arithmetic, as
has been stated above, a sexagesimal notation, which the Greeks
afterwards adopted for astronomical purposes. Herodotus[4] expressly states that the *polos* and *gnomon* (two kinds of sundials)
and the twelve parts of the day were made known to the Greeks
from Babylon. Much of the trigonometry and spherical geometry
of the later Greeks may also have been directly derived from
Babylonian sources.

Finally, it should be remembered that however scanty
geometrical theories may have been both in Egypt and Chaldea,
a very great variety of geometrical figures was used in both

---

[1] Strabo, XVII. 1. Meineke's ed.
p. 1124. Bretschneider (pp. 33, 34),
however, thinks that, before Plato's
time, Greek geometry had so far outstripped the Egyptian that no Greek,
after about 450 B.C., would have visited Egypt for the purpose of learning
geometry. He supposes therefore that
Plato and Eudoxus went to Egypt
to learn astronomy, as in fact the
passage of Strabo, above quoted, suggests.

[2] Diodorus I. 98 says also that
Telecles and Theodorus, the most

famous of the ancient Greek sculptors,
studied in Egypt, as did their father
Rhoecus, who designed the labyrinth
in Lemnos. (Bretschneider, p. 24.)

[3] See Cantor chap. III. pp. 67—94.

[4] II. 109. Pliny (*H. N.* II. 76) attributes the introduction of the gnomon
to Anaximenes, Suidas to Anaximander (*s. v.*). Diogenes L. (II. 1) and
Suidas both attribute a ὡροσκοπεῖον,
probably the *polos*, to the latter. On
the *gnomon* and *polos* see below p.
145 *n.*

countries for mural decoration and other ornamental purposes[1]. To a Greek, therefore, who had once acquired a taste for geometry, a visit to Egypt or Babylon would reveal a hundred geometrical constructions which, on inspection, suggested new theorems and invited scientific inquiry.

[1] See Cantor, pp. 58, 59, 89, 90.

CHAPTER VI.

81. An elaborate history of Greek geometry before Euclid was written by Eudemus[1], the pupil of Aristotle, who lived about 330 B.C. The book itself is lost but is very frequently cited by later historians and scholiasts, and it may be suspected also that many notices, not directly ascribed to it, were taken from its pages. Proclus, the scholiast to Euclid, who knew the work of Eudemus well, gives a short sketch of the early history of geometry, which seems unquestionably to be founded on the older book. The whole passage, which proceeds from a competent critic, and which determines approximately many dates of which we should otherwise be quite ignorant, may be here inserted verbatim by way of prologue. It will be cited hereafter as "the Eudemian summary." It runs as follows[2]:

"Geometry is said by many to have been invented among the Egyptians, its origin being due to the measurement of plots of land. This was necessary there because of the rising of the

---

[1] Diog. Laert. v. c. 2, n. 13 (ed. Huebner, I. pp. 347, 348), attributes to Theophrastus, another pupil of Aristotle, contemporary with Eudemus, a history of geometry in 4 books, of astronomy in 6, and of arithmetic in 1 book. Bretschneider (p. 27) is not inclined to the general opinion that Diogenes has here confused Theophrastus with Eudemus.

[2] Procli Diadochi *Comm. in primum Eucl. Elem. librum*, ed. Friedlein

(Leipzig, 1873) pp. 64 sqq. This work will be cited in future simply as 'Proclus.' Of the Eudemian summary, the original Greek is printed also by Bretschneider (pp. 27—31), with a (not very exact) German translation. A pretty close paraphrase is given by Prof. de Morgan in art. *Eucleides* of Smith's *Dic. of G. and R. Biography*, and another by Dr Allman in *Hermathena* (Dublin), no. v. for 1877, p. 160 sqq.

Nile, which obliterated the boundaries appertaining to separate owners. Nor is it marvellous that the discovery of this and the other sciences should have arisen from such an occasion, since everything which moves in development will advance from the imperfect to the perfect. From mere sense-perception to calculation, and from this to reasoning, is a natural transition[1]. Just as among the Phœnicians, through commerce and exchange, an accurate knowledge of numbers was originated, so also among the Egyptians geometry was invented for the reason above stated.

Thales first went to Egypt, and thence introduced this study into Greece. He discovered much himself, and suggested to his successors the sources of much more: some questions he attacked in their general form, others empirically[2]. After him Mamercus[3], the brother of the poet Stesichorus, is mentioned as having taken up the prevalent zeal for geometry: and Hippias of Elis relates that he obtained some fame as a geometer. But next Pythagoras changed the study of geometry into the form of a liberal education, for he examined its principles to the bottom and investigated its theorems in an immaterial and intellectual manner. It was he who discovered the subject of irrational quantities and the composition of the cosmical figures[4]. After him Anaxagoras of Clazomenae touched upon

---

[1] The text (ed. Friedlein) is ἐπειδὴ πᾶν τὸ ἐν γενέσει φερόμενον ἀπὸ τοῦ ἀτελοῦς εἰς τὸ τέλειον πρόεισιν. ἀπὸ αἰσθή- σεως οὖν εἰς λογισμὸν καὶ ἀπὸ τούτου ἐπὶ νοῦν ἡ μετάβασις γένοιτο ἂν εἰκότως. Both sentences are extremely obscure. The second, I should think, represents a chapter of Eudemus, in which the history of geometry was exhibited nearly as I have shown it in preceding pages. A pupil of Aristotle might well have adopted the evolutionary hypothesis here suggested. On the other hand, λογισμὸς does not necessarily mean 'arithmetical calculation' and νοῦς ought not to mean 'reasoning.' Dr Allman translates the first by 'reflection,' the second by 'know-

ledge,' which is even less permissible. Proclus, it should be remembered, was a neo-Platonist and addicted to hazy phraseology.

[2] Prof. de Morgan translates "attempting some in a general manner (καθολικώτερον), and some in a perceptive or sensible manner (αἰσθητικώ- τερον)." Dr Allman gives "in a more intuitional or sensible manner" for the last word.

[3] So Friedlein, other edd. have Ameristus or Mamertinus.

[4] That is, the five regular solids, the tetrahedron, cube, octahedron, eicosa- hedron, and dodecahedron, which were supposed by the Pythagoreans to be the primary forms of the matter of

many departments of geometry, as did Œnopides of Chios, who was a little younger than Anaxagoras. Plato mentions them both in his 'Rivals,' as having won fame in mathematics[1]. Hippocrates of Chios, next, who discovered the quadrature of the lune, and Theodorus of Cyrene became distinguished geometers, indeed Hippocrates was the first who is recorded to have written 'Elements.' Plato, who followed him, caused mathematics in general, and geometry in particular, to make great advances, by reason of his well-known zeal for the study, for he filled his writings with mathematical discourses, and on every occasion exhibited the remarkable connexion between mathematics and philosophy. To this time belong also Leodamas the Thasian and Archytas of Tarentum and Theaetetus of Athens, by whom mathematical inquiries were greatly extended, and improved into a more scientific system. Younger than Leodamas were Neocleides and his pupil Leon, who added much to the work of their predecessors: for Leon wrote an 'Elements' more carefully designed, both in the number and the utility of its proofs, and he invented also a *diorismus* (or test for determining) when the proposed problem is possible and when impossible. Eudoxus of Cnidus, a little later than Leon and a student of the Platonic school, first increased the number of general theorems, added to the three proportions three more, and raised to a considerable quantity the learning, begun by Plato, on the subject of the (golden) section[2], to which he applied the analytical method. Amyclas of Heraclea, one of Plato's companions, and Menaechmus, a pupil of Eudoxus and a contemporary of Plato, and also Deinostratus, the brother of Menaechmus, made the whole of geometry yet more perfect. Theudius of Magnesia made himself distinguished as well in other branches of philosophy as also in mathematics; composed a very good book of 'Elements,' and made more general propositions which were confined to particular cases[3]. Cyzicenus

---

which the universe is made. Timaeus (in Plato *Tim.* 53 c) says that fire consists of tetrahedrons, air of octahedrons, earth of cubes, water of eicosahedrons, and the dodecahedron is the shape of the universe.

[1] *Amatores*, c. 1, 132 A.

[2] The cutting of a line in extreme and mean ratio.

[3] πολλὰ τῶν ὁρικῶν καθολικώτερα ἐποίησεν.

of Athens also about the same time became famous in other branches of mathematics, but especially in geometry. All these consorted together in the Academy and conducted their investigations in common. Hermotimus of Colophon pursued further the lines opened up by Eudoxus and Theaetetus, and discovered many propositions of the 'Elements' and composed some on *Loci*. Philippus of Mende, a pupil of Plato and incited by him to mathematics, carried on his inquiries according to Plato's suggestions and proposed to himself such problems as, he thought, bore upon the Platonic philosophy."

"Those who have written the history of geometry," Proclus continues, "have thus far carried the development of this science. Not much later than these is Euclid, who wrote the 'Elements,' arranged much of Eudoxus' work, completed much of Theaetetus's, and brought to irrefragable proof propositions which had been less strictly proved by his predecessors."

82. To this extract should be added another, which supplies a very valuable criticism on the style of the early Greek geometers. Eutocius, at the beginning of his commentary on the Conics of Apollonius (p. 9, Halley's edn.), quotes from Geminus, an excellent mathematician of the first century B.C., the following remarks[1]:

"The ancients, defining a cone as the revolution of a right-angled triangle about one of the sides containing the right angle, naturally supposed also that all cones are *right* and there is only one kind of section in each—in the *right-angled* cone the section which we now call a parabola, in the *obtuse-angled* a hyperbola, and in the *acute-angled* an ellipse. You will find the sections so named among the ancients. Hence just as they considered the theorem of the two right angles for each kind of triangle, the equilateral first, then the isosceles, and lastly the scalene, whereas the later writers stated the theorem in a general form as follows, 'In every triangle the three interior angles are equal to two right angles[2],' so also with the conic

---

[1] The Greek is given also by Bretschneider, pp. 13, 14.

[2] Compare with this Aristotle, who says (*Anal. Post.* I. 5, p. 74, A. 17)

"The proposition that the terms of a proportion may be taken *alternando*, was formerly proved separately for numbers, lines, volumes, times, though

sections, they regarded the so-called 'section of a right-angled cone' in the right-angled cone only, supposed to be cut by a plane perpendicular to one side of the cone : and similarly the sections of the obtuse-angled and acute-angled cones they exhibited only in such cones respectively, applying to all cones cutting planes perpendicular to one side of the cone......But afterwards Apollonius of Perga discovered the general theorem that in every cone, whether right or scalene, all the sections may be obtained according to the different directions in which the cutting plane meets the cone." " This," adds Eutocius, "is what Geminus says in the 6th Book of his General View of Mathematics (μαθημάτων θεωρία)." The two extracts here quoted are our main clues to the history of geometry before Euclid. The first gives us the names of the leading geometers, the order of their appearance and a brief statement of their services. The second is valuable in enabling us to guess at the style in which a particular proposition would probably be treated at a given date. The sources from which further details may be obtained are generally very late in date and very meagre in information. They often ascribe the same proposition to different persons or different modes of proving the same proposition to the same person, or are silent altogether about modes of proof. The early history of Greek geometry must, therefore, be reconstructed largely by inference, and it is obvious that to this process the Eudemian summary and the authoritative statement of Geminus are of the greatest assistance.

(b)   Thales and the Ionic School.

83.   Thales, the acknowledged founder of Greek mathematics and philosophy, was born about B.C. 640 at Miletus, the chief city of the Ionian coast, and died at the same place

---

it might have been proved for all of them at once : but because these things are not called by one name and differ in kind, they were treated separately. But now it is proved generally " etc.

Hankel, pp. 114, 115. This is the passage cited above (p. 105 n.) as evidence that Aristotle knew the mathematical value of the alphabetical symbols which he introduced.

about B.C. 542[1].   He was apparently of Phœnician descent[2] but
probably not, as Diogenes relates, of Phœnician parentage, for
the names of his parents, Examius and Cleobuline, are good
enough Greek.   Many authorities concur in stating that he
was, in early life at least, engaged in commerce, for which he
seems to have had great aptitude[3].   Aristotle illustrates this by
a tale that one winter, when the stars promised an abundant
crop of olives, Thales at once secured by contract all the oil-
presses, and made, in the following autumn, a large profit by
lending these necessary implements.   It may be that he went to
Egypt for mercantile purposes, and there learnt in his leisure
the mathematical and other knowledge which he subsequently
introduced among the Greeks.   According to Plutarch, he was
somewhat advanced in years ($\pi\rho\epsilon\sigma\beta\upsilon\tau\epsilon\rho\sigma\varsigma$) when he returned
to Miletus.   According to other authorities[4] he was old, or had
given up an active share in political life, when he took to those
philosophical inquiries for which he is now remembered.   At
any rate the striking achievement which made his fame in his
own day did not occur till his later years.   He announced
beforehand a solar eclipse, which in fact took place at least in
the year predicted.   It happened on May 28th, 585 B.C. during
a great pitched battle between the Medes and the Lydians[5].

[1] The main facts of his life are given
by Diogenes Laertius (I. 1. nn. 1, 3, 6,
10, Huebner's ed. pp. 14, 16, 17, 24),
who cites Apollodorus, as authority for
the birth of Thales in the 35th Olym-
piad, and Socrates, for his death in
the 58th.

[2] Herod. I. c. 170.

[3] Plutarch, *Vita Solonis*, c. 2, Aristotle, *Pol.* I. c. 11, p. 1259 a. Plutarch
(*De Soll. Animal.* p. 45 of Reiske's
edition) says that Thales used mules
to carry his salt to market; one of
them, having slipped in fording a
stream, found its load considerably
lightened by the melting of the salt
and afterwards several times fell in
the water purposely.   To cure it of

this trick, Thales loaded it one time
with rags and sponges.

[4] Plut. *De plac. philos.* I. c. 3.
Themistius, *Orat.* XXVII. p. 317. Diog.
L. I. c. 1, n. 2.   Huebner's ed. p.
14.

[5] Herod. I. c. 74. Clem. Alex. *Strom.*
I. c. 14 (ed. Potter, p. 354).   The
latter quotes Eudemus as his authority.   The fact that Thales predicted
the eclipse is well attested, but we do
not know with what exactitude he
specified the time of its occurrence.
He may have learnt, from Egyptian
or Chaldæan registers, that a solar
eclipse occurs at intervals of 18 years
11 days. See Bretschneider, pp. 51,
52.

The circumstance gave additional *éclat* to the prophecy, and it was no doubt owing to this that, in the archonship of Damasias (B.C. 585—583 B.C.), Thales was added to the list of Wise Men[1]. "Thales apparently," says Plutarch[2], "was the only one of these whose wisdom stepped, in speculation, beyond the limits of practical utility: the rest acquired the name of wisdom in politics." It appears, nevertheless, that Thales possessed quite as much political shrewdness and knowledge of the world and had the same gift of epigrammatic counsel as his compeers among the famous Seven[3].

84. The well-known theory of Thales on the structure of the universe and the astronomical observations, to which he seems to have been chiefly devoted, do not fall within the scope of this history[4]. For the present purpose, it is necessary only to record that five geometrical theorems are expressly attributed to Thales and also two practical applications of geometry. The theorems are as follows[5]:

(1) The circle is bisected by its diameter.

(2) The angles at the base of an isosceles triangle are equal. (Euc. I. 5, part 1.)

---

[1] Diog. L. I. 1, n. 1, quoting Demetrius Phalereus.

[2] *Vit. Solonis*, c. 3.

[3] See, for instance, Herod. I. c. 170, and Diog. Laert.

[4] On the astronomy of Thales, see the authorities collected by Bretschn. pp. 47—49. The most copious of these is Plutarch, *De plac. philos.* II. cc. 12, 24, 28, III. cc. 10, 11. The chief extracts from Thales' astronomical teaching are: (1) that the year is 365 days: (2) that the intervals between the equinoxes are not equal: (3) that *Ursa Minor* was a better guide for mariners than *Ursa major*: (4) that the moon is illuminated by the sun: (5) that the earth is spherical.

[5] Of these (1) (2) (3) and (5) rest on the authority of Proclus (*Comm. in Eucl. I.* ed. Friedlein, pp. 157, 250, 299, 65), who cites Eudemus for (3)

and (5). The theorem (4) is attributed to Thales by inference from a passage of Diogenes Laertius (I. c. 1, n. 3) who says that Pamphila (*temp.* Nero) relates that Thales was the first person "to inscribe a right-angled triangle in a circle," and that he sacrificed an ox on performing this "problem." The same achievement was attributed by others to Pythagoras. Dr Allman (v. p. 170) has the excellent note: "It may be noticed that this remarkable property of the circle, with which, in fact, abstract geometry was inaugurated, struck the imagination of Dante:

'O se del mezzo cerchio far si puote
Triangol sì, ch' un retto non avesse'."

The lines (*Paradiso*, c. XIII. 101—2) are part of a description of the knowledge which Solomon did *not* choose from God.

(3) If two straight lines cut one another, the opposite angles are equal. (Euc. I. 15.)

(4) The angle in a semicircle is a right angle. (Euc. III. 31, part 1.)

(5) A triangle is determined if its base and base-angles be given (practically Euc. I. 26).

Of these the first and third are probably cases in which Thales relied on intuition, or as the Eudemian summary has it, attacked the question empirically ($αἰσθητικώτερον$), for, according to Proclus (p. 299), Euclid first thought (3) "worthy of proof," and he does not think (1) worthy of it at all, but leaves it to be inferred from definitions 17 and 18 to Book I. The language of Proclus also (p. 250) seems to hint that Thales proved the proposition (2), our old friend, the *Pons Asinorum*, by taking *two* equal isosceles triangles and applying them to one another as in Euc. I. 4, another case of experiment. But the two remaining theorems are obviously incapable of such treatment, and must have been supported either by deduction or at least by very wide induction. The last of them (Euc. I. 26) is attributed to Thales by Eudemus (Proclus, p. 65), apparently on the ground that Thales invented a mode of discovering the distance of a ship at sea, in which the proposition was used. In the application of this process, probably the given base was a tower of known altitude, and one of the given base-angles was the right angle which the tower forms with the shore. The other given angle was obtained by the observer who looked at the ship from the top of the tower[1]. It is hardly credible that, in order to ascertain the distance of the ship, the observer should have thought it necessary to reproduce and measure on land, in the horizontal plane, the enormous triangle which he constructed in imagination in a perpendicular plane over the sea. Such an undertaking would have been so inconvenient and wearisome as to deprive Thales' discovery of its practical value. It is therefore probable that Thales knew another geometrical proposition: viz. 'that the sides of equiangular triangles are proportional.' (Euc. VI. 4.) And here no doubt we have the

---

[1] Cantor, p. 122.

real import of those Egyptian calculations of *seqt*, which Ahmes introduces as exercises in arithmetic. The *seqt*, or ratio, between the distance of the ship and the height of the watchtower is the same as that between the corresponding sides of any small but similar triangle. The discovery, therefore, attributed to Thales is probably of Egyptian origin, for it is difficult to see what other use the Egyptians could have made of their *seqt*, when found. It may nevertheless be true that the proposition, Euc. VI. 4, was not known, as now stated, either to the Egyptians or to Thales. It would have been sufficient for their purposes to know, inductively, that the *seqts* of equiangular triangles were the same. The other practical application of geometry, attributed to Thales, depends upon the same proposition, but is described in two forms, the one very simple, the other more difficult. According to Pliny and Diogenes Laertius[1] (who quotes Hieronymus of Rhodes, a pupil of Aristotle, as his authority), Thales ascertained the height of pyramids and similar edifices by measuring their shadows at that hour of the day when a man's shadow is of the same length as himself. Plutarch[2], however, puts into the mouth of Niloxenus a different account of the process. "Placing your staff at the extremity of the shadow of the pyramid," says he to Thales, "you made, by the impact of the sun's rays, two triangles, and so showed that the pyramid was to the staff as its shadow to the staff's shadow." This is obviously only another calculation of *seqt*, though the proportion, as stated by Plutarch, is probably not exactly in its original form. There is no reason, now that Ahmes's book is well-known, to deny that Thales was acquainted with the simple process here attributed to him. It was, however, justifiable in Bretschneider, who knew Ahmes only from a brief abstract[3], which contained no mention of the *seqt* calculations, to question Plutarch's accuracy and to suppose that he was attributing to Thales the improved methods of his own day.

85. To infer from the knowledge which is expressly

[1] Pliny, *H. N.* xxxvi. 17. Diog. L.
I. c. 1, n. 3.

[2] *Sept. Sap. Conv.* 2.

[3] Dr Birch in Lepsius' *Zeitschrift*,
referred to *supra*, p. 16 *n*.

attributed to Thales what other geometrical knowledge he must have had is a peculiarly fascinating inquiry. It has been already suggested that he knew, in some form, the theorem Eucl. VI. 4. To this Dr Allman adds also two other inferences. If, he argues, Thales knew that the angle in a semicircle is a right angle, he must have known also that 'the interior angles of a triangle are equal to two right angles' (Euclid I. 32, pt. 2). He infers this, not from the fact that Euclid uses the proposition I. 32, in the proof of III. 31, pt. 1,[1] but in another way. Thales knew that the angle in a semicircle is a right angle: if he had then joined the apex of the triangle containing that right angle with the centre of the circle, he would have obtained two isosceles triangles, in which, as he also knew, the angles at the base are equal. Hence, he could not have failed to see that the interior angles of a right-angled triangle were equal to two right angles, and since any triangle may be divided into two right angled triangles, the same proposition is true of every triangle. It is justifiable, no doubt, to ascribe so much *intelligence* to Thales, but it is another matter to attribute to him a particular piece of knowledge and a particular method of proof: on the same plan, Thales might be held to have known the first six books of Euclid. It will be remembered that Geminus, in the extract quoted above, attributes to "the ancients" (οἱ παλαιοί) the knowledge of the proposition that the interior angles of a triangle are equal to two right angles. It may be conceded that he alludes here to Thales among others, but it is also to be borne in mind that he says that this proposition was separately proved for the different classes of triangles. Hence Dr Allman suggests, as an alternative, that the theorem was arrived at from inspection of Egyptian floors paved with tiles of the form of equilateral triangles, or squares, or hexagons[2].

---

[1] There would be two objections at least to such an inference, viz. that Euclid I. 32 contains two propositions, of which only the first, which is not the prop. in question, is used in III. 31: and also that Euclid I. 32 is said by Proclus (p. 379) to have been proved almost as it stands by the Py-

thagoreans. Cantor, however (p. 120), is inclined to attribute to Thales Euclid's proof (or something very like it) of III. 31.

[2] Proclus, p. 305, attributes to the Pythagoreans the *theorem* that only three regular polygons, the equilateral triangle, the square and the hexagon,

If, for instance, Thales observed that six equilateral triangles could be placed round a common vertex, he would also notice that six equal angles make up four right angles, and therefore the angles of each equilateral triangle are equal to two right-angles. Hankel (pp. 95, 96) suggests a similar theory, which is adopted also by Cantor (pp. 120—121), with the addition that the scalene triangle was divided into two right-angled triangles, each of which was considered as half a rectangle. It seems needless to dwell further on this proposition.

86. Dr Allman, however, makes a second inference of a far bolder character. He converts the theorem that the angle in a semicircle is a right angle into a theorem that, if on a given straight line as base, there be described any number of triangles each having a right angle at the vertex, then the *locus* of their vertices is the circumference of a circle described on the given base as diameter, and attributes to Thales, therefore, the conception of *geometrical loci*. If Thales proved the first theorem empirically, by constructing a great number of right-angled triangles on the same base, no doubt the notion of a *locus* may have occurred to him : but what becomes then of that deductive, that essentially Greek character which Thales is always said to have imparted to Egyptian geometry ?[1] There will not be left a single theorem, attributed to Thales, which he is not likely to have discovered by inspection or inductively. He may, no doubt, have arrived at any theorem in two ways, at first inductively or by inspection, and later also by a formal deductive process, but there is no available evidence on this matter. If he used deduction only for this particular theorem, he would probably not have conceived a *locus*. If he used induction only, he might have conceived a *locus*, but there would have been no great merit in the conception.

Of speculation in this style there is no end, and there is hardly a single Greek geometer who is not the subject of it. A

can be placed about a point so as to fill a space, but Dr Allman (p. 169 note) supposes, no doubt rightly, that the Egyptians habitually used these figures for tiles.

[1] The Eudemian summary expressly says that Thales "attacked some questions in their general form" (καθο-λικώτερον).

mathematician, writing for mathematicians, is perhaps entitled, and may even be required, to fill up with his own opinions the gaps in his evidence. But his theories, however ingenious, are necessarily of such a kind that even a non-mathematical reader can see that they are, for the most part, imaginary, and a mathematician will think he can make better for himself. A history, like this, of which the utility will no doubt vary as the brevity, had best omit long and inconclusive discussions. Suffice it then to say, of Thales, that he certainly introduced geometry to the Greeks, that he probably improved upon Egyptian geometry by teaching more particularly of *lines* than of areas, and by giving deductive instead of inductive proofs, and that at any rate he formed a school which derived from him its subjects and methods of inquiry, its belief in the stability of natural laws, its tradition of the beauty and utility of the intellectual life[1].

87. The Eudemian summary names, immediately after Thales, **Mamercus,** the brother of the poet Stesichorus, as one of the founders of Greek geometry. Nothing more is known of this person, and his name itself is exceedingly doubtful. Stesichorus lived in Sicily, and died about 560 B.C. Mamercus nevertheless may have been a pupil of Thales, for it is difficult to imagine how he could have learnt any geometry in Sicily at that time. However this may be, Thales undoubtedly had some pupils (e.g. Mandryatus of Priene[2]) whom the Eudemian summary does not mention. Another pupil of Thales, **Anaximander** of Miletus, became very famous. He was born about 611 B.C., and died about 545 B.C.[3]. He also, like Thales, devoted himself mainly to physical speculations and to astronomy. It has been already mentioned that he first introduced the *gnomon* and the *polos* or sundial into Greece[4].

---

[1] Thales apparently composed some astronomical treatise in verse, but the authorities on his writings are conflicting. See Bretschneider § 39, pp. 54, 55.

[2] Apuleius, *Florida*, IV. n. 18, ed. Hildebr. p. 88, ed. Delphin. p. 817. Bretschneider, pp. 53, 56.

[3] Diog. Laert. II. c. 1.

[4] The *gnomon* was an upright staff placed in the centre of three concentric circles, so that at the summer solstice its shadow at noon just reached the inner circle, at the equinoxes the middle, at the winter solstice the outer. Afterwards in places, of which the meridian

Simplicius also relates (*in Ar. de Coelo,* ed. Brandis, p. 497 *a*), on the authority of Eudemus, that Anaximander ascertained the relative sizes and distances of the planets: and Diogenes states that he first constructed terrestrial and celestial globes[1]. These facts favour a presumption that Anaximander also was greatly interested in geometry, and Suidas, in particular, attributes to him a work entitled ὑποτύπωσις τῆς γεωμετρίας, which would seem to mean 'a collection of figures illustrative of geometry.' Pliny (*H. N.* II. c. 76) as was mentioned above (p. 67 *n.*), attributes the introduction of the gnomon to the younger philosopher, **Anaximenes,** who lived B.C. 570—499, and there may be some confusion between him and Anaximander. Nothing is known of any geometrical work by Anaximenes, and the same might be said of the more famous **Anaxagoras** of Clazomenae[2], (B. C. 500—428) were it not that the Eudemian summary expressly mentions him as a geometer; that Plutarch (*de exilio,* c. 17), relates that when in prison he wrote a treatise on quadrature of the circle, and that Vitruvius (vii. *praef.*), ascribes to him a work on perspective.

88.   We may add finally to the Ionic school, with which he seems to have had most affinity, **Œnopides** of Chios, a contemporary perhaps of Anaxagoras, or according to the Eudemian summary, a little later. Of him Diodorus, as quoted above (p. 131), relates that he studied in Egypt. He was certainly devoted chiefly to astronomy ; and Ælian (*Var. Hist.* x. 7), says that he invented a "great year" of 59 years, that is, a period at the end of which, according to his observations, the lunar and solar years would exactly coincide[3]. He was however interested in geometry, and Proclus[4] attributes to him the

---

was known, the circles were omitted and three spots, marked on the meridian line, were substituted. The *polos* can hardly have been similar to our sundials, but was probably a staff placed in the centre of six concentric circles, such that every two hours the shadow of the staff passed from one circle to the next. Bretschneider, p. 60. Cantor, p. 92.

[1] His fellow-townsman, Hecataeus, made about the same time the first *map.*

[2] Anaxagoras lived, in his later years, with Pericles at Athens.

[3] Censorinus c. 18, says that a "great year" of this length was attributed also to Philolaus, the Pythagorean. See the note to Ælian in Gronovius' ed. II. p. 655.

[4] Ed. Friedlein, pp. 283 and 333. Eudemus is cited in the latter passage.

solution of two problems, 'To draw a straight line perpendicular to a given straight line of unlimited length, from a given point without it' (Euclid I. 12.) and 'At a given point in a given straight line to make a rectilineal angle equal to a given rectilineal angle' (Euclid I. 23). On the first of these, Proclus' note is curious and worth quoting. He says, " Œnopides first invented this problem, thinking it useful for astronomy. He calls the perpendicular (κάθετος) in the antique manner a 'gnomon,' because the gnomon is at right angles (πρὸς ὀρθάς) to the horizon, and the line drawn is at right angles to the given line, differing in plane only (τῇ σχέσει), but not in principle (κατὰ τὸ ὑποκείμενον)."

It is plain enough from these scanty facts and from their scantiness, that the Ionic school did not, in nearly two hundred years, do anything like what might have been expected for the advancement of geometry. It introduced the study, kept it alive, and by working at astronomy, opened up a vast field of research, to which geometry soon became essential. The progress of geometry itself, however, was due mainly to the Pythagoreans in Italy.

(c.) *The Pythagoreans.*

89. Pythagoras, the son of Mnesarchus, was born in Samos, probably about 580 B.C. The date of his birth, however, and the other facts of his biography are the subject of disputes, which, owing to the nature of the evidence, can never be satisfactorily settled. The following summary statement perhaps excludes most of the very doubtful matter. Pythagoras was at first the pupil of Pherecydes of Syros[1], but afterwards visited Thales[2], and was by him incited to study in Egypt, particularly at Memphis or Diospolis (Thebes). In pursuance of this

---

[1] Pherecydes is said (Suidas, *s. v.* Pliny *H. N.* VII. 56) to have been the first writer of prose. He is also said to have introduced the doctrine of metempsychosis, which Pythagoras a-dopted. See Ritter and Preller, *Hist. Philos.* c. II. § 92.

[2] Iamblichus (*Vita Pyth.* c. 2) is the authority for this statement, which is not intrinsically improbable.

10—2

advice, Pythagoras went to Egypt and stayed there a long time, perhaps 22 years. He may subsequently have visited Babylon[1]. He returned ultimately to Samos and attempted to found a school there, but without success. For this reason or because of some political disturbance he emigrated to Croton in S. Italy[2]. The colonies in Magna Graecia, of which Sybaris was the chief, were at this time more wealthy and important than the mother country, and a very considerable commerce was carried on between them and the Ionian coast. Pythagoras, therefore, did not arrive at Croton among a strange and uncouth people, and was able soon to gain a leading position among his fellow townsmen. Among the noblest and best of these he formed a brotherhood, the members of which were united by common philosophical beliefs and pursuits. They were, however, bound by oath not to divulge the tenets and discoveries of their school, and it is due to this fact that the historian of philosophy is now obliged to speak of 'the Pythagoreans' as a body and is unable to identify the author of any particular portion of their creed. This Masonic society[3], so to say, soon spread into other cities of Magna Graecia, and as it was capable of taking united action on political questions, especially on the side of the aristocrats, from whom its members were chiefly drawn, it became the object of popular suspicion and hatred[4]. Ultimately, the Pythagoreans of Croton, their leader with them, were attacked by the plebeian party: Pythagoras fled first to Tarentum and then to Metapontum, and was there murdered in another popular outbreak about 500 B.C.

90. It has been already stated that, by writers of other schools, Pythagorean doctrines are generally attributed to "the Pythagoreans" and not to Pythagoras himself. On the other hand, the Pythagoreans were wont to attribute all their tenets to their master. Αὐτὸς ἔφα, *ipse dixit*, was the formula which secured acceptance for any doctrine however remote it may

---

[1] Strabo, xiv. i. 16.
[2] Diog. Laert. viii. 3. Cicero, *De Rep.* ii. 15.
[3] It contained two orders, the μαθηματικοί and the ἀκουσματικοί, the 're-searchers' or 'mathematicians' and the 'listeners.' The former apparently were communists. Iamblich. *V. P.* 81; Porph. *V. P.* 37.
[4] Polybius, *Hist.* ii. 29.

have been from the teaching of Pythagoras himself. Further, the teaching of the school seems to have been traditional and founded on no text-books, until the wide dispersion of its members made it desirable that some record should be secured. Philolaus, a contemporary of Plato, is generally credited with the first publication of a detailed Pythagorean philosophy[1]. His work is lost, save a few very brief fragments (not undoubted) preserved by Stobaeus and similar compilers. In default of this, we are compelled to rely on incidental remarks or mere allusions of the earlier Greek writers, or else on histories obviously uncritical of a very late date. Now these are precisely the kind of authorities who would naturally omit to mention discoveries of Pythagoras and his school in geometry. Aristotle, for instance, had no occasion to discuss geometrical details to which he did not, though the Pythagoreans did, attach any profound significance. To Iamblichus, on the other hand, geometry was not in itself interesting, or, if it was, the geometry of his day had so far outstripped the Pythagorean that the latter would have seemed childish by comparison. Hence it is that, though the evidence is abundant that Pythagoras really made geometry the Greek science *par excellence*, yet very few particular inventions can be attributed to him or his immediate followers.

91. It has been already stated (see above pp. 67—72) how it was that Pythagoras came to attach so much importance to geometry, and how closely he connected it with arithmetic. It will be remembered also that the geometry of Ahmes is exhibited only as leading to arithmetical problems, and we may suppose therefore that Pythagoras was profoundly influenced by his Egyptian teaching. We shall also be prepared to find that the Pythagorean geometry, like the Egyptian, is concerned, more than that of Thales, with the relations of areas and volumes, and is not largely concerned with those relations of lines which do not admit of, or do not readily suggest, arith-

---

[1] Lucian, *Pro Lapsu in Salut.* c. 5, mentions that Pythagoras had not thought fit to leave any authoritative writings (μηδὲν ἴδιον καταλιπεῖν τῶν αὑτοῦ). Diog. Laert. viii. 15, says "before Philolaus it was impossible to learn any Pythagorean dogma."

metical expression. This being premised, it remains only to set out in order such doctrines, of geometrical interest, and such special discoveries in geometry as are attributed to Pythagoras or the Pythagoreans.

According to Aristotle[1] "the Pythagoreans first applied themselves to mathematics, a science which they improved; and penetrated with it, they fancied that the principles of mathematics were the principles of all things." Proclus[2] says expressly that the specialised meaning of 'mathematics' ($\mu a\theta\acute{\eta}\mu a\tau a$) was first used by the Pythagoreans. The Eudemian summary says that Pythagoras changed the study of geometry into the form of a liberal education, for he examined its principles to the bottom, and investigated its theorems in an immaterial and intellectual manner ($\acute{a}\ddot{v}\lambda\omega\varsigma$ $\kappa a\grave{\iota}$ $vo\epsilon\rho\hat{\omega}\varsigma$). Diogenes Laertius[3] states, on the authority of Favorinus, that Pythagoras "used definitions, on account of the mathematical matter of his subject." This perhaps was the first step towards that systematization of geometry which Eudemus ascribes to him. The following details are also preserved[4]:

(1) The Pythagoreans define a point ($\sigma\eta\mu\epsilon\hat{\iota}ov$) as "unity having position." (Proclus, ed. Friedlein, p. 95.)

(2) They considered a point as analogous to the monad, a line to the duad, a superficies to the triad, and a body to the tetrad. (*Ib.* p. 97.)

(3) They showed that the plane about a point is completely filled by six equilateral triangles, four squares or three regular hexagons. (*Ib.* p. 305.)

(4) They first, according to Eudemus, proved generally that the interior angles of a triangle are equal to two right-angles. (*Ib.* p. 379.)[5]

---

[1] *Metaph.* i. 5, 985.

[2] Friedlein's ed. p. 45.

[3] viii. 25.

[4] All the following quotations are in Bretschneider, pp. 67—91. They are more neatly arranged by Dr Allman.

[5] The Pythagorean proof, according to Eudemus, is as follows. Let ABC be a triangle. Through A draw DE

parallel to BC. Then the alternate

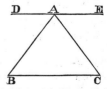

angles ($a\grave{\iota}$ $\acute{\epsilon}va\lambda\lambda\acute{a}\xi$) are equal, DAB to

(5) They also, according to Eudemus, invented the problems concerning the application of areas, including the cases of defect (ἐλλεῖψις) and excess (ὑπερβολή) as in Eucl. VI. 28, 29. (*Ib.* p. 419, Comm. on Eucl. I. 44[1]. See above p. 84 *n.*)

(6) Pythagoras sacrificed an ox on solving the problem how to construct a figure equal to one and similar to another given figure. (Euclid II. 14, VI. 25)[2].

(7) Pythagoras, according to Eudemus, discovered the construction of the regular solids[3]. (Proclus, p. 65).

(8) The triple interwoven triangle, the pentagram-star, (τὸ τριπλοῦν τρίγωνον, τὸ δι' ἀλλήλων, τὸ πεντάγραμμον) was used as a badge or symbol of recognition by the Pythagoreans, and was called by them Health (ὑγίεια). (Lucian, *Pro Lapsu,* c. 5, Schol. in Ar. *Nub.* 611)[4].

(9) Pythagoras discovered the theorem of the three squares, Euclid I. 47. (Proclus, p. 426)[5].

ABC, EAC to *ACB.* Add the angle *BAC.* Then the three angles *DAB, BAC, CAE,* that is, *DAB, BAE,* that is, two right angles are equal to the three angles of the triangle. The redundant explanation in the last sentence is curious. The text is given by Bretschneider, p. 78.

[1] This passage will, for the sake of some other matter contained in it, be quoted later on. The statement is confirmed by Plutarch (*Non posse suav. vivi sec. Epicur.* c. 11), who says, " Pythagoras, according to Apollodorus, sacrificed an ox on completing the figure...either for the proposition concerning the hypotenuse, that its square is equal to those of the sides containing the right angle, or else the problem about application of an area." The texts have περὶ τοῦ χωρίου τῆς παραβολῆς, for which Bretschneider (p. 79, *n.*) proposes, evidently rightly, περὶ τῆς τοῦ χωρίου παραβολῆς. The text is

sometimes translated "on the area of the parabola," which involves a gross anachronism.

[2] Plutarch, *Quaest. Conv.* VIII. 2, c. 4.

[3] According to Iamblichus (*Vita Pyth.* c. 18, s. 88) Hippasus was drowned for divulging the knowledge of "the sphere with the twelve pentagons" (i.e. the inscribed ordinate dodecahedron) " for he took the glory as discoverer, whereas everything belonged to Him (εἶναι δὲ πάντα ἐκείνου) for so they call Pythagoras."

[4] See Chasles, p. 477, sqq. This *Pythagorae figura* was used through the middle ages, and was regarded even by Paracelsus as a symbol of health. It is the *drudenfuss* of Goethe's *Faust,* sc. iii.

[5] The oldest authority for this is Vitruvius, IX. pref. 5, 6, 7. It is attested also by Plutarch (*supra, n.* 1). Diog. L. VIII. 11.

(10)   Pythagoras used to say that of all solids the sphere was the most beautiful ; of all plane figures, the circle.   (Diog. Laert. VIII. 19.)

(11)   The Pythagoreans are said to have solved the quadrature of the circle.   (Iamblichus quoted by Simplicius *in Ar. Phys.* 185, *a*, 16.   Ed. Brandis, p. 327, *b*.)

(12)   The Pythagoreans, as has been already stated (*supra* p. 70) were largely occupied with the study of proportion, doubtless not in arithmetic only but in geometry[1].

(13)   From the Pythagorean use of 'gnomon' as a designation of those numbers, which, when added to a square number, make a square total, it is evident that the Pythagoreans were accustomed to consider and use the gnomon in geometry[2].

92.   It will be seen at once that all this knowledge can by no means be attributed to Pythagoras himself or to his earliest successors.   There must have been, notwithstanding the enthusiasm and ability of the school, a slow progress from empirical to reasoned solutions, from the diffuse treatment of special cases to the concise treatment of one general case.   But we are hopelessly in the dark as to when and how this progress was effected.   It is probable, indeed, that much of it was not effected inside the Pythagorean school at all, but that later writers ascribe to the Pythagoreans theorems which they first proved for one special case but which some Academic geometer afterwards proved generally.   A Pythagorean, for instance, may very well have solved Eucl. II. 14, without going so far as VI. 25.   Some statements also, in themselves beyond doubt, may lead to very plausible but erroneous inferences.   For instance, if Pythagoras was

---

[1] Proclus (ed. Friedlein, p. 43) says that Eratosthenes regarded proportion as 'the bond of mathematics,' and says elsewhere that the 5th Book of Euclid is common to geometry, arithmetic, music and, in a word, to all mathematics.   See Knoche, *Untersuch. über die Schol. des Proklus zu Eucl. Elem.* Herford, 1865, p. 10.

[2] It should be mentioned before leaving this enumeration, that Bret-

schneider p. 89, § 71, conclusively shews that Montucla (I. p. 117) is wrong in attributing to the Pythagoreans any investigations in isoperimetry.   What Diogenes Laertius (VIII. c. 1. n. 19) says is stated above (10).   He does not say that Pythagoras taught that the circle is the greatest among figures of equal periphery, and the sphere among solids of equal superficies.

acquainted (as no doubt he was) with the regular solids, he was acquainted also with the regular pentagon. This fact, together with the form of the pentagram and together with the directions as to dividing figures into triangles which Plato puts into the mouth of the Pythagorean Timaeus[1], suggests that Pythagoras constructed the regular pentagon in the manner of Euclid IV. 11. But Euclid IV. 11 is founded on IV. 10, which is founded on II. 11, and the Eudemian summary, the most authoritative of all our historical accounts of ancient geometry, says that *Plato* invented the learning on the subject of cutting a line in extreme and mean ratio. It can hardly, therefore, serve any useful purpose to criticise minutely a whole body of geometrical teaching much of which is not properly authenticated, and which, if it be correctly ascribed to the Pythagorean school, must belong to very different dates[2]. It is sufficient to say, generally, that the Pythagoreans seem at a very early time to have been masters of most of the geometry contained in the first two books of Euclid, and that they knew some propositions of the 5th and 6th books. To them also is probably due the introduction of definitions of some kind and the use of orderly deductive proofs in geometry. Further, just as Aristoxenus tells us that they raised arithmetic above the needs of merchants, so the Eudemian summary tells us that they made geometry 'a liberal education'; and other writers record as one of their proverbial maxims, "A figure and a stride : not a figure and sixpence gained[3]".

93. There are, however, two portions of the Pythagorean geometry which have provoked interesting comments. One is the construction of the five regular solids, the other is the Pythagorean theorem, Euclid I. 47.

Timaeus, in the dialogue of Plato above cited, explains that every rectilineal figure is made up of triangles, and that every

---

[1] *Tim.* c. 20, 107. See next par.

[2] A very curious instance of the distracting nature of the evidence about the Pythagoreans is furnished by Diogenes Laertius (VIII. 83), who says that Archytas, one of the last of the school, "first found the cube"!

[3] σχᾶμα καὶ βᾶμα, ἀλλ' οὐ σχᾶμα καὶ τριώβολον. Proclus, ed Friedlein, p. 84. Iamblichus, *Adhort. ad Philos. Symb.* xxxvi. c. 21, quoted by Dr Allman, v. p. 206.

triangle may be divided into two right-angled triangles, either isosceles or scalene. "Of such scalene triangles the most beautiful is that out of the doubling of which an equilateral arises, or in which the square of the greater perpendicular is three times that of the less, or in which the less is half the hypotenuse. But two or four right-angled isosceles triangles, properly put together, form the square : two or six of the most beautiful scalene right-angled triangles form the equilateral triangle, and out of these two figures arise the solids which correspond with the four elements of the real world, the tetrahedron, octahedron, icosahedron and the cube[1]." Of these solids, the tetrahedron, octahedron and cube must have been familiar to a traveller who had lived in Egypt; on the construction of the other two, Dr Allman has the following remarks: "In the formation of the tetrahedron, three, and in that of the octahedron, four, equal equilateral triangles had been placed with a common vertex and adjacent sides coincident, and it was known too that if six such triangles were placed round a common vertex with their adjacent sides coincident, they would lie in a plane, and that, therefore, no solid could be formed in that manner from them. It remained then to try whether five such equilateral triangles could be placed at a common vertex in like manner: on trial it would be found that they could be so placed and that their bases would form a regular pentagon. The existence of a regular pentagon would thus be known (sic). It was also known from the formation of the cube that three squares could be placed in a similar way with a common vertex; and that, further, if three equal and regular hexagons were placed round a point as common vertex with adjacent sides coincident, they would form a plane. It remained then only to try whether three equal regular pentagons could be placed with a common vertex and in a similar way: this on trial would be found possible and would lead to the construction of the regular dodecahedron which was the regular solid last arrived at." It should be added that there is no reason to suppose that the Pythagoreans knew that there are, in fact, no other regular solids save these.

[1] The dodecahedron represented the universe itself.

94. The famous proposition, Euclid I. 47, has always been known as the *theorem of Pythagoras*. It will be remembered that the converse of this (Eucl. I. 48) was known to the Egyptians and to other nations, at a very early date, in the case in which the sides of the triangle are to one another as $3:4:5$, and that Pythagoras extended it to cases in which the sides are to one another as $2n+1 : 2n^2+2n : 2n^2+2n+1$[1]. The first proposition also may have been known to the Egyptians in the particular case where the right-angled triangle is isosceles. It would of course be at once suggested by a floor paved with tiles in the form of isosceles right-angled triangles. But the general proof is attributed to Pythagoras; and Proclus says expressly (p. 426) that the form of Euclid I. 47 (as well as Euclid VI. 31) is due to Euclid himself. Hence Bretschneider (p. 82), after Camerer[2], proposes as a possible restoration of the original proof, the following. If a straight line

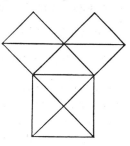

be divided into any two parts a and b, then the square on the whole line is equal to a^2+b^2 with the two complementary rectangles ab. Draw the diagonals c of these rectangles, and dispose the four triangles so formed about the square in the

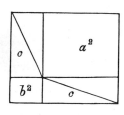
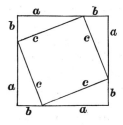

manner shown in the second figure. There is thus left, in the middle of the square, a figure c^2 which is obviously equal to a^2+b^2. Upon this Hankel (p. 98) remarks that "it has no specifically Greek colouring, and reminds us rather of the

---

[1] See above p. 71.   [2] *Euclidis Elem.* I. p. 444, and reff. there given.

Indian style[1]." This criticism will serve to introduce an
Indian proof of the same theorem, taken from the *Víja-ganita*
('Root-calculations') of Bhâskara, who was born A. D. 1114[2].

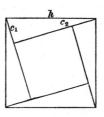

Here a square is constructed on the
hypotenuse and the original triangle
repeated four times, is disposed round
it as in Bretschneider's proof. The
square left in the middle is that of the
difference between the two perpendiculars. Bhâskara merely draws the figure
and adds 'Look!' without thinking it
necessary to add that if $h^2 = 4 \left(\dfrac{c_1 c_2}{2} \right) + (c_2 - c_1)^2$, then $h^2 = c_1^2 + c_2^2$.

A proof of precisely the same kind is given, two hundred years
earlier, by the Arab Abû 'l Wafâ (A.D. 940—998), who translated Diophantus[3]. It would seem also that the Chinese had
a similar proof. The passage on which this presumption is
founded occurs in a book called the *Tcheou pei*[4] or 'signal in a
circle,' of which the first part, containing the passage, is
attributed to 1100 B.C. It may not be so early as this, but
it certainly existed and was the subject of a commentary in the
2nd century after Christ. Here, apparently, the same figure
as Bhâskara's is drawn and is named 'the Rope figure,' as
though it were intended to ex-
plain the practice of some Chinese
Harpedonaptes. Another proof
given by Bhâskara, in the same
place, is also worth quoting. A
perpendicular is drawn from the
vertex to the hypotenuse, dividing the triangle into two

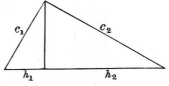

---

[1] Dr Allman (p. 193) adopts this
criticism, but accepts Bretschneider's
proof and attributes it to the Egyptians.
See, however, the passage from the
Meno, cited below p. 174.

[2] Colebrooke, *Algebra etc. of Brahmegupta and Bhâskara*, 1817, p. 220—
222, § 146. Hankel, p. 209. Cantor
p. 557.

[3] See Cantor, pp. 637, 639, quoting
Woepcke in *Journal Asiat.* 1855, pp.
346, 350—351 (Feb. and March).

[4] Cantor, pp. 579—581, quoting E.
Biot in *Journ. Asiat.* pp. 593—639, for
June 1841. Cantor's restoration of
the figure is founded, conjecturally, on
Biot's description of it, which is by no
means clear.

others similar to it. Then in the figure (if h be the original hypotenuse) $\dfrac{c_1}{h} = \dfrac{h_1}{c_1}$ and $\dfrac{c_2}{h} = \dfrac{h_2}{c_2}$, whence it follows that $h\,(h_1 + h_2) = h^2 = c_1^{\,2} + c_2^{\,2}$. Hankel remarks that this proof was revived in the West by Wallis[1].

95. It is not to be supposed that when the Pythagorean brother-hood was for political reasons broken up, the Pythagorean philosophy disappeared also. On the contrary, the school continued to flourish for at least two centuries more. Tarentum seems to have been its head-quarters, but it seems also to have sent out occasional missionary expeditions into Greece. Its habit of secrecy prevents the possibility of naming its earliest leaders. The most celebrated of its earliest disciples was Epicharmus, the founder in Sicily of Greek comedy (*cir.* B.C. 480). Considerably later Philolaus wrote his book, and thus the Pythagorean doctrines became accessible to the public. Two other disciples, Archippus and Lysis, are also reported to have written text-books of their philosophy, and by the time of Plato the Pythagorean teaching seems to have been well known. Simmias of Thebes, the companion of Socrates, says in Plato's *Phaedo* (61 D) that he had himself heard Philolaus in Thebes. Most of the Sophists also, who introduced geometry into Athens, came from Sicily and it is said of some of them and may be presumed of others that they acquired their knowledge of the science from Pythagorean sources. So also, no doubt, did Plato himself, who bought a copy of Philolaus, and who, in Sicily, studied with Archytas and Timaeus of Locri[2]. This **Archytas** was a mathematician of great celebrity. The Eudemian summary mentions him without attributing to him any particular discovery, but a good deal is known of him from other sources[3]. He was a leading politician and chief of the Pythagorean school in Tarentum. According to Diogenes

---

[1] *De Sect. Ang.* c. VI. in Wallis *Op. Math.* (1693), Vol. II. Hankel, p. 209. Cantor, p. 557. A collection of proofs of Euclid I. 47 was made by J. J. Hofmann. *Der Pythagorische Lehrsatz.* Mainz (2nd ed.) 1821. See the

note in Todhunter's *Euclid*.

[2] Cic. *De Rep.* I. 10, 16.

[3] See the authorities collected in Ritter and Preller, *Hist. Philos.* c. II. sec. 100, n. *d*. Cantor, p. 203.

Laertius (VIII. 83) he first applied geometry to mechanics and handled the latter subject methodically, and used mechanical contrivances in the construction of geometrical figures. A very ingenious solution of the problem 'to double a cube' is attributed to him[1] and will be cited below (pp. 181, 182). He also is said to have defined the three chief kinds of proportion. Horace in a well-known ode (I. 28) describes him as *maris ac terrae numeroque carentis arenae mensor*, from which it might be inferred that he attempted some of the problems which Archimedes treats in his *Arenarius*. Gellius (x. 12) ascribes to Archytas also the invention of a mechanical dove[2].

From this it will be seen that the later Pythagoreans were worthily maintaining the traditions which they had received from their master. But in the meantime the Persian Wars had made Athens by far the wealthiest and most brilliant city of Greece. To her resorted, from all parts, those men who had something to teach and were not too proud to make a living by teaching it. Among such there were no Pythagoreans, and thus it is that the history of geometry must leave Italy and the Pythagorean school for Athens and the Sophists and the Academy.

96. Two schools of Greek philosophy, founded early in the 5th century B.C. remain yet to be mentioned. In Sicily Xenophanes of Colophon had formed a school which afterwards made its head-quarters at Elea in Italy. Here Parmenides, his pupil Zeno and Melissus of Samos instituted their famous inquiries into the inconceivable. They denied the infinite divisibility of time and space and illustrated their position by such paradoxes as that concerning Achilles and the tortoise[3],

---

[1] Eutocius, *Comm. in Archim. de Sph. et Cyl.* and Diog. L. *loc. cit.* The method of Archytas contains the first example of a curve of double curvature.

[2] Poggendorff, *Gesch. der Physik*, p. 12, compares this with the automatic eagle made by Regiomontanus in 1489 to greet the Emperor Maximilian I. on his entry into Nuremberg. Young, in his lectures (xx. p. 182) ascribes to

Archytas the invention of the pulley and screw, but I have seen no authority for this statement.

[3] It will be remembered that Zeno maintained that Achilles could never overtake a tortoise, if the tortoise had any start. For, supposing the tortoise to take 100 yards start and Achilles to run 10 times as fast as the tortoise, when the former has covered the 100 yards, the tortoise has run 10 yards,

the heap of corn etc. The doubt and difficulty into which their arguments led the early mathematicians were, no doubt, the cause of the banishment of *infinity* from Greek mathematical terms and conceptions[1]. Both Parmenides and Zeno came to Athens at a Panathenaic festival about 450 B.C. and were heard by the youthful Socrates.

Somewhat later Leucippus of Miletus, a disciple of Zeno, founded the Atomistic school, of which, before Epicurus, Democritus, who lived at Abdera in Thrace, became the most famous professor. Democritus at least was a very ardent and successful geometer. His boast that he was the equal of the Egyptian *harpedonaptae* has been already cited from Clemens Alexandrinus. Diogenes Laertius[2] says that he was a pupil of Anaxagoras as well as of Leucippus, that he was an admirer of the Pythagoreans and intimate with Philolaus, and that he wrote mathematical works on geometry, on numbers, on perspective ($\dot{a}\kappa\tau\iota\nu o\gamma\rho a\phi\acute{\iota}\eta$)[3], another in two books on incommensurable lines and (?) solids ($\pi\epsilon\rho\grave{\iota}\ \dot{a}\lambda\acute{o}\gamma\omega\nu\ \gamma\rho a\mu\mu\hat{\omega}\nu\ \kappa a\grave{\iota}\ \nu a\sigma\tau\hat{\omega}\nu\ \beta'$), and another 'on the difference of the gnomon or the contact of the circle and sphere' ($\pi\epsilon\rho\grave{\iota}\ \delta\iota a\phi o\rho\hat{\eta}\varsigma\ \gamma\nu\acute{\omega}\mu o\nu o\varsigma\ \mathring{\eta}\ \pi\epsilon\rho\grave{\iota}\ \psi a\acute{\upsilon}\sigma\iota o\varsigma\ \kappa\acute{\upsilon}\kappa\lambda o\upsilon\ \kappa a\grave{\iota}\ \sigma\phi a\acute{\iota}\rho\eta\varsigma$). It is impossible to say what these titles mean. It appears also from Plutarch[4], that Democritus was interested in the cone and raised a question, of the Eleatic kind, as to the infinitesimal gradations in its slope. The life of Democritus is generally said to fall between 460 and 370 B.C.

Now though the history of geometry, after about 450 B.C., can be traced with any definiteness only at Athens, yet it is plain that the progress of the science was due to contributions from many other places. Throughout Hellas, in Ionia, in Sicily, Italy, at Athens, at Abdera far away in Thrace, there were men who were working earnestly at the formation of rules for exact thinking or at the exemplification of such rules in

---

and when Achilles has covered this, the tortoise is a yard ahead and so on. Coleridge's answer to this paradox is discussed by Mr S. Hodgson in *Mind*, XIX. (July, 1880).

[1] See esp. Hankel, p. 115 sqq.

[2] Diog. L. IX. 7, 47.

[3] Comp. Vitruvius *Arch.* VII. *praef.*

[4] *De Comm. Not. adv. Stoic.* 39, § 3. Allman in *Hermathena.* VII. p. 208. The question was whether, if a cone be cut by a plane parallel and infinitely near to its base, the conic section so exposed was equal to the base or not.

geometry. It is sufficient only to remark the birthplaces of the philosophers and teachers of this time to see how close an intellectual communion was maintained between the most widely separated cities.

(d.) The Sophists.

97. It was in the Second Persian War, and at the battle of Salamis in particular (B.C. 480), that Athens discovered her vocation to be a maritime power, and that Hellas perceived that a strong fleet was the best protection against any future invasion. For this reason, a joint fleet was during many years kept up by Athens and the islands and cities of the Archipelago, but as Persia showed no sign of moving and the islands found the fleet a serious burden on their resources, a league was formed on the terms that the islands should pay tribute to Athens and Athens should find the ships. It soon followed that the tribute was rigorously exacted but the fleet was not maintained. Immense sums were poured yearly into the Athenian treasury and were spent by Pericles in the adornment of Athens. In the meantime also Athens was become a great commercial city with a large carrying trade, and petty wars in various parts of the Levant filled her streets and markets with captive slaves. Thus she became the richest and most beautiful city in the world. Her citizens were, for the most part, well-to-do and enjoyed an unexampled amount of leisure. The constitution of Athens, moreover, compelled every man to be more or less a politician, and opened a splendid career to any citizen who could but once make a successful speech in the *ecclesia.* Litigation was rife and actions were conducted always by the parties in person. Hence there arose, among the wealthy and ambitious youth, a strong desire to cultivate rhetoric and any other branches of knowledge which could conduce to correct reasoning or successful disputation. The demand was met by the necessary supply. Corax and Tisias in Sicily, had laid the foundations of the rhetorical art and from Sicily and elsewhere there came to Athens a multitude of teachers, calling themselves and called by others, "Sophists." Their business was to teach, for pay, rhetoric principally, but some of them added also geometry, astronomy, philosophy as necessary ingredients of a liberal education. The most famous of them were Protagoras of Abdera, Hippias of Elis, Polus of Agrigentum, Gorgias of Leontini, Prodicus of Ceos, Licymnius of Sicily, Alcidamas of Elaea in Aeolis, Theodorus of Byzantium, Thrasymachus of Chalcedon, Hippocrates of Chios. Physicists of the old school, Anaxagoras, Diogenes of Apollonia in Crete, Diagoras of Melos, came also but were persecuted by charges of impiety and driven away. Of the whole army of Sophists, two only seem to have been Athenians born, namely Antiphon and Meton, the astronomer[1] who introduced the Metonic cycle which the Church still uses. The dates of these teachers cannot be more precisely determined than this, that they were all teaching in Athens between 440 and 400 B.C. A few of them, as has been said, were geometers, but the merits of these (as of the rest) have been greatly obscured by Plato's well-known hostility to their class. Hence perhaps it is that Proclus, an ardent Platonist, in his Eudemian summary, names only Hippocrates as a good geometer.

98. By the Pythagoreans, it will be remembered, the geometry of the circle was practically neglected. This part of the science was revived in the Athenian schools, which occupied themselves chiefly with three famous problems (1) Quadrature of the circle (2) Trisection of an angle (3) Duplication of the Cube. It was mainly through a thousand attempts to solve these problems that new propositions and new processes were discovered and geometry made daily progress. It is not surprising that the first two of the three should have invited attention. Quadrature of the circle was a problem almost as old as geometry itself, and the Pythagoreans, who were so busy with symmetrical divisions of all kinds, would have been led

---

[1] I add him here because Aristophanes (*Birds*, 992—1020) seems to treat him as a sophist. He is there introduced carrying a machine for squaring the circle. The Metonic cycle is said to have been adopted from B.C. 432. The year, according to it, is stated by Ptolemy to have been 365¼ days + $\frac{1}{76}$th of a day. This is more than half an hour too long.

naturally enough to trisection. But the duplication-problem is
not so easily accounted for. Eratosthenes, in a letter[1] describing the solutions of this problem, addressed to Ptolemy III,
(Euergetes), says that an old tragic poet had represented King
Minos as wishing to erect a tomb for his son Glaucus: but
being dissatisfied with the dimensions (100 feet each way)
proposed by his architect, the King exclaims: "The enclosure is
too small for a royal tomb : double it, but fail not in the cubical
form[2]." A little further on, Eratosthenes says that the
Delians, who were suffering under a pestilence, were ordered by
the oracle to double a certain cubical altar and, being in
a difficulty, consulted Plato on the matter. Both these
statements, perhaps, contain a minute portion of truth.
The problem was certainly called 'the Delian,' and it may have
originated in an architectural difficulty. But for this evidence,
one would have been inclined to say, with Bretschneider, that
the problem was suggested in the investigation of incommensurables. It was at least well known in Athens before Plato's
time[3].

99.   Hippias of Elis is mentioned in the Eudemian summary
as authority for the geometrical performances of the brother of
Stesichorus, but is not named as the author of any original
work himself. A certain Hippias, however, who can hardly be

[1] Quoted by Eutocius in Archimed.
De Sph. et Cyl. Torelli's ed. p. 144.
Bretschneider, p. 97, suggests that the
duplication-problem is due merely to
this: the Pythagoreans had found that
the diagonal of a square is the side of
a square twice as large as that of
which it is the diagonal, and they
wished to find a similar law for the
cube.

[2] Valckenaer (*Diatribe de fragm.
Eurip.* p. 203), suggests that the lines
are from the *Polyidus* of Euripides
and ran

μικρὸν γ᾽ ἔλεξας βασιλικοῦ σηκὸν τάφου·
διπλάσιος ἔστω, τοῦ κύβου δὲ μὴ σφαλῆς.

[3] On the Greek circle-squarers the
chief authority is Simplicius in *Ar.*

Phys. printed, from the Aldine edition
(1526), with many corrections, by Bretschneider, pp. 100 sqq. On the duplication-problem, Eutocius in Archimed.
De Sph. et Cyl. Bk II. is most copious,
but very little is said by any ancient
writer about trisection. The modern
commentators (Bretschneider, pp. 94
—134. Hankel, pp. 115—127, 150—
156. Allman in Hermathena VII. pp.
180—228. Cantor, *Vorles.* pp. 172—
176, 180—182, 194—201, etc.) present
an *embarras de richesses*. I shall in
the main follow Cantor, whose arrangement, though it does not offer the
same opportunities for brilliant and
comprehensive criticism as Allman's
or Hankel's, is better suited to my plan.

anybody else than Hippias of Elis[1], is mentioned elsewhere by Proclus and the mathematical learning of this sophist is directly attested by Plato himself[2]. It is true that he is mentioned by Plato with a certain sarcasm. Protagoras, for instance, in his long and eloquent plea for his own teaching, is made to say "The others injure the young : for they drag them back against their will into arts which they would fain avoid, teaching them arithmetic and astronomy and geometry and music (and here he glanced at Hippias), but he who comes to me shall learn only that for which he comes." Hippias evidently was the polymath of his time and had high notions of a liberal curriculum. Proclus mentions him twice[3]. In the first passage, he says that Nicomedes had solved the *trisection* problem by means of the conchoid curve, which he himself invented : that others had used for the same purpose the mixed curve called the *quadratrix* of Hippias and Nicomedes and that others divided an angle in any given proportion by using the spirals of Archimedes. In the second passage, he says that mathematicians have described the properties of various curves, Apollonius of the conic sections, Nicomedes of the conchoids, Hippias of the *quadratrix* (τετραγωνίζουσα) and Perseus of the spirals. Pappus[4], however, says that the quadrature of the circle was effected by Dinostratus, Nicomedes and other later geometers by means of a line which, from this use, was called the *quadratrix*. Here Hippias is ignored. Now Dinostratus belongs to the end of the 4th century B.C. and Nicomedes seems to be a century later. Cantor, therefore, proposes[5] to reconcile the statements of Proclus and Pappus by supposing that Hippias, i.e. Hippias of Elis, invented a curve which was found useful for both the quadrature- and the trisection-problems, and that this curve was, by Dinostratus or Nicomedes or later, called

---

[1] Allman (VII. p. 220) and Hankel (p. 151) deny this. Bretschneider (p. 94) and Cantor (p. 165) affirm it. The latter shows, by many instances, that Proclus was always careful to distinguish writers of the same name.

[2] *Hippias Maj.* 285, C D. *Hippias Minor*, 367, 368. *Protagoras*, 318 E.

[3] Ed. Friedlein, pp. 272, 356.

[4] IV. c. XXX. ed Hultsch, p. 251. So also Simplicius *loc. cit.* quoting Iamblichus, names Nicomedes only in connexion with the *quadratix*. Bretschneider, p. 108.

[5] p. 167.

the *quadratrix*, τετραγωνίζουσα. Originally, it may have been intended only for the trisection.

The construction of the *quadratrix* is thus described by Pappus (*loc. cit.*). "In the square

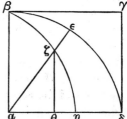

αβγδ, from *a* as centre with *aβ* as radius, describe a quarter of a circle *βεδ*. The straight line *aβ* moves evenly about its end *a* so that the other end *β* moves in a given time along the whole arc *βεδ*. The line *βγ* moves evenly in the same time, remaining always parallel to itself from the position *βγ* to the position *aδ*. The *locus* of intersection of this straight line with the moving radius *aβ* in the curve *βζη*, which is the *quadratrix*." The property of this curve consists in this, that any straight line *aζε* drawn to the circumference of the circle, makes the ratio of the quadrant to the arc *εδ* equal to the ratio of the straight lines *βa* : *ζθ*. And since the straight line *βa* can be divided into any number of parts, in any given ratio to one another, so also can the quadrant or the arc *εδ*, and the trisection or any other section of an angle is performed. The quadrature of the circle is given by this curve, since the straight line which is equal to the quadrant *βεδ* is a third proportional to *aη, ηδ*[1].

100. Theodorus of Cyrene, whom the Eudemian summary names with praise, is known to us only as the mathematical teacher of Plato[2]. Iamblichus says he was a Pythagorean and Plato introduces, in the *Theaetetus*, his discovery in effect that the square roots of numbers between 3 and 17 (except 4, 9, 16) are irrational. He does not seem to have visited Athens.

101. Hippocrates of Chios, who is mentioned with Theodorus in the summary, was one of the greatest geometers of antiquity. Like Thales, he began life as a merchant but lost his property either by piracy or through the chicanery of the

---

[1] Pappus, iv. 26. Bretschneider, p. 96. Hankel, p. 151. Cantor, p. 168, 213, (*sub* Dinostratus).

[2] Diog. Laert. ii. 104. Iamblichus *Vita Pyth.* 267. Plato, *Theaet.* 147 D.

Byzantine custom house[1]. He came to Athens to prosecute the offenders, employed his leisure in attending lectures[2] and ultimately himself became a teacher of geometry. Aristotle says he had a talent for the science but was in other respects slow and stupid (βλάξ καὶ ἄφρων). The Greeks, however, would naturally call any man a fool who was cheated of his property and Aristotle seems to have no other evidence for his criticism of Hippocrates. He may, of course, have been right. There are still extant mathematicians who are singularly deficient in ability for any studies but their own.

The most celebrated achievement of Hippocrates was that 'squaring of the lune' which the Eudemian summary attributes to him. He was, however, ardently engaged on both the quadrature and the duplication-problems and added enormously, in the course of his researches, to the geometry of the circle. He wrote also the first textbook of 'Elements,' a sufficient service in itself to the cause of the science.

The first step[3] in Hippocrates' attempts at quadrature was the squaring of a particular lune as follows. On a given straight line AB, he described a semi-circle, and inscribed in this an isosceles triangle AΓB. On the equal sides of this triangle he described two other semicircles. Now in the right-angled triangle AΓB, $AB^2 = A\Gamma^2 + \Gamma B^2$, and (since circles or semicircles are to one another as the squares of their diameters)[4] the semicircle AΓB is equal to both the smaller

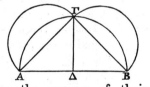

[1] Aristotle, *Eth. Eudem.* vii. 14. Joh. Philoponus *in Ar. Phys.* ed. Brandis, p. 327.

[2] Iamblichus (*De Philos. Pyth.* lib. iii., Villoison, *Anecdota Gr.* ii. p. 216) says that Hippocrates and Theodorus divulged the Pythagorean geometry. Fabricius, *Bibl. Gr.* i. p. 505 (Hamburg. 1718), referring to this passage of Iamblichus, says wrongly that Hippocrates and Theodorus were expelled from the Pythagorean school for making money by teaching geometry. See All-

man, *Herm.* vii. pp. 188, 189.

[3] Simplicius in Bretschneider, pp. 102—103. Vieta (*Opera*, p. 386), quotes these two proofs of Hippocrates from Simplicius, and Montucla follows Vieta (Bretschn. pp. 122, 123).

[4] This proposition (Euclid xii. 2) is expressly attributed to Hippocrates by Eudemus "in the second book of his History of Geometry," as quoted by Simplicius shortly afterwards (Bretschn. p. 110 top). The proposition as stated by Hippocrates seems to have

semicircles on ΑΓ, ΓΒ or is double of either of them. But the
semicircle ΑΓΒ is also double of the quadrant ΑΓΔ, which,
therefore, is equal to the semicircle on ΑΓ. Take away from
both the common part and it is seen that the triangle ΑΓΔ
is equal to the lune (μηνίσκος) which lies outside the semicircle
ΑΓΒ.

The next step[1] was as follows. In a semicircle he inscribed
half of a regular hexagon, and on
the three sides of this as diameters
he described the semicircles ΓΗΕ,
ΕΘΖ, ΖΚΔ. Then, since the
sides ΓΕ, ΕΖ, ΖΔ are equal to
the radius ΓΑ of the large semi-
circle and the semicircle on a radius is a quarter of that on a
diameter of the same circle, it follows that each of the three
smaller semicircles is a quarter of the large one. It follows that
the three smaller semicircles together with that on the radius
ΓΑ is equal to the larger semicircle. Deduct the common
parts. Then the external lunes, together with the semi-circle
on ΓΑ, are equal to the trapezium ΓΕΖΔ. But the lune
has been shown, in the first step, to be equal to a rectilineal
figure. Deduct therefore from ΓΕΖΔ the three rectilineal
figures equal to the three external lunes, and the remainder is
a rectilineal figure equal to the semicircle on ΓΑ, and twice
this rectilineal figure is equal to the circle on ΓΑ and thus the
circle is squared.

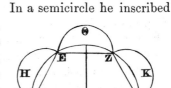

The fallacy[2] here lies, as Simplicius rightly points out, in

been (see Bretschneider, p. 120, n. 1),
that "*similar* circles are to one another as the squares of their diameters,"
from which it would appear that he
was not quite sure that all circles are
similar to one another.

[1] Simplic. in Bretschn. pp. 103, 104.

[2] ψευδογράφημα in Simplicius, i. e. a
false delineation, a fallacy founded on
a faulty diagram. The errors of Hippocrates, Antiphon and Bryson, in
their attempts to square the circle are
referred to and contrasted with one an-

other by Aristotle, *Soph. Elench.* pp. 171
b. 172: *Phys.* 185, a. and also (as well
as by Simplicius) by the commentators Themistius and Joh. Philoponus
(*Schol. in Ar.* ed. Brandis, p. 327 b.
33, 211 b. 19, 30, 41, 212 a. 16).
Bretschneider (p. 122) thinks that Hippocrates was too good a geometer
to make the mistake here attributed to
him and supposes that, in his second
step, he merely said "*If* the lune on
the side of a hexagon can be squared,
so can the circle."

assuming that the lunes in the second step are the same as those in the first step, which they are not. The first step squares the lune formed on the side of an inscribed *square* in a circle : the second step deals with lunes formed on the sides of an inscribed *hexagon*. Hippocrates seems to have felt this difficulty, for he proceeded to examine other lunes which might lead to a quadrature of the circle. Simplicius quotes from Eudemus, with some additions of his own, these further attempts. It appears that Hippocrates made some important additions to his proposition that circles are to one another as the squares of their diameters. He proved[1] that similar segments of a circle are to one another as the squares of their chords (βάσεις) ; that similar segments contain equal angles, and that in a segment less than a semicircle the angle is obtuse, in a segment greater than a semicircle the angle is acute[2]. Using these propositions he squared a lune of which the exterior arc is greater than a semicircle[3] and again a lune of which the exterior arc is less than a semicircle[4]. Lastly, he squared a lune and a circle together in the following manner[5]. Describe two circles about a common centre K, and let the square on the diameter of the exterior circle be six times the square on that of the interior. Inscribe in the inner circle a hexagon ΑΒΓΔΕΖ and draw the radii ΚΑ, ΚΒ, ΚΓ and

---

[1] Bretschneider, p. 110. Allman, *Herm.* VII. p. 197. Hippocrates defined similar segments as those which contained the same *quotum* of their respective circles, e.g. a semicircle is similar to a semicircle, a quadrant to a quadrant.

[2] He uses also the props. Euclid II. 12 and 13, but it does not appear that he invented these.

[3] Bretschneider, pp. 111, 112, fig. 8. Allman, VII. pp. 198, 199. This lune is obtained by the following construction. Hippocrates draws a trapezium having three equal sides and the fourth such that the square on it is three times the square on any other side. About this trapezium he described a circle, and

on its greater side he described a segment of a circle similar to those of which the three equal sides are the chords. The exterior arc of the lune so obtained is greater than a semicircle.

[4] Bretschneider, pp. 114—119, fig. 9. Allman, VII. pp. 199—201 (with additions and corrections to Bretschneider). The proof and even the construction are too long and complicated to be given here. The proposition is remarkable as involving the consideration of a pentagon with a re-entrant angle. This is described however as "a rectilineal figure composed of three triangles."

[5] Bretschneider, pp. 119—121, fig. 10. Allman, VII. pp. 201, 202.

produce them to meet the circumference of the outer circle in
H, Θ, I and join HΘ, ΘI, HI.
Then it is evident that HΘ, ΘI
are sides of a hexagon inscribed
in the outer circle. On HI describe a segment similar to that
cut off by HΘ. Then since the
square on HI is three times the
square on HΘ, the side of the
hexagon[1], and the square on
HΘ is six times the square
on AB, it is evident that the

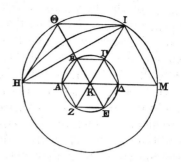

segment on HI must be equal to the sum of the segments of
the outer circle on HΘ, ΘI, together with those cut off in the
inner circle by all the sides of the hexagon. If we add to both
equals the part of the triangle HΘI lying over the segment HI,
then the triangle HΘI is equal to the lune HΘI together with
the segments of the inner circle cut off by the sides of the
hexagon: and if we add to both equals the hexagon itself, the
triangle together with the hexagon is equal to the lune HΘI
together with the interior circle.

These demonstrations, though they do not lead to quadrature
of the circle, must have greatly stimulated the study of that
problem, since they indisputably prove that some curvilinear
figures are capable of quadrature[2]. They are given here almost
verbally as they are reported by Simplicius who found them
in Eudemus who must have had them from Hippocrates' own

---

[1] The proof of this is inserted by
Simplicius. Join IM. Then the square
of the diameter HM is equal to the
squares of HI, IM, and is also equal to
four times the square of IM or any
other side of the hexagon.

[2] These performances of Hippocrates
are very neatly described by Hankel,
(p. 127) after Bretschneider. He says
that Hippocrates squared "lunes which
are contained by two arcs standing on
the same chord, the central angles of
the arcs being to one another as 1 : 2

or 1 : 3 or 2 : 3. To these surprising
discoveries he attached great hopes
and shewed that if in the same way
certain other lunes could be squared,
the quadrature of the circle would also
be solved." He adds in a note from
Clausen (Crelle's *Journ.* XXI. p. 375):
"It is interesting that the lunes squared
by Hippocrates are in fact the only
ones whose area can be constructed in
the elementary manner, with the aid
only of ruler and compasses."

work. They are interesting, apart from their intrinsic ability, as being the oldest specimens of reasoned geometrical proofs in existence. They appear to be in part taken verbally from Hippocrates, for the matter is rather confused and full of repetitions and the diction is in places archaic, witness such expressions as "the line on which AB is marked," "the point on which K stands" ($\dot{\eta}$ $\dot{\epsilon}\phi$' $o\tilde{\upsilon}$ AB, $\tau\dot{o}$ $\dot{\epsilon}\phi$' $o\tilde{\upsilon}$ K). From this it seems that letters, for the purpose of describing a geometrical figure, were of recent introduction[1]. It is to be observed also that Hippocrates does not, like Euclid, omit I from the geometrical alphabet. Another fact of great interest, if Simplicius is really citing Hippocrates, is this, that we have here the first use of the word $\delta\acute{\upsilon}\nu\alpha\mu\iota\varsigma$ in the sense of 'square,' from which the Latin translation *potentia* and the English 'power' have passed into algebraical nomenclature[2].

102. Beside the quadrature of the circle, Hippocrates was busy also with the duplication-problem. He observed that in the proportion $a : x :: x : y :: y : 2a$, since $x^2 = ay$ and $y^2 = 2ax$ and $x^4 = a^2y^2$, then $x^4 = 2a^3x$ and $x^3 = 2a^3$. Consequently, the problem of doubling a cube may be reduced to finding two mean proportionals between one straight line and another twice as long. The problem thus ceases to be one of solid and becomes one of plane geometry[3]. The same ill-luck however attended Hippocrates with this problem as with the other. He merely, as Eratosthenes in his letter above-quoted remarks, exchanged one difficulty for another. Nevertheless the duplication-problem was afterwards treated always in the form in which Hippocrates stated it and thus stereometry, as Plato complains[4], went entirely out of fashion.

103. In connexion with this recasting of the duplication-problem, Proclus (*loc. cit.*) ascribes to Hippocrates the invention of $\dot{\alpha}\pi\alpha\gamma\omega\gamma\acute{\eta}$, or geometrical *reduction*, which he defines as a

---

[1] Cantor, p. 177, surmising that letters were used with diagrams in the Pythagorean schools, points out that the letters $\dot{\upsilon}\gamma\iota\epsilon\alpha$ (Health) seem to have been placed on the vertices of the pentagram. For the Greek expressions

cf. Aristotle, *An. Priora*, p. 69 *a*.

[2] See *supra*, p. 78 *n*.

[3] Proclus, ed. Friedlein, p. 212. Eratosthenes in Eutocius *uti sup.* (ed. Torelli, p. 144).

[4] See, for instance, *Rep.* VII. 528 D.

transition from one problem or theorem to another, which being
solved or proved, the thing proposed necessarily follows[1]. The
reductio ad absurdum[2] is a particular and the commonest kind
of ἀπαγωγή, in which the substituted contrary theorem is
disproved by *analysis*. The introduction of the analytical
method of proof is attributed to Plato but it must have been
constantly resorted to before[3]. The proposition Euclid XII. 2,
which is attributed to Hippocrates by Eudemus may therefore
have been proved as it stands by Hippocrates. This style of
proof was regularly used by the Greek geometers in their
"method of *exhaustion*," i.e. the method of exhausting, by
means of inscribed and circumscribed polygons, the area of a
curvilinear figure.

104. The *process* of exhaustion was introduced, for the purposes of the quadrature-problem, about the time of Hippocrates,
by Antiphon and Bryson. **Antiphon,** a sophist who is said to
have often had disputes with Socrates[4], inscribed in a circle a
square: on the sides of this he constructed in the segments
isosceles triangles, on the sides of these other isosceles triangles
and so on, exhausting the circle: or, according to Themistius,
he began with an equilateral triangle, on the sides of which he
constructed isosceles triangles and so on. **Bryson** of Heraclea,
a Pythagorean of the same time, attacked the quadrature of
the circle by inscribing a polygon and circumscribing another[5].
He then assumed that the circle was an arithmetical mean
between the inscribed and the circumscribed polygons. Of
these two methods, the latter was more consonant with the

---

[1] In Aristotle, *Anal. Prior.* II. p. 69 a.
c. 25 (27) ἀπαγωγή is a syllogistic proof
which involves a probable assumption.
The example chosen is as follows ;
Δ is *capable of quadrature: E* a *rectilineal figure: Z* a *circle*. All *E* is Δ, but
that *Z* is *E* is one step short of certainty, since we know only that a circle
with a lune is equal to a rectilineal
figure.

[2] ἀπαγωγὴ εἰs ἀδύνατον.

[3] There are signs of it in Hippo-

crates. See Bretschneider, § 89, p.
114, *n.* Hankel, p. 149.

[4] Diog. Laert. II. 46. Bretschneider,
p. 101 (quoting Simplicius *uti supra*),
and p. 125 (quoting Themistius *in
Ar. Physica*, ed. Brandis, p. 327).

[5] Bretschneider. p. 126 (quoting Joh.
Philoponus *in Ar. Anal. Post.* ed.
Brandis, p. 211), and p. 127 (quoting
Alex. Aphrod. in *Ar. Soph. Elench.* ed.
Brandis, 306 *b*).

ancient, the former with modern notions. Upon Antiphon, Simplicius remarks, and quotes Eudemus to the same effect, that the inscribed polygon will never coincide with the circumference of the circle, else a geometrical principle would be set aside which lays down that magnitudes are divisible *ad infinitum*[1]. Antiphon, indeed, would seem to be the sole exception to the rule that the ancients never considered a circle as a polygon with an infinite number of sides. "This principle" says Chasles[2], not adverting to Antiphon "has never appeared in their writings : it would not have suited the rigour of their demonstrations. It was the moderns[3] who introduced it into geometry and simplified thereby the ancient demonstrations. This happy idea was the passage from the method of exhaustion to the infinitesimal method."

105. It being admitted that Antiphon and Bryson introduced the practice of exhaustion and that Hippocrates shows signs of using analysis in geometry, the question arises whether he did in fact prove Euclid XII. 2. as it stands. If he did, then he invented that rigorous mode of proof, called "the method of exhaustion," which is generally attributed to Eudoxus. This method may be considered as contained in two propositions[4], as follows.

(1) If A and B be two magnitudes of the same kind, of which A is the greater, and there be taken from A more than its half (or any other fraction) and from the remainder more than its half (or any other fraction) and so on, the ultimate remainder will be less than B. (This is the prop. Eucl. X. 1. now prefixed as a *Lemma* to the 12th Book.)

(2) Let there be two magnitudes P and Q, both of the same kind and let a succession of magnitudes X_1, X_2, X_3 etc.

---

[1] Bretschneider, p. 102. This is the proposition which Zeno denied and which Aristotle is always supporting. See especially his treatise περὶ ἀτόμων γραμμῶν, shewing that there are no *indivisible* lines, and compare Hankel, pp. 117—120.

[2] *Aperçu*, p. 16.

[3] Kepler (in his *Nova Stereometria Doliorum*) and Descartes. Cf. Hallam, *Hist. of Lit*. Pt. III. c. 8, secs. 9, 14.

[4] See De Morgan's article 'Geometry of the Greeks' in the *Penny Cyclop*. Cantor, pp. 233, 234. Hankel, pp. 122—124.

be each nearer and nearer to P, so that any one X_r, shall differ from P less than half as much as its predecessor differed. Let Y_1, Y_2, Y_3 etc. be a succession of magnitudes similarly related to Q, and let the ratios $X_1 : Y_1$, $X_2 : Y_2$, etc. be all the same with each other and the same as $A : B$. Then $P : Q :: A : B$. (Suppose X_1, X_2 etc. all *less* than P and Y_1, Y_2 etc. *less* than Q.) Now if $A : B$ is not $P : Q$, then A is to B as P is to some other quantity S, greater or less than Q. Take S less than Q. Then by the hypothesis and prop. I. we can find one of the series Y_1, Y_2 (say Y_n) which is nearer to Q than S is, and is therefore greater than S. Then, since $X_n : Y_n :: A : B :: P : S$, it follows that $X_n : P :: Y_n : S$. But X_n is less than P, therefore Y_n is less than S. But Y_n is also greater than S, which is absurd. In like manner, it may be proved that, if S be taken greater than Q, then the proportion $A : B :: P : S$ is an absurdity. Therefore $A : B :: P : Q$. (*Vide* Euclid XII. 2.)

106. The discussion of the question whether Hippocrates or Eudoxus was the author of this method proceeds on the following lines. The opening *lemma* was the mathematicians' evasion of the difficulty which Zeno had found in *infinite* division. They avoided the expression "infinitely small magnitudes" by substituting for it "magnitudes as small as we please." Now Archimedes[1] says that this *lemma* (in a different form) was used by "former geometers" for the theorem *Eucl.* XII. 2. Eudemus attributes this theorem to Hippocrates and there is in fact no other way of proving it save by the method of exhaustion, which Euclid adopts. Dr Allman[2] replies that Archimedes mentions this theorem not with particular emphasis but along with three others, two of which were beyond question proved by Eudoxus[3], who is said also to be the author of the theory of proportion contained in Euclid V. He does not, however, suggest any proof of the theorem which Hippocrates might possibly have arrived at without using the *lemma*. Here Cantor[4] is more satisfactory, for he points out that the Egyptians had long ago adopted a fixed arithmetical ratio

---

[1] *Pref.* to *Quadr. Parab.* (Torelli's ed. p. 18). Hankel, p. 120 *sqq.*
[2] *Hermathena* VII. pp. 222—223.
[3] Archim. *Pref.* to *Sph. et Cyl.* Torelli's ed. p. 64.
[4] p. 178.

between a circle and the square of its diameter and Hippocrates may have known this through the Pythagoreans. It is probable, moreover, as will be seen presently, that Plato first raised analysis to the dignity of a legitimate geometrical method. The evidence, therefore, inclines to the opinion that the method of exhaustion is to be ascribed not to Hippocrates, but to Eudoxus, who lived nearly a century later, but we know, in truth, so little about the Greek geometry of the period that no man is entitled to hold this opinion very confidently.

(e.) The Academy.

107. Plato was born of wealthy and distinguished parentage, at Athens in 429 B.C. the year of the great plague. He was a pupil of Socrates, who was executed 399 B.C., but he did not derive from this teacher his enthusiasm for mathematics, since Socrates was of opinion that it was no use learning more geometry or astronomy than would suffice for daily wants, such as to measure a field or tell the time of day[1]. But Plato, after the death of Socrates, went away from Athens and consorted in many places with Pythagoreans who no doubt indoctrinated him with a passion for their favourite science. He went certainly to Egypt, also to Cyrene where he studied with Theodorus, and lastly to Magna Graecia and Sicily (in B.C. 389) where he became a close friend of Archytas and Timæus of Locri. He returned to Athens and formed about himself a school of students who heard his discourses in the grove of the *Academia*, a suburban gymnasium. He died, at the age of 81, in 348 B.C.

The physical philosophy of Plato, being partly founded on the Pythagorean, is partly, like the latter, an attempt to find in arithmetic and geometry the key to the universe. He held that God was a great geometer[2] and therefore made a knowledge of geometry an indispensable preliminary to the study of philosophy. It is said that he inscribed over his porch "Let

[1] Xenophon, *Memorab.* IV. 7. Diog. Laert. II. 32.

[2] According to Plutarch, *Quaest. Conv.* VIII. 2, Πῶς Πλάτων ἔλεγε τὸν θεὸν ἀεὶ γεωμετρεῖν; The expression does not occur in any extant work of Plato's, but he does say (*Rep.* 527 B) that geometry, rightly treated, is the knowledge of the eternal.

none that is ignorant of geometry enter my doors"[1] and Xenocrates also, who after Speusippus succeeded to the professorial
chair, so to say, of the Academy, is reported to have turned
away an applicant for admission who knew no geometry, saying
"Depart, for thou hast not the grip of philosophy."[2] But
it was not really with a view to physical speculation that Plato
thus glorified geometry. He was interested, no doubt, in the
inanimate world but he was interested far more in man. The
nature and laws of thought and the rules of conduct were his
especial subject, and he valued geometry mainly as a means of
education in right seeing and thinking and in the conception of
imaginary processes. Hence it was that, as the Eudemian
summary says, "he filled his writings with mathematical discourses, and exhibited on every occasion the remarkable connexion between mathematics and philosophy." This statement
may be illustrated by two interesting passages in the *Meno*, a
dialogue on Virtue which Socrates is supposed to hold with
Meno, a pupil of the sophist Gorgias. In the first of these
passages[3], Socrates has just suggested that the knowledge
which we seem to have by intuition, is really recollected from
a former state of existence, that in fact "our birth is but a
sleep and a forgetting." In illus-
tration of this theory, he calls up
one of Meno's slaves and draws
before him, by several steps, the
accompanying figure, a square of
4 feet. The boy apprehends the
steps perfectly well and correctly
answers Socrates' questions[4], until
at length Socrates, having induced
him to say that the square obliquely

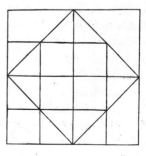

placed is double of the square of 2 feet with which the diagram

---

[1] μηδεὶς ἀγεωμέτρητος εἰσίτω μου τὴν
στέγην. Tzetzes. *Chil.* VIII. 972.

[2] πορεύου, λαβὰς γὰρ οὐκ ἔχεις τῆς
φιλοσοφίας. Diog. Laert. IV. 10.

[3] *Meno*, 82 B—85 B.

[4] The conversation begins as follows.

Socr. 'Tell me, boy, do you recognise
this for a square?' *Boy*, 'Yes.' *Socr.*
'Is not it a square that has all these
four lines equals?' *Boy*, 'Of course.'
Socr. 'And these cross-lines equal too.'
Boy, 'Yes,' etc. etc.

was commenced, elicits the incorrect answer that the *side* of the oblique square is twice 2 feet. Upon this Socrates retraces his steps and, by judicious questions, leads the boy on to say that the side of the oblique square is more than 2 and less than 4 and yet is not 3 feet. Here it is obvious that Plato is interested not in the Pythagorean theorem or incommensurable lines, but in the chain of reasoning. Similarly, a little later in the dialogue[1] Meno asks Socrates whether virtue may be imparted by teaching. To this Socrates replies "Let me argue this upon hypothesis. A geometer, if he were asked 'Can this area (i.e. the square of 2 feet) be inscribed in this circle[2]?' might say 'I don't know, but I think I can suggest a useful hypothesis. If this area is such that, when applied to the given diameter, it is deficient ($\dot{\epsilon}\lambda\lambda\epsilon\dot{\iota}\pi\epsilon\iota$) by an area equal to itself, then one consequence follows, but if this be impossible, then another[3].' So in the case of virtue, we must assume virtue to be or not to be (*ex hypothesi*) one of some class of goods etc." Here also it is the logical procedure and not the problem which is intended to be observed. The reader therefore is prepared to find, as the fact is, that Plato was rather a maker of mathematicians than himself distinguished for original discoveries and that his contributions to geometry are rather improvements in its method than additions to its matter. It was he who turned the instinctive logic of the earlier geometers into a method to be used consciously and without misgiving[4]. With him, apparently, begin those careful definitions of geometrical terms, that distinct

---

[1] *Meno*, 86 D—87 A.

[2] Socrates may, early in the dialogue (73 E), have drawn a circle on the ground.

[3] The text of this passage (which is absurdly translated by Jowett) is extremely obscure, but it seems certainly to refer to the previous figure. A square of 2 ft. is there shown to be equal to an isosceles right-angled triangle of which the base is 4 ft. If the diameter of the given circle is 4 ft. a triangle equal to the given square of 2 feet can certainly be inscribed in it.

A square of 2 ft. applied to a line of 4 ft. is *deficient* by a square of 2 ft. (See *supra* p. 84 *n.*) This explanation seems to be Benecke's (*Ueber die Geom. Hypoth. in Plato's Menon*, 1867). Hankel (p. 134 *n.*) says that the text is unnecessarily difficult for describing so simple a fact, but I am inclined to think that Plato was fond of 'showing off' his mathematics. The famous 'Nuptial number' in *Rep.* 546 B.C. is an instance in point.

[4] See a brilliant chapter of Hankel, pp. 127—150.

statement of postulates and axioms, which Euclid has adopted. The Pythagorean so-called definitions, such as "A point is unity in position," are not explanations of terms but statements of a philosophical theory. But the Academics, as became the pupils of Socrates, desired explicit determination of the meanings of words. Thus Aristotle[1] says that Plato objected to calling a point a 'geometrical fiction' ($\delta\acute{o}\gamma\mu a$), and defined it as 'the beginning of a straight line' or 'an indivisible line.' Aristotle gives also as definitions customary in his time the following: 'the point, the line, the surface are respectively the boundaries of the line, the surface and the solid:' 'a line is length without breadth:' 'a straight line is one of which the middle point covers both ends' (the eye being placed at either end of the line): the surface arises from 'the broad and the narrow:' 'a solid is that which has three dimensions[2].' So also Aristotle refers to 'mathematical axioms' and often quotes one of them, viz. 'If equals be taken from equals the remainders are equal[3].' Although probably not all of these definitions and axioms are due to Plato himself, yet one great invention in geometrical methods is expressly attributed to him. Both Proclus (ed. Friedlein p. 211) and Diogenes Laertius (III. 24) state that Plato invented the method of proof by *analysis*[4]. It is not, indeed, to be supposed from this that analysis was new to Greek geometers for Hippocrates uses it, as was above-stated, and most of the early geometers probably were led, by the contemplation of constructions, to the invention of theorems, and were thus using analysis without knowing it. But Plato may very well have introduced analysis as a legitimate method

---

[1] *Metaphys.* I. 9. 992, a. 20.

[2] The passages here quoted are in order Aristotle, *Top.* VI. 141 *b.* 19, 143 *b.* 12, 148 *b.* 29, *Metaphys.* I. 9, 992 *a.* 12: *Top.* VI. 5, 142 *b.* 24, obs. Aristotle calls a point $\sigma\tau\iota\gamma\mu\acute{\eta}$, the later word being $\sigma\eta\mu\epsilon\hat{\iota}o\nu$: and a surface $\dot{\epsilon}\pi\acute{\iota}$- $\pi\epsilon\delta o\nu$, later $\dot{\epsilon}\pi\iota\phi\acute{a}\nu\epsilon\iota a$, the former word being later appropriated to 'plane.'

[3] Metaphys. IV. 3, 1005 *a.* 20, XI. 4, 1061 *b.* 17: *An. Post.* I. 11, 77 *a.* 31

(with the addition $\mathring{\eta}$ $\tau\hat{\omega}\nu$ $\tau o\iota o\acute{v}\tau\omega\nu$ $\mathring{a}\lambda\lambda a$ 'or any such axioms'). Hankel, p. 136 *nn.*

[4] Both state also that he "gave it to Leodamas of Thasos, which probably means that Plato orally described the method to Leodamas and the latter wrote or lectured upon it, describing it as Plato's but giving his own geometrical illustrations.

in geometry to be consciously employed, may have given rules
for its conduct and pointed out under what conditions it was
satisfactory or not so.

108. The oldest definition of *Analysis* as opposed to
Synthesis is that appended to Euclid XIII. 5. It was possibly
framed by Eudoxus [1]. It states that "Analysis is the obtaining
of the thing sought by assuming it and so reasoning up to an
admitted truth: synthesis is the obtaining of the thing sought
by reasoning up to the inference and proof of it." In other
words, the synthetic proof proceeds by shewing that certain
admitted truths involve the proposed new truth: the analytic
proof proceeds by shewing that the proposed new truth involves
certain admitted truths. An analytic proof begins by an
assumption, upon which a synthetic reasoning is founded. The
Greeks distinguished *theoretic* from *problematic* analysis. A
theoretic analysis is of the following kind. To *prove* that A is
B, *assume* first that A is B. If so, then, since B is C and C is
D and D is E, therefore A is E. If this be a known falsity [2], A
is not B. But if this be a known truth and all the intermediate propositions be convertible, then the reverse process,
A is E, E is D, D is C, C is B, therefore A is B, constitutes a
synthetic proof of the original theorem. Problematic analysis
is applied in cases where it is proposed to construct a figure
which shall satisfy a given condition. Hence the process consists in constructing a figure which is assumed to satisfy the
given condition. The problem is then converted into some
theorem which is involved in the condition and which is proved
synthetically, and the steps of this synthetic proof taken

[1] Bretschneider, p. 168. Pappus
(*Math. Coll.* VII. ed. Hultsch, p. 635) has
Euclid's definition. Chasles (p. 5),
takes a definition from Vieta, *Isagoge
in Artem Analyticen, ad init.* "Il est en
mathématiques une méthode pour la
recherche de la vérité que Platon passe
pour avoir inventée, que Théon a nommée
analyse et qu'il définit ainsi : Regarder
la chose cherchée comme si elle était
donnée et marcher de conséquences en

conséquences jusqu'à ce que l'on recon-
naisse comme vraie la chose cherchée.
Au contraire la synthèse se définit :
Partir d'une chose donnée pour arriver
de conséquences en conséquences à
trouver une chose cherchée." See also
a note in Todhunter's *Euclid.* App.
§§ 35 sqq.

[2] Thus the *reductio ad absurdum* is a
kind of theoretic analysis. This is the
only analysis which Euclid admits.

backwards are a synthetic solution of the problem. Suppose there is only one condition : e.g. *To describe a triangle having each of the angles at the base double of the third angle.* Draw an isosceles triangle *ABC* and assume that the base angles are each double of the vertical angle. An addition must be made to the figure. Bisect the angle *ACB* by the straight line *CD*. There thus arises a theorem that *AB* is cut, at *D*, in extreme and mean ratio, and that *BC = AD*, from which a synthesis (Eucl. IV. 10) is obtained. (It will be seen that the whole aim of problematic analysis is

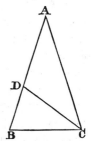

to find a synthetic solution of the problem, and therefore the ancient geometers never omitted to add the synthetic solution so found.) If there are more conditions than one, the procedure is just the same. A figure is drawn which is assumed to satisfy all the conditions, but the subsequent analysis is directed to shewing what each condition, in turn, involves.

A very good authentic example of this more complicated analysis is given by Hankel (p. 143) from Pappus[1]. The problem is : "Given the position of a circle *ABC* and of two points *D, E*, outside it, it is required to draw from *D, E* the straight lines *DB, EB* cutting the circle in *B* and produced to *A, C*, so that *AC* shall be parallel to *DE*." The analysis is as follows. "Let the figure be drawn and also the tangent *FA*. Then, since *AC* is parallel to *DE*, the angle at *C* is equal to the angle *CDE*. It is also equal to the angle *FAE* (Euclid III. 32). Therefore the angle *FAE* is equal to the angle *CDE*, and the points *ABDF* lie on the circum-

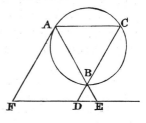

ference of a circle, and the rectangle *AE, EB* is equal to the rectangle *FE, ED*. But the rectangle *AE, EB* is given[2], because it is equal to the square of the tangent ; therefore

---

[1] *Coll. Math.* VII. prop. 105.

[2] δεδομένον, *datum*, in Euclid's sense. Euclid's *Data* are a collection of theorems stated in an abridged form, e.g. Prop. 92 (95 in Simson's *ed.*) is "If a straight line be drawn from a

the rectangle FE, ED is given, and since ED is given, so also is FE, both in length and in position. And since FA is a tangent to the given circle and F is given, so also A is given. And since A is given, so also is AE and the point B." Then follows the synthesis. "Join ED and produce it to F, so that the rectangle ED, EF is equal to the square of the tangent. From F draw the tangent FA and join AE," etc. The reader will see that here the analysis is directed to two facts involved in the conditions, the condition that AE cuts the circle in B, involves the fact that the rectangle $AE \cdot EB$, wherever A may be, is equal to a certain square which can be found. The condition that AC must be parallel to ED, involves the further fact that the rectangle $AE \cdot EB$ is equal to the rectangle under ED and EF, where F is that point in which ED produced meets the tangent at A. The addition of a synthetic solution is made *ex majori cautela*, lest a condition should not have been examined in the analysis or lest a proposition reached in the analysis should not be convertible (e.g. all A may be B, but not all B need be A). Further, the problem may be under some conditions impossible, and this fact is likely to be overlooked in the analysis. Hence, to the synthetic solution, the Greeks appended, if necessary, a *diorismus* (*determinatio*) or statement of the conditions in which the given problem is or is not soluble. The Eudemian summary ascribes the invention of the *diorismus* to Leon the Platonist, but it is observable that the passage above quoted from the *Meno* (p. 175) contains a partial *diorismus* which is undoubtedly Plato's. It is probable therefore that the whole systematization of analysis is due to Plato. "The conjunction of philosophical and mathematical productivity" says Hankel, "such as we find, beside Plato, only in Pythagoras, Descartes and Leibnitz, has always borne the finest fruits for mathematics. To the first we owe scientific mathematics in general. Plato discovered the analytic method,

given point without a circle given in position, the rectangle contained by the segments betwixt the point and the circumference of the circle is given." This is an abridgement of III. 36. One sense of the word "given" is determined by Def. I. "Spaces, lines, and angles are said to be *given in magnitude* when equals to them can be found."

through which mathematics were raised above the standpoint of the Elements; Descartes created analytical geometry; our own celebrated countryman (Leibnitz) the infinitesimal calculus— and these are the four greatest steps in the development of mathematics." It must be admitted, however, that the introduction of analysis is just the sort of service which might be ascribed, by a vague exaggeration, to a philosopher who certainly had a great influence on mathematics but left no mathematical work.

109. The one respectable solution which is attributed to Plato seems to have been obtained through analysis in the first instance. It will be remembered that Hippocrates had recast the duplication-problem into one of plane geometry, to find two mean proportionals to two straight lines. Let a, b be the given straight lines, x and y the mean proportionals, so that $a : x :: x : y :: y : b$. Take $CA = a : CX = x : CY = y : CB = b$

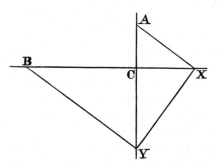

and place these lines in a right-angled cross about the common extremity C. Then the triangles ACX, XCY and YCB are similar, and the angles AXY, XYB are both right-angles (Euclid VI. 8 and *Cor.*). Hence a synthetic solution would be obtained if a straight line XY could be so placed between two arms of the cross that the perpendiculars to it at the points X and Y would pass through A and B respectively. For this purpose Plato[1] is said to have invented a little apparatus consisting of a rectangular frame, one side of which would slide up

[1] Eutocius in Torelli, p. 135. Hankel, pp. 154, 155. Bretschneider, pp. 141—143. Cantor, p. 195.

and down so as to diminish or enlarge the rectangle at pleasure. On the other hand, Plutarch relates[1] that Plato blamed Archytas and Eudoxus and Menaechmus for using such instruments for the purpose of solving the duplication-problem, and said that the good of geometry was spoilt and destroyed thereby, for it was reduced again to the world of sense and prevented from soaring among the unseen and incorporeal figures. Elsewhere[2] Plutarch repeats the same story, and adds that, owing to this remonstrance of Plato, mechanics were wholly dissociated from geometry and reduced to a mere department of strategy. These statements of Plutarch are much more likely to be true than the other, which rests on the authority of Eutocius, and it may be said therefore that to Plato we owe the strict limitation of geometrical instruments to the ruler and compasses. It will be remembered also that Plato deplored the decay of stereometry, and we shall find this department of geometry reopened with great zeal by Plato's immediate pupils. In short, however we discount the evidence, it is plain that Plato was almost as important as Pythagoras himself to the advance of Greek geometry.

110. It is desirable for two or three reasons to insert here the solution of the duplication-problem which is attributed to the Pythagorean Archytas. It could hardly be given before, because it solves the problem in that form in which Hippocrates recast it. Further, it is the kind of solution which Plato blames and it involves some stereometrical considerations, which Plato is thought to have revived. It will serve also to remind us that, side by side with the Athenian mathematical school, there was still the older Pythagorean at Tarentum, to which Plato was probably under very great obligations. The solution of Archytas[4] is reported by Eutocius[3] from Eudemus. It is as

---

[1] *Quaest. Conv.* VIII. 9, 2, c. 1. Bretschneider, p. 143.

[2] *Vita Marcelli*, c. 14, § 5.

[3] In Torelli, p. 143. The form given in the text is Cantor's (p. 196), but is only very slightly abridged from the original. The latter gives the synthesis only. Bretschneider, p. 152, sug-

gests an analysis by which Archytas may have been led to his solution. The figure on the next page is awkward and defective, for MI and KΔ' should be joined, but it serves its purpose sufficiently well and is a little better than Cantor's.

[4] See p. 318.

follows.   Let AΔ, AB be the two straight lines, between which two mean proportionals are required, and of these let AΔ be the greater.   Describe a semicircle on AΔ as diameter, and let AB be a chord of this semicircle.   Describe on AΔ, in the perpendicular plane, another semicircle which can be moved round

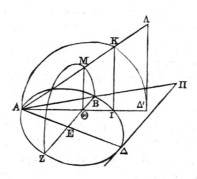

from Δ towards B, the extremity of the diameter A remaining fixed.   This revolving semicircle will describe a curve on a half-cylinder supposed to stand on ABΔ.   Draw the tangent at Δ and produce AB to meet it in Π.   The triangle AΔΠ turning about AΔ as axis, produces a cone, which penetrates the half-cylinder and cuts the curve thereon in the point K.   This point K being on the half-cylinder, the line KI drawn from it perpendicular to the plane of the semicircle ABΔ meets the circumference of that semicircle in I.   While AΠ is describing this cone, the point B moves through a circle, BMZ, which is perpendicular to the circle ABΔZ of which ABΔ is half.   Since AKΔ′ is perpendicular to the same plane, the line MΘ (which is the line of section of BMZ and AKΔ′) is also perpendicular to it, and is likewise perpendicular to the line BZ, which is the line of section of the plane BMZ with the plane ABΔZ.   Then since BMZ is a semicircle and BZ its diameter, $MΘ^2 = BΘ . ΘZ$.   But BΘ . ΘZ = AΘ . ΘI (BZ and AI being two chords cutting one another in Θ).   Therefore $MΘ^2 = AΘ . ΘI$.   Therefore the angle AMI is a right angle and is equal to AKΔ′ (which is an angle in a semicircle), and therefore MI is parallel to KΔ′.   Therefore the triangles Δ′AK, IAM, KAI are similar to one another, and

AM : AI :: AI : AK :: AK : Δ'A. Take AM = AB = a, and Δ'A
= AΔ = b, and we then get two middle proportionals between a
and b. It will be seen that this solution uses Euclid III. 18 and
35, and XI. 19, and discloses also very clear notions on the origin
of cylinders and cones, the section of a surface by a surface and
the curves thence arising. It must be remembered however that
Archytas probably used a mechanical apparatus in the solution
(*supra*, pp. 158, 181).

111. It was said above that Plato made many mathematicians, and the observation is fully borne out by the discoveries
which are attributed to his immediate pupils. Of **Leodamas**
of Thasos, for whom Plato invented the method of analysis,
nothing more is known, save what the Eudemian summary says
of him, viz. that he and Archytas and Theaetetus greatly
extended mathematical inquiries, and improved them into a
more scientific system. This **Theaetetus** is the same who gives
a name to one of Plato's dialogues and who was chiefly occupied
with the study of incommensurables. Suidas (*s. v.*) attributes
to him a treatise on the five regular solids, but to what effect
this treatise was is not known. Of **Neocleides** and his pupil
Leon also, we know no more than the Eudemian summary tells
us, in which the only important fact is that Leon wrote an
improved ' Elements ' and treated particularly of *diorismus*.

112. But **Eudoxus**, who is mentioned next, was one of
the most brilliant mathematicians of antiquity. He was born
about B. C. 408 in Cnidus, was a pupil of Archytas, and subsequently, for a few months, of Plato. He then went to Egypt
(with Plato, according to Strabo), thence to Cyzicus, where
he founded a school, and came from Cyzicus with his pupils
to Athens, where he met Plato again not on very friendly
terms. He returned finally to Cnidus and died there at the
age of 53 (B. C. 355).[1] He is described by Diogenes Laertius
as astronomer, physician and legislator, as well as geometer.
In the first capacity he is said by Aristotle[2] to have made
a kind of orrery, and various discoveries are attributed to

---

[1] Diogenes Laert. VIII. 86—90. Bretschneider, pp. 163—164. Cantor, pp.
205, 206.

[2] *Metaphys.* VII. c. 8. See Schiaparelli (trans. Horn) in *Suppl.* to
Zeitschr. Math. Phys. vol. XXII.

him.   He wrote also a work on practical astronomy, Φαινόμενα,
on which the extant poem of Aratus is founded[1]. The
Eudemian summary states that he added three kinds of
proportion to those introduced by Pythagoras, and increased
by the analytical method the learning, begun by Plato, on the
subject of the *section*.   This must mean the so-called 'Golden
Section,' the cutting of a line in extreme and mean ratio[2],
and the Eudemian summary is very well explained by supposing
that Eudoxus was in fact the author of the first five propositions
of Euclid XIII., which deals with the regular solids (see below
p. 197*n*).   A scholiast on Euclid, thought to be Proclus[3], says
further that Eudoxus invented practically the whole of
Euclid's Fifth Book.   Beside this work in proportion,
Archimedes expressly says (in the passages above cited upon
which the method of exhaustion is attributed to Hippocrates)
that Eudoxus proved by means of the Lemma, Euclid X. (XII),
1, the propositions that every pyramid is a third of a prism on
the same base and with the same altitude (Euclid XII. 7. *Cor.* i.),
and that every cone is the third part of a cylinder on the same
base and with the same altitude (Eucl. XII. 10).   It is on the
strength of this perfectly clear evidence that Eudoxus is
supposed to have invented the method of exhaustion.   Lastly,
Eudoxus is reported[4] to have invented a curve which he
called ἱπποπέδη, or 'horse-fetter,' and which resembled those
hobbles which Xenophon describes as used in the riding-school.
They were of the form

Proclus calls this curve a 'spiral,' and has some interesting
remarks on its origin[5].   The word σπεῖρα means a so-called

---

[1] Aratus is criticised by Hipparchus, who preserves some of the original statements of Eudoxus. These are criticised, as usual, with the utmost contempt by Delambre. *Astron. Ancienne*, Vol. I. p. 107.

[2] The possible meanings are discussed by Bretschneider, pp. 167—168. Cantor, p. 208.

[3] Knoche, *Untersuch. über Schol. des Proclus*, pp. 10—13.

[4] Simplicius, in *Ar. De Coelo*, ed. Brandis, p. 500, 10.

[5] Cantor pp. 209, 210, quoting Schiaparelli, *uti sup.* section v. Proclus, ed Friedlein, pp. 112, 119, 127, 128. Heron Alexandrinus, ed. Hultsch, p. 27, *def.* 98.

'tore,' a ring-shaped solid of revolution which is produced
by the revolution of a circle about a straight line which lies
in the same plane with but does not cut the circle[1]. If this
solid be cut by a plane, there arises a "spiral" line, which
may assume three forms, according as the cutting plane is
more or less near the axis. If it is further from the axis than
the centre of the circle, we get an oval[2]: if nearer, we get a
curve "narrower in the middle and broader at the ends; but if
the plane is still nearer to the axis, so that it
touches the tore at an inner point, which is in
fact the double-point of the curve, we get the
ἱπποπέδη." Eudoxus somehow used this curve
in his description of planetary motions, but
nothing more is known of his treatment of it.
Eutocius, however, in the passage so often quoted
on the duplication-problem, says that Eudoxus used certain
curves (καμπύλαι γραμμαί) for his solution of this problem, but
he disdains to give his solution, because it had nothing to do
with these curves after all and contained an absurd mistake in
proportion. Eratosthenes, however, whose letter on the subject
Eutocius has himself previously quoted, mentions Eudoxus in
the same breath with Archytas, and calls him, in an epigram
appended to the letter, "godlike." The probability, therefore,
is that Eutocius was himself mistaken.

113. Amyclas of Heraclea, a Platonist, is unknown save
from the Eudemian summary, but **Menaechmus**, "a pupil of
Eudoxus and a contemporary of Plato," is well known to fame.
It was he who invented the geometry of the conic sections[3],
which, after him, were sometimes called "the Menaechmian
triads." Democritus, indeed, had cut a cone by a plane parallel to
the base (*supra*, p. 159*n*.) but it was Menaechmus who took three
kinds of right cones, the "right-angled," "acute-angled" and
"obtuse-angled" (as Geminus describes them in the passage

---

[1] An anchor ring is the common example of this solid.

[2] Proclus (p. 112) describes this as
a "παραμηκής, broad in the middle
and narrow at the ends."

[3] See the often-quoted letter of Eratosthenes in Eutocius, ed. Torelli, p.
146, and Proclus, *ed.* Friedlein, p. 111,
(citing Geminus as well as Eratosthenes).

quoted above, p. 137) and cutting these by planes at right-angles to one of their sides, exposed the parabola, ellipse and hyperbola[1], which he called the section of the right-angled, acute-angled and obtuse-angled cone respectively. He seems, however, not to have regarded these curves always in the cone itself, but to have used some mechanical apparatus for drawing them[2]. How far he proceeded in the geometry of conics can be guessed only from two very neat solutions of the duplication problem, which are attributed to him by Eutocius[3]. Menaech-

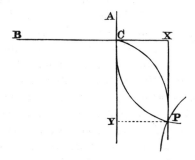

mus observed that if $a : x :: x : y :: y : b$, then $ay = x^2, y^2 = bx$. If through the point C there be described two parabolas, one

[1] These *names* were invented not by Menaechmus but by Apollonius of Perga, a century later. That Menaechmus used the names "section of right-angled cone" etc. is attested by Pappus, vii. (ed. Hultsch), p. 672.

[2] Compare Eratosthenes again in Eutocius, *ed.* Torelli, p. 144, and the reproach of Plato against Menaechmus. Eutocius (a little earlier) says that his own master, Isodorus of Miletus, had invented a compass (διαβήτης) for drawing parabolas. Bretschneider (p. 170), says that a modern geometer would suppose that the notion of *loci* preceded the conic sections, but that in fact the ancients always regarded conics in the cone itself: the *foci* of the ellipse and hyperbola are only just mentioned by Apollonius and charac-

terised by two of their simplest properties: the *focus* of the parabola is not mentioned at all. But Mr Taylor (*Ancient and Modern Conics*, pp. xxxi —xxxiii. and xliii.) suggests that the conic sections may have been discovered as plane loci in investigations of the duplication problem. In support of this he urges that Menaechmus used a machine for drawing conics, that in his solutions of the duplication problem he uses only the parabola and hyperbola, and that the ellipse, the most obvious of the sections, is treated last by Apollonius. He admits, however, that the conception of a conic as a plane locus was immediately lost.

[3] Archimed. *ed.* Torelli, p. 141. Bretschneider, pp. 159—161. Cantor, pp. 198, 199. Hankel, p. 155. Eutocius

with parameter a and axis AC, the other with parameter b and axis BC, they will cut one another in a point P of which the co-ordinates CX, CY are the desired mean proportionals, x and y. Secondly, since $xy = ab$, the same point P may be found as the point of section between one of these parabolas and a hyperbola, of which CX and CY are asymptotes, subject to the condition that the rectangle contained by the straight lines drawn from any point on the hyperbola parallel to one asymptote and meeting the other, shall be equal to ab. The learning of Menaechmus, indicated by these solutions, is very considerable, and it is not surprising to find that before the century was out **Aristaeus** "the elder"[1] (about B.C. 320) wrote an "Elements of Conic Sections" in five books, which, according to Pappus, Euclid highly approved[1]. Menaechmus is said to have been the teacher of Alexander the Great, who asked him whether he could not make his instructions somewhat shorter. To this Menaechmus replied in the famous words, "There is no royal road to geometry[2]." The brother of Menaechmus, **Dinostratus**, was also a great geometer. It was he who as stated above (p. 163) used the *quadratrix* of Hippias for the solution both of the trisection and the duplication problems. Nothing more than this, however, is known of him,

gives both analysis and synthesis. The form in the text is Hankel's, and contains of course (as indeed does that of Eutocius) technical terms which Menaechmus knew nothing of. It gives the solutions also in a different order from Eutocius.

[1] Pappus VII. *Praef.* (ed. Hultsch), pp. 672, 676. This Aristaeus wrote also on the regular solids. Cantor, p. 212. Curves of all kinds were at this time called τόποι διεξοδικοί or "running loci." The straight line and circle were called "plane loci" (τόποι ἐπίπεδοι) the conic sections "solid loci" (στερεοί) and all other curves beside these were called "linear loci" (γραμμικοί) or, from the manner of their construction, "mechanical loci" (μη-χανικοί). Hankel, p. 152. Cantor, p. 214, quoting Pappus *uti sup.* pp. 662, 672. Pappus seems to say that Aristaeus wrote two books, one on Conics, the other on Solid Loci. Cantor (*loc. cit.*) suggests that the second was a series of solutions in which the conic sections were used. Viviani restored it conjecturally (*pub.* 1701). Chasles, p. 7n. Cantor (p. 197) says that the word τόπος first occurs, in its geometrical sense, in Eutocius' report of Archytas on duplication, but this is erroneous. The word τόπος there means only "place."

[2] Bretschneider, pp. 162—163. Stobaeus *Florileg.* ed. Meineke, IV. p. 205. The same story, however, is told of Euclid and King Ptolemy.

save what the Eudemian summary relates, viz. that he and Menaechmus and Amyclas "made the whole of geometry yet more perfect." After these names the same catalogue gives others, for which also it is the sole authority. **Theudius** of Magnesia wrote yet another and further improved 'Elements': he and **Cyzicenus** of Athens were Platonists, like Amyclas. **Hermotimus** of Colophon added yet more to the Elements and wrote on *Loci*. **Philippus** of Mende, another Platonist, tried to find in geometry illustrations of the Academic philosophy. This name finally brings the history of geometry down to the time of Eudemus[1] himself, and with it the Eudemian summary closes. The next great name in Greek geometry is Euclid but with him the scene is shifted and mathematics desert Athens for Alexandria.

114. The Athenian school, however, should not be left, without a word on that great philosopher who, for nearly 2000 years, was in all subjects "the master of those who know." **Aristotle** (B.C. 384—322) was not, any more than Plato, a professed mathematician, but, like Plato, he was learned in the mathematics of his day and was above all things interested in correct reasoning on every subject. The man who systematized deductive logic must be admitted to have performed a great service to geometry. But Aristotle's benefits are not confined to this. He is the author or the improver of many of the most difficult geometrical definitions (*vide supra*, p. 176*nn.*). One of these, which has not yet been quoted, may be here given. He defined *continuity* as follows. "A thing is *continuous* ($\sigma\upsilon\nu\epsilon\chi\acute{\epsilon}\varsigma$) when of any two successive parts the limits, at which they touch, are one and the same, and are, as the word implies, held together[2]." Hence, he said in answer to Zeno, motion is not, like counting, a discrete operation, a series of jerks: the moved thing does not stop at the stages which the calculator chooses to make. The interest which Aristotle took in these inquiries accounts for the fact that the sole extant Greek work in which,

---

[1] Eudemus was a native of Rhodes, but a pupil of Aristotle. Proclus says he wrote περὶ γωνίας, 'on the angle,' so it may be inferred that he was a com-petent historian of geometry.

[2] *Phys.* III. c. 3. 227, *a*. 10. *Insec. Lin.* 969, *a*. 30.

before Archimedes, geometry is applied to mechanics is attributed to him. The passage in the *Physics* which contains a hint of the principle of virtual velocity has been cited above (p. 105n)[1]. The *Mechanica,* which is perhaps not Aristotle's, though certainly of his time, is a series of 35 questions, mostly on the performances of various levers[2]. The explanation here given of the lever is founded on the observation that, in a revolving circle, the circumference moves faster than the parts near the centre: the power, therefore, at the end of a lever overcomes the weight by its superior velocity. This explanation leads to the correct inference (Qu. 3) that if two weights keep a lever in equilibrium, they are to one another in the inverse proportion of the arms of the lever. Nearly all the questions are answered by reference to one or other of these facts. The book contains many errors but is worth noticing as evidence that, about this time, questions were asked which ultimately lead to a correct theory of mechanics[4].

115. There remains still to be mentioned another writer, who lived about 330 B.C. and whose works, still unpublished, are the oldest of extant Greek mathematical treatises. This is **Autolycus** of Pitane in Æolis, an astronomer of whom nothing is known save that he wrote two elementary works on the apparent motion of the sun and stars[3]. The first (in 12

---

[1] Too much stress should not be laid on this passage, for Aristotle goes on immediately to say that if A moves B a distance Γ in time Δ, it does *not* follow that $\dfrac{A}{2}$ will move B a distance $\dfrac{\Gamma}{2}$ in the same time, for $\dfrac{A}{2}$ may not be able to move B at all. A hundred men may drag a ship a hundred yards, but it does not follow that one man can drag it one yard.

[2] The questions are of great variety, both in subject and in merit: e.g. 'Why are carriages with large wheels easier to move than those with small?' (Q. 11). 'Why are pebbles on the seashore round?' (Q. 15) 'Why is it easier to extract teeth with the forceps than with the fingers?' (Q. 21). 'Why in rising from a seat do we lean the body forward at an acute angle with the thigh?' (Q. 30). 'Why does a missile, once thrown, ever stop?' (Q. 33). 'Why do objects in a whirlpool move towards the centre?' (Q. 36).

[3] An account of both is given by Delambre. *Hist. Astron.* I. c. II. pp. 19—48. They exist in 3 MSS. at Oxford, but are published only in a Latin translation by Auria (Rome, 1587. 1588). Delambre used Dasypodius' *Sphaericae Doctrinae Propositiones* (Strasburg, 1572), which contains only the enunciations of Autolycus.

[4] See p. 318.

propositions) is called "on the moving sphere" (περὶ κινουμένης σφαίρας). The sphere is supposed to revolve uniformly and to be divided by a great circle (not called the "horizon" but drawn obliquely to the axis) so that half of the sphere is always invisible. The propositions which are of an excessively simple character, relate to the appearance, disappearance and reappearance, of various points on the sphere. The same subject is dealt with more particularly in the other work, "On Risings and Settings" (περὶ ἐπιτολῶν καὶ δύσεων), which is in two books, the first of 13, the second of 18 props. Here Autolycus premises that the rising or setting of any star is invisible unless the sun be at least 15⁰ (measured on the ecliptic) below the horizon[1]. The propositions, which are very obscurely worded, consist mostly of deductions from this fact as to the time both of the night and of the year at which or during which a particular star will be visible. The results are of the most general character, to the effect that after a given phenomenon certain others will happen at certain times or in a certain order[2].

116. A brief summary may here be added of the progress of Greek geometry up to 300 B.C., the date at which the Alexandrian school may be taken to arise, and which begins the most brilliant century in the history of Greek mathematics.

It will be remembered that Thales about 580 B.C. introduced the Egyptian geometry into Ionia: Pythagoras about 530 B.C. introduced it into Magna Graecia. In these places, the extreme Eastern and Western limits of Hellas, mathematical schools survived for nearly 200 years, but the Ionian was by far the less meritorious of the two. This school seems to have been concerned chiefly with the geometry of angles, the Pythagorean chiefly with the geometry of areas and the theory of proportion. To the former we owe much of Euclid's 1st Book, to the latter, no doubt, the 2nd, and the foundations of the 4th, 5th, and 6th,

---

[1] Autolycus divides the ecliptic not into degrees but into 12 parts (δωδεκα-τημόριαι) of 30⁰ each.

[2] Delambre proves many of the propositions for Arcturus and Aldebaran supposed to be observed from Thebes in 360 B.C. He also, in his usual scornful manner, reduces the whole book to a few trigonometrical formulae.

Books.   Before these schools were extinct, the Sophists about
450 B.C. introduced geometry to Athens; and here, under the
stimulus of three insoluble problems, arises the geometry of the
circle and other curves (with Hippias and Hippocrates); still
later the geometry of conics in particular, and of *loci*, stereometry,
mechanics and astronomy.   But it is not to be supposed that
during this time any department of geometry was the peculiar
study of any place.   Intercommunication was so frequent and
rapid that the Pythagoreans of Italy and Sicily and the
Atomists of Abdera, no doubt, were acquainted soon with the
last geometrical discovery of Athens, and *vice versa*.   The systematization of geometrical methods and the orderly arrangement
of elementary text-books, since Hippocrates wrote the first, had
specially occupied the attention of the Platonic school.   Thus it
was that the substance of nearly all the geometry of Euclid's
elements was known before Euclid's time, the forms of geometrical proof were settled, and the arrangement of at least large
fragments of geometry was practically determined.   To collect
these fragments and connect them where necessary, and to
embellish the proofs, was the chief work which was left for ὁ
στοιχειωτής, the writer of *the* Elements *par excellence*.

CHAPTER VII.

EUCLID, ARCHIMEDES AND APOLLONIUS.

117. IT has been already pointed out that the conditions of life in Athens were unfavourable to the growth of any "natural" science. Her practical men were absorbed in politics, her philosophers in metaphysical speculation. Neither of these classes objected to deductive science, for deduction is the chief instrument of rhetoric and is also the most interesting part of logic: but the patient and unrewarded industry, which leads to inductive science, was not to the Athenian taste. The practical men thought it profane, the philosophers vulgar. The schools of inductive science remained therefore far away from the turmoil of Athens: the observatories of the astronomers were at Cyzicus on the Hellespont or at Cnidus on the south coast of Asia Minor: the school of medicine was maintained by one illustrious family in the island of Cos. If it be objected that Aristotle lived in Athens, the answer is that Aristotle was the son of a physician, was not born or bred in Athens, never became an Athenian citizen, disliked Athens and left it, and was not able to command in Athens an audience for anything but metaphysics. The Peripatetic school was as unscientific as the Platonic. There was not yet a "university," to which all the world might come and learn all the knowledge that was in existence. Alexandria was the first city to deserve that name. Athens might have won it, but when Athenian politics were no more and the field was free for other pursuits, Alexandria had forestalled her.

118. The political supremacy of Athens was first broken by the Peloponnesian War (B.C. 431—404). During the next 50 years she was slowly recovering, but in the meantime a more powerful enemy was growing in the North and the Macedonian phalanx was in training for the subjugation of Greece. It came down at last (352 B.C.) under the command of Philip, the father of Alexander the Great. In the struggle that followed, Athens once more took the lead; but she was beaten at Chaeronea (338 B.C.) and never held up her head again. Alexander succeeded his father in 336 B.C. and, after securing his power over Greece, started on his unparalleled career of conquest. In thirteen years he scoured the earth from Macedonia to the Indus, from the Caspian to the cataracts of the Nile, and left behind him, wherever he went, a monument of his visit in the shape of a new city, founded on some aptly chosen position, to be at once a fortress and a centre of commerce[1]. In this way Alexandria was founded in B.C. 332, when Alexander turned from Palestine into Egypt. The site was chosen, the ground-plan drawn and the mode of colonization directed, by Alexander himself, but the building of the city, which was entrusted to Dinocrates, the architect of the temple of Diana at Ephesus, was not completed till many years afterwards. The structure, when finished, was worthy of the site and Alexandria seemed to Ammianus Marcellinus still, in the 4th century after Christ, "vertex omnium civitatum," the noblest of all cities. It was divided into three districts, Greek, Jewish and Egyptian, for Alexander was above all things cosmopolitan and deliberately attempted, on many occasions, to break down the barriers of race and

---

[1] Besides Alexandria in Egypt, he founded at least 17 other Alexandrias (not to mention other cities) in different parts of central Asia. Herat, Candahar and probably also Merv, attest the excellence of his judgment. The abundance of Greek coins which are still current in the bazaars of Afghanistan and the signs of the influence of Greek sculpture and architecture which everywhere abound in those regions, show how successfully they were opened up to Greek commerce and civilization. Alexander's example in this respect was followed by his successors. Ten cities of *Antiochia*, six of *Seleucia*, six of *Apamea*, and six of *Laodicea* were built in a short time by the kings of Syria: and similarly in Egypt the Ptolemies studded the coasts with cities of *Ptolemais*, *Arsinoe*, *Berenice*.

creed[1]. It became at once the meeting-place of all the most important trade-routes; Greek, Egyptian, Arabian and Indian produce passed through it and brought with them a motley host of new settlers. The travels of Alexander had excited throughout the civilized races a new and burning curiosity to see and know more of one another and of the world, and the place where of all others this curiosity could best be satisfied, was Alexandria.

To the sovereignty of this magnificent city Ptolemy, the son of Lagus, succeeded on the partition of Alexander's empire after his death in 322 B.C. Ptolemy was a man who had caught much of Alexander's own enthusiasm, and it was he who created the university of Alexandria. The university buildings stood near the palace and were provided with lecture-rooms, laboratories, museums, a zoological garden, promenades and other accommodations, all clustered near the great Library. This contained in the time of Ptolemy Philadelphus (about B.C. 260) 400,000 rolls, representing about 90,000 distinct works, and there was another library in the Egyptian quarter, containing about 40,000 works. The collections were afterwards greatly increased and were always under the care of some distinguished scholar. So equipped for the pursuit of learning, Alexandria had yet another advantage, in that she enjoyed under the Ptolemies for nearly 200 years a profound peace both internal and external. A short period of conflict followed and then again the 'majestas Romanae pacis' settled upon her and she was free to pursue her old callings, of commerce on the one side, of learning on the other. It is no wonder that to this haven every student resorted and that to Alexandria we owe whatever is best in the science of antiquity. Criticism, mathematics, astronomy, geography, medicine, natural history, jurisprudence

---

[1] E.g. at Susa, in B.C. 325, he himself married Statira, daughter of Darius, and compelled 100 of his generals and 10,000 of his soldiers to marry Persian wives. At Babylon, in B.C. 323, he corporated 20,000 Persians in his army and mixed them with Macedonians in the same phalanx. To the cities which he founded he imported colonists of all nations, and after his death there was found, in his written orders to Craterus, a plan for the wholesale transportation of inhabitants from Asia to Europe and *vice versa*.

were for nearly a thousand years taught in her schools. Other schools arose elsewhere (notably at Pergamum) on the model of these, but none were so complete or so long-lived. Almost all the mathematicians who remain to be mentioned in this history were professors or had been students in the University of Alexandria.

119. A distinguished Athenian, Demetrius Phalereus, was invited to take charge of the Alexandrian Library and it is probable that **Euclid** was invited, with him, to open the mathematical school. That Euclid lived and taught in Alexandria is certain[1], but in fact nothing more is known of him save what Proclus has added to the Eudemian summary, viz. that he lived in the time of Ptolemy I. and was junior to Plato, senior to Archimedes and Eratosthenes. The first Ptolemy reigned B.C. 306—283, and these dates must be taken to determine the period of Euclid's greatest activity. Proclus[2] says he was a Platonist but adds immediately the obviously untrue state-ment that the whole aim of the Elements was to show the construction of the five regular solids, "the Platonic figures." It is true that the XIIIth Book concludes with the construction of these solids, but it is not true that the whole of the preceding books are designed purely for this purpose. It may, neverthe-less, be that Euclid was a Platonist, for most of the geometers who could have taught him, were of that school. All the other historical notices of Euclid are either trivial or un-trustworthy or false. Pappus[3] says that he was gentle and amiable to all those who could in the least degree advance mathematical science, but the context shows that Pappus here refers not to Euclid's personal conduct but to his criticism of his predecessors. A little story related by Stobaeus[4] is perhaps authentic and is at least *ben trovato.* "A youth who had begun to read geometry with Euclid, when he had learnt the first proposition, inquired 'What do I get by learning these

---

[1] Pappus, VII. 35, p. 678 (Hultsch's ed.).

[2] Ed. Friedlein, p. 68.

[3] Ed. Hultsch, VII. 34, pp. 676—678. Hultsch, however, thinks the passage is not genuine. It is given *à propos* of some disparaging remarks on Euclid by Apollonius of Perga. See the pre-face to his *Conics,* quoted *infra,* p. 248.

[4] *Floril.* IV. p. 250.

things ?' So Euclid called his slave and said 'Give him three-pence, since he must make gain out of what he learns.'" These are almost the only personal details which Greek writers have preserved, for in truth 'Euclid' soon became with them, as it is with us, synonymous with 'Elementary Geometry[1].' Syrian and Arabian writers however know a great deal more. They tell us that Euclid's father was Naucrates, his grandfather Zenarchus: that he was a Greek who was born in Tyre and lived in Damascus: that he was much later than Apollonius, whose Elements he edited: that the name *Euclides* is derived from two Greek words, *ucli* a 'key,' and *dis* 'geometry,' so that Euclides means 'key to geometry[2].' Much of this information is pure invention[3], the rest is founded on the preface to the xivth Book of the Elements, which was written not by Euclid but by Hypsicles. In the middle ages some new statements appear, for Euclid was then always confused with Euclides of Megara[4], a pupil of Socrates who founded a small philosophical school which Plato greatly disliked. Dismissing these errors, we can retain only the meagre biography that Euclid was a Greek who lived and taught at Alexandria about 300 B.C.

120. The fame of Euclid, both in antiquity and in modern times, has always rested mainly on his *Elements* ($\sigma\tau o\iota\chi\hat{\epsilon}\iota a$). From this work he acquired among Greeks the special title of

---

[1] So Aelian, *Hist. Anim.* vi. 57, says that spiders can draw a circle and "need not Euclid" (Εὐκλείδου δέονται οὐδέν). So an Arabian, Ibn Abbad, quoted by Hadji Kalfa, maintained that Euclid was the name of a book.

[2] Casiri, *Biblioth. Arab.* i. 339, Abulpharagius, *Hist. Dynast.* p. 41, Hadji Kalfa, *Lexic. Bibliogr.* i. p. 380 sqq. etc. The Arabian authorities on Euclid's life and writings are most carefully collected, discussed and rejected by Heiberg, *Litterar. geschichtliche Studien über Euklid.* Leipzig, 1882, pp. 1—21. This work will hereafter be referred to as 'Heiberg.' Compare also Cantor, p. 224, Hankel, p. 383.

[3] The Arabs tried to claim Oriental origin or education for all the great Greek mathematicians. So Nasir Eddin, who was born at Tus in Khorassan, says that Euclid was born there also.

[4] E.g. Campano's translation is described in the colophon *Opus Elementorum Euclidis Megarensis* (Venice, 1482). Many more examples of this error are collected by Heiberg, pp. 23, 24. Diog. Laertius (ii. 106) says that Euclides Megarensis was sometimes said to have been a Sicilian from Gela. Hence arose the very frequent statement, that Euclid the geometer was a Sicilian.

ὁ στοιχειωτής, 'the author of the Elements[1],' for he so completely superseded his predecessors Hippocrates, Leon and Theudius, that not a trace of their works survives, and so completely satisfied posterity that, until recent times, no attempt seems ever to have been made to supersede the Elements as Euclid left them. His success, moreover, was evidently immediate, for Archimedes and Apollonius and "all the rest," as Proclus says[2], "treat the Elements as perfectly well known and start from them."

It is needless, in England, to describe this book with any detail[3], or to criticise it. Every schoolboy possesses the greater part of it and every one, who is likely to read these pages, is able to recognise both its merits and some at least of its defects[4]. The space which is saved by omitting matter which is so well known, may be better utilised by remarks which do not find a place in English editions of Euclid.

It should be said then, what the preceding chapter has

---

[1] See for instance, the beginning of Heron's *Definitions*, Hultsch's ed. p. 7, s. 1, and the last section of Marinus' *Pref. to Euclid's Data*, printed in Gregory's ed. of Euclid, pp. 453—459. More reff. in Heiberg, pp. 29, 30.

[2] Ed. Friedlein, p. 71, 16.

[3] It may be useful, perhaps, to add a short statement of the contents of the XIIIth Book, which is now seldom seen. It is composed of 18 propositions. The first five (attributed to Eudoxus) relate to lines cut in extreme and mean ratio. Suppose the whole line a, the segments b, c, of which b is the greater. Then, prop. 1, $\left(b + \dfrac{a}{2}\right)^2 = 5\left(\dfrac{a}{2}\right)^2$ and conversely, (prop. 2) if this equation is assumed, b is the greater segment. Prop. 3, $\left(c + \dfrac{b}{2}\right)^2 = 5\left(\dfrac{b}{2}\right)^2$. Prop. 4, $a^2 + c^2 = 3b^2$. Prop. 5, If b be added to a, the line $(a+b)$ is divided in extreme and mean ratio: $b(a+b) = a^2$. Prop. 6, If a *rational* line be cut in extreme and mean ratio, each segment is the

irrational line called ἀποτομή (*supra*, p. 81). These props. are then applied in an investigation chiefly of the relations between the sides of a pentagon, hexagon and decagon inscribed in the same circle with one another and with the diameter (props. 7—11). Prop. 12, The square of the side of an equilateral triangle inscribed in a circle is three times the square of the radius. Then follow five problems, to inscribe in the same sphere (13) a pyramid, (14) an octahedron, (15) a cube, (16) an eicosahedron, (17) a dodecahedron and to show the relations of their sides to the diameter. In prop. 18 the sides of these inscribed figures are compared together. The reader will here see for what purpose Book x. is inserted in the Elements. Book XIV. was added by Hypsicles, xv. probably by Damascius of Damascus about A.D. 510.

[4] A very neat and comprehensive criticism is given by Prof. de Morgan in the article *Eucleides* in Smith's *Dict. of Gk. and Rom. Biogr.*

abundantly proved, that Euclid is certainly not the *author* of
all the propositions which are contained in the Elements. In
the whole collection there is only one proof (I. 47) which is
directly ascribed to him. Many more, no doubt, are his or
partly his[1], but his merit, as Proclus (p. 69 of Friedlein's ed.)
expressly says, lies chiefly in the selection and arrangement of
the propositions. The word 'selection' (ἐκλογή) implies that
some matter is omitted and Proclus again (pp. 72—74) expressly says that much which was not in itself generally useful,
or followed very easily from inserted propositions, was discarded:
e. g. (p. 72, 13) the prop. that the perpendiculars drawn from
the angular points of a triangle to the opposite sides meet in a
point: (p. 74, 18) the construction of an isosceles or scalene
triangle, or propositions on unclassed irrational lines. Hence
not only Archimedes and Apollonius, but Euclid himself, refer
to and use, as well-known truths, propositions which are not
included in the Elements at all. Thus, to take only an instance or two, in the *Sectio Canonis*, prop. 2, Euclid says
" I have learnt that if any series of numbers be in continued
proportion, and the first is a measure of the last, it is also
a measure of all the rest," which is not stated in the Elements.
In the *de Divisionibus*, prop. 23[2], he cites the fact that if
$a : b > c : d$, then $a - b : b > c - d : d$. In the *Data*, prop. 67,
(76 of Simson's ed.) Euclid uses a proposition that, if in an
isosceles triangle a straight line be drawn from the vertex
to the base, then the square of one of the equal sides is equal to
the square of the straight line so drawn + the rectangle under
the segments of the base[3]. (Simson adds a *lemma* to prove this.)
Evidence of this kind, which shows that Euclid used his dis-
cretion in rejecting available matter, which was unquestionably
useful for some purposes, shows also that he had a definite

---

[1] Proclus, at the end of the Eudemian summary, says that Euclid
brought to irrefragable proof propositions which had been less strictly
proved by his predecessors (τὰ μαλα-
κώτερον δεικνύμενα τοῖς ἔμπροσθεν).

[2] So in prop. 22 if $ad < bc$, then

$a : b < c : d$, which Archimedes also
uses. *Sph. et Cyl.* II. 9, ed. Torelli, p.
186. 12.

[3] For many other examples see
Heiberg, pp. 15, 31, 32, 53, n. and reff.
there given.

design in the composition of the Elements. We may therefore, perhaps, attribute to his deliberate choice all the characteristics of the book. With Euclid the word στοιχεῖα[1] no longer means "easiest" or "preliminary" propositions in geometry, but means the whole of geometry, exclusive of certain subjects (the geometry of conics and other higher curves), treated by one method (that of synthesis) only. To him also perhaps may be attributed that orderly method of proof by the regular stages of general *enunciation* (πρότασις), particular *statement* (ἔκθεσις), *construction* (κατασκευή), *proof* (ἀπόδειξις), *conclusion* (συμπέρασμα) and the addition of the final Q. E. D. (ὅπερ ἔδει δεῖξαι) or Q. E. F. (ὅπερ ἔδει ποιῆσαι)[2]. At any rate, Autolycus, just before Euclid, knows only πρότασις and ἀπόδειξις, Archimedes, just after, often dispenses with πρότασις or ἔκθεσις[3]. The design of the whole book, viz. to proceed from a few definitions and axioms, by sure steps which are always of precisely the same kind, to the furthest limits of the subject, is certainly Euclid's, and the pattern of each particular proof is of a piece with the pattern of the whole book[4].

121. Secondly, some remarks will not be out of place on the text of Euclid as we have it. Theon of Alexandria, the

---

[1] Etymologically, στοιχεῖον means any one of *a series* (στοῖχος), one thing of a number of similar things placed in a row. Hence it comes to mean the *elements* of which composite things are compounded, e.g. the single sounds which go to make a word or the parts of speech (Arist. *Poet.* 20, 1 & 2), or the four elements of which the universe was supposed to be made. With Euclid the etymological meaning seems to be uppermost. He calls his book τὰ στοιχεῖα because it is a connected whole and each proposition leads to another.

[2] Proclus, pp. 203, 210.

[3] Bretschneider (p. 21), Cantor (pp. 236, 237) and Heiberg (pp. 35, 36) deny that Euclid invented this form of proof, on the ground that Proclus does not expressly attribute it to him,

and that large portions of Euclid (e.g. Book v.) are attributed to Eudoxus and other predecessors.

[4] A few of Euclid's Greek terms may be here added from the definitions (ὅροι). σημεῖον = a point: εὐθεῖα γραμμή = straight line: ἐπιφάνεια = superficies: ἐπίπεδος = plane: γωνία = angle: εὐθύγραμμος = rectilineal: ὀρθός = right: κάθετος = perpendicular: ἀμβλύς = obtuse: ὀξύς = acute: σχῆμα = figure: περιφέρεια = circumference. There is no word for *radius*, which is called ἡ ἐκ τοῦ κέντρου (γραμμή). τετράγωνον = square: ἑτερομήκης = oblong: ἐκβάλλεσθαι = to be produced: ἐγγράφεσθαι = to be inscribed: περιγράφεσθαι = to be circumscribed: ὅμοια σχήματα = similar figures: ἀντιπεπονθότα = reciprocals: ἄκρον καὶ μέσον λόγον τετμῆσθαι = to be cut in extreme and mean ratio.

father of Hypatia, says, in his commentary on the Almagest
(ed. Halma, I. p. 201), "that the sectors of equal circles are to
one another as the angles which they span (ἐφ᾽ ὧν βεβήκα-
σιν) has been proved by *me* (ἡμῖν) in my edition of the Elements
at the end of Book VI." (VI. 33, pt. 2).   From this it is evident
that Theon edited the Elements, and in fact all the MSS. which
first came to light are entitled 'after Theon's edition' or 'after
Theon's lectures' (ἀπὸ συνουσιῶν τῶν Θέωνος)[1].   For this reason,
on the one hand most commentators of the 16th century
supposed that Euclid had left only the enunciations but Theon
added the proofs[2], and, at a still later time, when this notion
was exploded, other commentators, especially Robert Simson[3],
attributed to Theon all the defects which they could not fail to
perceive in the Elements as they knew them.   But at the
beginning of this century, among various other MSS. which
Napoleon sent to Paris from the Vatican library and which were
restored after the peace in 1815, there was found one (Vat. 190)
of the 10th century, in which the second part of Euclid VI. 33
was written not with the text but in the margin.   Many other
variations from the received text were also perceived in it
(e. g. the useless definition of compound ratio, VI. def. 5,
was omitted[4]), and from these facts F. Peyrard, who printed
it (Paris, 1814—1818), concluded that he had here a copy
of Euclid anterior to Theon's recension.   Nevertheless the
variations between this MS. and the others, which give Theon's

---

[1] E.g. for the first title Cod. Flor.
Laur. xxviii. 3 of the 10th or 11th
century, for the second Laur. xxviii. 1
of the 13th century.   More in Heiberg,
p. 174.

[2] Heiberg, p. 175, gives a great many
instances: e g. Xylander (Holtzmann)
in his German translation (Basil, 1562)
warns the reader that the demonstrations were added 'nit von jme dem
Euclid selbs' but by other learned men,
Theon, Hypsicles, Campano etc.   For
the contrary opinion see the quotations
from Sir Henry Savile in the Preface
to Gregory's *Euclid* (1703).

[3] See the conclusion of his notes.
"From the preceding notes, it is sufficiently evident how much the Elements of Euclid, who was a most
accurate geometer, have been vitiated
and mutilated by ignorant editors," etc.

[4] Simson had, on his own authority,
rejected it.   See his note to VI. 23.   It
does not occur, nor does VI. 33, pt. 2,
in Campano's translation (Venice, 1482)
from the Arabic, but though Campano's
Arabic original was not the Theonic
text, it is not a close enough version of
Euclid to be useful for critical purposes.   Heiberg, p. 178.

text, are not at all important and show that Theon, in the main, confined himself to trifling verbal alterations.

It appears, however, from the citations contained in Proclus' commentary on Eucl. I. that Proclus, though he did not use the Theonic text, did not use the Vatican either or, if he did, was sometimes dissatisfied with it: and it appears also, from quotations in other authors, that the text of Euclid had for many centuries been subject to criticism[1]. This criticism, it is true, was, for the most part, of a verbal kind, but some real discussion seems to have taken place over the definitions ($ὅροι$), postulates ($αἰτήματα$), and axioms ($κοιναὶ ἔννοιαι$, 'common notions'[2]) to Book I. Thus in our MSS. the definition now printed as III. def. 6 (segment of a circle) is appended to the def. I. 18, but Proclus did not have it in that place[3]. This is not an important matter, and, in fact, Heron, who lived about 100 B. C., quotes in his 'Definitions' all the definitions of Euclid, save the arithmetical, in practically the same form, though not in the same order, as that in which we now have them[4]. But the postulates and axioms were the subject of more serious controversy. Our editions have three postulates and twelve axioms, of which the last three are **10.** Two straight lines cannot enclose a space: **11.** All right angles are equal: **12.** If a straight line meet two straight lines, so as to make the two interior angles on the same side of it together less than two right angles, these straight lines will meet if produced on that side. Of these three, the first (*Ax.* 10) appears in many ancient MSS. as *Ax.* 12, but in the Vatican as *Postulate* 6[5]. Proclus (p. 184, 8), however, who omits it altogether, says that Geminus (*cir.* 60 B.C.) would reject it from the *Axioms*, as a proposition requiring proof, and himself (p. 239) gives a

---

[1] Alexander Aphrod. in *Arist. Anal. Prior.* (Venice, 1530) 87. *a*, quotes as Euclid x. 4, the proposition which is now Euclid x. 5: and Eutocius in *Apollonii Conic.* p. 44, quotes as Eucl. III. 15 the prop. which is now Eucl. III. 16. Both these cases may be mere slips.

[2] Euclid does not use the name 'axioms,' $ἀξιώματα$, which Proclus has (e.g. p. 193).

[3] See Proclus, p. 158. The definition is quoted in Heron (ed. Hultsch), *Deff.* no. 33, but, curiously enough, Heron's no. 31 is Eucl. I. 18 and his no. 34 is Eucl. III. 8, so that no inference can be founded on this arrangement.

[4] Heiberg, pp. 186—192.

[5] Heiberg, p. 182, *nn.*

proof in his commentary on I. 4[1]. The last two (*Axx.* 11, 12)
are given by Proclus as *Postulates* 4, 5, and so also in the
Vatican MS. and the older MSS. of the Theonic recension.
But as to *Post.* 4 (*Ax.* 11), Proclus says (p. 188, 2) that
Geminus wished to take it from the postulates and add it
to the axioms, and as to *Post.* 5 (*Ax.* 12), he says (p. 191, 21)
that it ought to be struck out of the postulates and proved as a
theorem, like its converse, and for this opinion he again cites
Geminus as an authority[2]. With regard to Axioms 1—9,
Proclus says (p. 196, 15) that Heron wanted to admit only the
first three, and in fact Martianus Capella in the 4th century
(*Nupt.* VI. 723) quotes only these three as 'communes animi
conceptiones.' Proclus himself quotes only five (viz. 1, 2, 3, 9, 8,
in this order), says (p. 197, 6) that Pappus added *Axx.* 4, 5,
though not in their present form, and himself expressly rejects
Axx. 6 and 7 (p. 196, 25), which stand in the Vatican and are
therefore older than Theon. The evidence, therefore, on the
whole, shows that Euclid originally wrote *five* postulates, of
which the fourth and fifth were those which are now printed as
Axx. 11 and 12, and perhaps *four* axioms, of which the first
three were the present *Axx.* 1—3, and the fourth was the
present *Ax.* 10. The number of the postulates is clearly
attested by Geminus, Proclus and the oldest MSS.: but of the
axioms we can only say, with certainty, that Nos. 4 and 5 are
due to Pappus and 11 and 12 are transferred from the postulates.

But though some reasonable doubt remains as to the
axioms, there is none at all as to the *proofs* of the propositions.
These are very seldom mentioned by ancient writers with
an exact reference to the number of the proposition nor are
whole proofs ever quoted, but there is no trace of any controversy as to any Euclidean proof: the extracts of Proclus show
that he had Book I. almost word for word as it stands now, and
the Vatican MS. agrees, in all but trifling details, with the

---

[1] Nevertheless, p. 193, 22 he rejects
it from the axioms as unnecessary, on
the ground that it merely describes 'a
characteristic of the subject-matter of
geometry' (ἴδιά ἐστι τῆς γεωμετρικῆς
ὕλης).

[2] Ptolemy (in Proclus, pp. 362—368)
attempted to prove it as theorem. See
post.

copies of Theon's recension. From this it is evident that the defects of Euclid are of his own making and not, as Simson would have it, the fault of bungling editors.

122. Lastly, it will perhaps be interesting to show what have been the fortunes of Euclid's Elements and how they have come into the possession of English schoolboys and been made the staple of our mathematical education[1]. In Alexandria this book occupied the same place as with us, and Theon's edition of it was made, nearly 700 years after Euclid, for the benefit of the students who attended the editor's lectures. It does not, however, seem to have been at all known in Italy, for Boethius, who (about A. D. 500) wrote a Latin geometry, contents himself with giving merely the *enunciations* of Book I. and of some propositions in Books III. and IV. of Euclid and adding at the end, as a stimulus to the mind, the whole proofs of the first three propositions of Book I. He then proceeds (in Book II.) to the calculation of areas etc. of given dimensions, the practical geometry for which alone the Romans had any desire. Euclid was the Greek text-book and was confined to Greek schools, or to those which were founded on the model of Alexandria, such as the Syrian schools of Antioch and Emesa and Damascus, and in particular, the school of Nestorian Christians at Edessa. These latter, after the terrible sack and ruin of Alexandria in 640, became the chief repositories in the East of all Greek learning. To them belonged the chief physicians of that time, who were invited to Bagdad to attend upon the Abbasid Caliphs[2]. The Arabs did not fail to remark that these Jewish and Christian doctors relied upon the writings

[1] Most of the facts given in this section are taken from various chapters of Hankel, pp. 231—237 (on Arabic translations): pp. 307—317 (Gerbert and his predecessors): pp. 334—348 (translations from the Arabic, etc.): pp. 354—359 (Mathematics in foreign Universities). I have added some trifling details from Cantor, whose *Vorlesungen* Vol. I. only go as far as the year 1200. Hankel's account of the Arabs is taken chiefly from Wenrich,

De Auctorum Graec. Versionibus Arab. Syr. et Pers. Leipzig, 1842. Cantor has a more recent authority, Kremer, *Kulturgesch. des Orients unter den Chalifen* (Vienna, 1877). English literature is ridiculously deficient in such monographs.

[2] It is said that the Arabs, when they gave up the nomad life and settled in Bagdad, became subject to various disorders, which their native physicians were unable to cure.

of Hippocrates, Aristotle and Galen, and the medical books
of these three Greeks were therefore translated into Arabic from
the Syriac in the time of Harun al Raschid (786—809). An
intense interest in Greek science of every kind was thus aroused,
and in a few years translations of all the principal mathematical
books of Alexandria were secured. The Caliph Al Mamun[1]
(813—833) was especially zealous in this cause. He obtained
from the Byzantine empire, through his ambassadors, copies of
the Greek MSS. and established in Bagdad a college of Syrian
Christians who were nominally his physicians but were chiefly
engaged in translating the Greek books into Arabic. A little
earlier than this, in the time of Al Mansur (754—775), the
Arabian commerce with India had brought to the knowledge of
Bagdad the *Siddhanta* or 'System' of Brahmagupta[2]. This also
was translated and thus the Arabs acquired the Indian numerical symbols. The interest of Al Mamun in foreign science
was more than rivalled by his successors. The most famous of
the translators was one Honein ibn Ishak, a Syrian physician,
who was acquainted with both Greek and Arabic. He was
appointed, by the Caliph Mutawakkil (847—861), president of
the college of translators some of whom were busy in rendering
Greek books into Syriac, the rest in rendering the Syriac
into Arabic. Honein and his son, Ishak ibn Honein, revised
the final Arabic translations, but as they were both ill-versed in
mathematics, Tabit ibn Korra (836—901), another Syrian,
edited their texts with the knowledge of a competent mathematician. It was in this way that the works of Euclid, Archimedes, Apollonius, Theodosius, Ptolemy and other Greeks received a new lease of life among a strange people[3]. Ptolemy
seems to have been the first of these to be translated. A
portion of the Elements of Euclid was translated in the time of

---

[1] See Gibbon's Chapter LII.

[2] Cantor, pp. 597, 598.

[3] Diophantus was not translated till
the end of the 10th century (by Abul
Wefa). Arabian algebra however begins in 820 with the *Al gebr w' al Muka-
bala* of Mohammed ibn Musa Alchwar-
izmi, who cites no authorities but has
an advanced knowledge of his subject.
Hankel (p. 263) suggests that the Dio-
phantine analysis had become traditional in the Syrian schools. Cantor
(pp. 619, 620) leaves it an open question whence Mohammed obtained his
Algebra. See p. 318.

Harun al Raschid : Al Mamun ordered another and a complete translation, Honein or his son Ishak prepared a third, Tabit ibn Korra published the final redaction.

But Euclid did not come into Europe from Bagdad. Some fifty years before the Abbasid Caliphs settled in that city, the Arabs had penetrated into Spain and taken possession of the ancient city of Cordova. Here, in 747, the Emir Abdarrahman founded a separate kingdom and the Arabs of Spain were thenceforth wholly dissociated from their kinsmen in the East. Both nations had the same intellectual tastes : each was as enthusiastic as the other for medicine, mathematics and astronomy, but each pursued its studies in its own way and with some considerable jealousy of the other. Nevertheless, by some means which has not been explained[1], the Arabs of Spain acquired the same books which were used in Bagdad and had also their Indian numerals, their Ptolemy, Euclid and Aristotle. In the meantime, among the Christians of the West, learning was at its lowest ebb. Their mathematical interest was confined almost entirely to arithmetic and, as to geometry, "we find in the whole literature of that time hardly the slightest sign that any one had gone further in this department of the Quadrivium than the definitions of a triangle, square, circle or of a pyramid or cone, as Martianus Capella and Isidor (Hispalensis, bishop of Seville in 636) left them[2]." The study was revived by the great Gerbert, a native of Auvergne, born in the first half of the 10th century. He, after a visit to Barcelona, where perhaps he acquired somehow an inkling of the Arabic sciences[3], became the teacher of the Cathedral school at Rheims and acquired the greatest renown by his mathematical ability. He was, after many other promotions, elected Pope in 1003, under the name of Sylvester II. Gerbert, while abbot of Bobbio on the Trebbia (about 980), came across the *Codex Arcerius* con-

---

[1] Possibly Jewish pedlars of books had something to do with it.

[2] Hankel, pp. 307, 311, 312. Compare Hallam, *Middle Ages*, III. chap. ix. pt. 2, p. 420 (12th ed.).

[3] Gerbert certainly knew no Arabic, and the common statement that he went to Cordova is for many reasons incredible. Nevertheless, it was he who introduced the Arabic numerals (as *apices*) into the Western schools. Hankel, pp. 327, 328, *supra*, pp. 37—39.

taining the works of the old Roman surveyors (*gromatici*)[1]. He
studied them with avidity and founded on them his own
Geometry. A little later, he found at Mantua a copy of the
Geometry of Boethius. In this way the study of practical
geometry was renewed and some small portions of Euclid
became the common property of the Christian schools. But it
was not for 100 years yet that men began to seek the Arabic
text-books. The Moorish Universities of Cordova and Seville
and Granada were dangerous resorts for Christians and, though
it was known that all manner of learning was to be had there,
no student ventured to steal it. An Englishman was the first,
or one of the first two, to undertake the enterprise. In 1120,
Adelhard of Bath obtained in Spain a copy of Euclid's Elements
and translated them into Latin. Translations from the Arabic
of other Greek works, especially those of Aristotle, soon followed[2].
About 1186 Gherardo of Cremona made another translation
of the Elements and, again in 1260, Giovanni Campano reproduced Adelhard's translation under his own name[3] and obtained with it a wide celebrity. The fruit of these translations
soon followed. In 1220, Leonardo of Pisa, a mathematician of
great power and originality, published his *Practica Geometriae*,
which though it deals with the calculation of areas and numerical ratios of spaces, is founded on Euclid and Archimedes and
Ptolemy[4], and contains some trigonometry and conics. A little
later Roger Bacon (1214—1294) was urging the claims of
experimental science as taught by Aristotle. But the greatest
result of the inflow of Arabian learning was the organisation of
study in Universities. At Paris[5], indeed, the study of geometry

---

[1] See Cantor, pp. 467, 734, 738—743.

[2] The Jews, who were tolerated by
both Arabs and Christians, assisted
largely in this movement. See Jour-
dain, *Rech. sur les Trad. Lat. d'Aris-
tote*.

[3] Prof. de Morgan first suspected
this. For a full bibliography of Euclid
see his art. *Eucleides* in Smith's *Dic.
of Gr. and Rom. Biogr.*

[4] Hankel, pp. 344—346.

[5] Unofficial lectures of some kind

seem to have been given in Paris all
through the 11th century and even
earlier. But it was Abelard (1079—
1142) who made the University of
Paris famous. Similarly, all the other
Universities seem to have been at an
early time centres of instruction. But
it is in the 13th and 14th centuries
that they first receive charters of incorporation. Paris received its charter
in 1200. See Hallam *supra cit.* pp.
420—427.

was neglected and Aristotle's logic was the favourite subject. But at the reformation of the University in 1336 a rule was introduced that no student should take a degree without attending lectures in mathematics and from the preface to a commentary on the first six Books of Euclid, dated 1536, it appears that a candidate for the degree of M.A. was then required to take an oath that he had attended lectures on the said books. In Leipzig (founded 1389), the daughter of Prague, a similar rule was made, but it is doubted whether the rule was enforced, since in the lists of lectures for the years 1437, 1438, none on Euclid are mentioned[1]. But in Prague itself (founded 1350) mathematics were more regarded. Candidates for the Baccalaureat were required to take up the treatise of the Globe by Johannes de Sacrobosco (i.e. of Holywood in Yorkshire) and, for the Master's degree, the first six Books of Euclid and many subjects of applied mathematics were required. At Oxford, in the middle of the 15th century, the first two books of Euclid were read[2] and no doubt the Cambridge curriculum was similar. It will be seen, however, that though the study of geometry was maintained (indeed it was part of the ancient Quadrivium) it was maintained only in a half-hearted manner and did not produce a tithe of the results which might have been expected from the brilliant commencement of Leonardo of Pisa. It was, in fact, driven out of the field by Aristotelian logic and the stupid subtleties on which that logic was employed by the schoolmen. Another *Renaissance* was still wanted. This came after Constantinople was taken by the Turks in 1453 and a crowd of Greeks fled into Italy bringing with them precious manuscripts of Greek literature. About this time also printing was invented and books became comparatively cheap and common. Campano's (stolen) translation of Euclid was printed in 1482 by

---

[1] Cologne, founded 1389, was equally behindhand and so were the Italian Universities of Bologna, Padua and Pisa, where astrology was the favourite subject. As late as 1598, the professor of mathematics in Pisa was required to lecture, not on the Almagest, but on the *Quadripartitum*, an astrological work attributed to Ptolemy. Hankel, p. 357.

[2] Churton's *Life of Smyth*, p. 151, quoted by Hallam *Lit. of Eur.* Pt. I. ch. 2, *s.* 34, *n.* I have been unable to find any statement of the Cambridge course.

Ernest Ratdolt at Venice, and many times afterwards. The Greek text was printed in 1533 by Simon Grynæus at Basle. Even still it cannot be pretended that Euclid or any other mathematician occupied anything like the same amount of attention as the writers of *belles lettres*. Nevertheless a con- siderable number of commentaries were produced in the 16th century and in 1570 an English translation from the Latin was published (by Henry Billingsley). About the same time Sir Henry Savile began to give *unpaid* lectures on the Greek geometers at Oxford. In 1619, the Savilian professorships were founded in that University, but it was not till 1663 that a professorship of mathematics (the Lowndean) was given to Cambridge[1]. The 70 years or so, from 1660 to 1730, when Wallis and Halley were professors at Oxford, Barrow and Newton at Cambridge, were the period during which the study of Greek geometry was at its height in England. After Newton's time the whole field of mathematics and natural philo- sophy was so rapidly enlarged that the Greeks, all except Euclid, fell into neglect. But as modern learning advanced, so also it became necessary that boys leaving school for the Universities should take with them some preliminary knowledge of mathe- matics and should stay at school longer to acquire this[2]. For this purpose Euclid's *Elements* was especially suited, but it may be safely guessed that its place among our schoolbooks dates only from the middle of the last century at the earliest. To

[1] Sir Thomas Gresham founded a professorship of geometry in London in 1596. Briggs, a Cambridge man, was the first professor but afterwards became the first Savilian professor of geometry at Oxford. At the latter place, he began lecturing on Eucl. i. prop. 9, at which Savile had himself left off. The mathematicians of this time were more interested in algebra than geometry. Lord Herbert of Cher- bury (1581—1648), in his Autobiogra- phy, says that he sees little use in geometry for gentlemen, though it may perhaps help them to understand for- tification.

[2] This statement and the next are made without much authority. I have looked through all manner of biogra- phies and "memorials" without finding any useful information on the curricu- lum of a public school before 1750. The evidence is abundant that, during the last century, the average age of fresh- men was gradually increasing. It may be gathered (e.g. from Wordsworth's *Scholae Academ.* ch. vii. and app. iii.) that, during the same time, Euclid was gradually passing from the Universities to the schools. There is obviously some connexion between the two facts. When boys stayed longer at school,

this time belong also all the famous editions of Euclid in England, from Gregory's Greek text (pub. 1703) to Simson's translation and commentary (pub. 1756) upon which all subsequent editions have been more or less founded. Attempts have recently been made to depose Euclid from his place in the English educational system, but they are not likely to be successful. No modern text-book can acquire an equal prestige and the advantage to teachers, in knowing that all their pupils possess and have studied the same rudimentary treatise, is not lightly to be foregone.

123. The extant works of Euclid comprise, beside the Elements, books of *Data* (Δεδομένα), Φαινόμενα ('appearances of the heavens'), 'Οπτικά, Κατοπτρικά (' Reflections'), Κατατομὴ Κάνονος ('Division of the Scale'), a probably spurious Εἰσαγωγὴ 'Αρμονική (' Introduction to Harmony'), and a work *De Divisionibus*, known only in the Arabic and in a Latin translation from another Arabic edition.

The *Data*, the authenticity of which is attested by Pappus[1], consists of 95 propositions (Pappus knew only 90), preceded now by an explanatory introduction written by Marinus of Neapolis, a pupil of Proclus, at the end of the 5th century. The book, which is printed in Simson's *Euclid* with many alterations, begins with some definitions declaring the meaning of the word δεδομένον in various cases[2]: e.g. 1. Spaces, lines and angles, are said to be *given in magnitude* when equals to them can be found : 4. Points, lines and spaces are said to be *given in position*, which have always the same situation [and which are either actually exhibited or can be found, Simson] : 6. A circle is said to be given *in position and in magnitude* when the centre is given in position, the radius in magnitude. The propositions which follow deal with magni-

they would necessarily begin to learn higher subjects. But why did they stay longer at school? The answer suggested in the text is inadequate but is no doubt correct. Classical studies at the Universities are not, and never were, much different from those of schools.

[1] VII. ed. Hultsch, pp. 638—640.

[2] Marinus says that Euclid ought to have started with a general definition of "given" and, after discussing many such himself, concludes with the opinion that the best definition is "knowable and obtainable" (γνώριμον καὶ πόριμον). Gregory, pp. 457, 458.

tudes, lines, rectilineal figures and circles, in this order. The
following specimens will sufficiently show their character. Prop.
VIII. (Simson, 9): Magnitudes which have a given ratio to the
same magnitude have also a given ratio to one another. Prop.
XXXII. (35): If a straight line be drawn between two parallel
straight lines given in position, and make given angles with them,
the straight line is given in magnitude. Prop. XXXIX. (42):
If each of the sides of a triangle be given in magnitude,
the triangle is given in species. Prop. LII. (56): If a rectilineal figure, given in species, be described on a straight line
given in magnitude, the figure is given in magnitude. Prop.
LXXXIX. (92): If a straight line, given in magnitude, be drawn
within a circle given in magnitude, it shall cut off a segment
containing a given angle. The word 'given,' it will be seen, is
employed in two significations. It means first 'actually given'
and secondly, 'given by implication,' and the propositions are
all to this effect, that a certain partial description of a certain
magnitude, or of a certain geometrical figure, involves a more
complete description, just as the description of a triangle as
equilateral involves its description as equiangular. The book,
in fact, is a series of easy riders on the Elements. The proof
of the prop. LXXXIX. stated above, will serve well enough as a
specimen. By def. 1 the angle is 'given,' if equals to it can be
found. Now let the straight line AC, given in magnitude, be drawn
within the circle ABC given in magnitude. It shall cut off a
segment containing a given angle. Draw
AE, passing through the centre, and join
EC. Then because each of the straight
lines AC, AE is given, their ratio is
given: and the angle ACE is a right
angle, therefore the triangle ACE is
given in species and consequently the
angle AEC is given (i.e. can always be

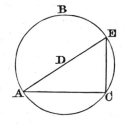

reproduced). But the *Data* had a special use in Greek
geometry. They are described by Pappus and Marinus[1] as
forming part of the τόπος ἀναλυόμενος. This was the name of
a special department of geometry, "matter prepared for those

---

[1] Marinus in Gregory's *Euclid*, p. 458.

who, after going through the Elements, wish to acquire the power of solving problems proposed to them and useful for this purpose only[1]," a course of practice in analysis. The way in which the *Data* were found serviceable in analysis will be seen at once by reference to the specimen of analysis given above (p. 178) from Pappus. Analysis begins with a construction which is assumed to satisfy the proposed conditions. These conditions being thus converted into given elements of the figure, involve others which are *given by implication* in the Euclidean sense, and these again involve more, until by steps, every one of which is legitimate, we reach a construction from which a synthesis is obtainable. The *Data* are hints upon the most usual steps in analysis.

The *Phaenomena* is a book of 18 propositions with a preface. The authenticity of this also is attested by Pappus[2], who gives some *lemmas*, or explanatory propositions to it. The preface is a statement of the considerations which show that the universe is a sphere, followed by some definitions of technical terms. Among these ὁρίζων, as a substantive, and μεσημβρινὸς κύκλος, *meridian* circle, occur for the first time. The book consists of geometrical proofs of propositions which are established by observation, to the effect chiefly that stars situate in given positions rise or set together or one after another in a certain order. It is beyond question founded on the *Moving Sphere* of

---

[1] Pappus, vii. ed. Hultsch, p. 634. In the same place it is said that the τόπος ἀναλυόμενος was written entirely by Euclid, Apollonius and Aristaeus the elder. The word τόπος here does not mean *locus*, but has its Aristotelian meaning of 'store-house.' So, at the beginning of Book vi. of Pappus τόπος ἀστρονομούμενος means 'the astronomical treasury,' consisting of books which he afterwards discusses. Τόπος ἀναλυόμενος means "the treasury of analysis," just as in Aristotle's rhetoric τόποι, or κοινοὶ τόποι are collections of "commonplaces," remarks and criticisms to which the rhetorician may always resort. The translation of τόπος ἀναλυόμενος as 'locus resolutus,' 'lieu résolu' or 'aufgelöster Ort' is therefore misleading and has led, I believe, to some misconception. See the translation in Chasles, *Les Porismes* etc. p. 16.

[2] vi. (Hultsch), pp. 594—632. The text which Pappus used was not quite the same as that of Gregory's edition, which has a great many evident interpolations. These are discussed by Heiberg (pp. 47—52), who has found at Vienna a better MS. On the *Phaenomena* see also Delambre, *Astr. Anc.* i. ch. 3, pp. 48—60.

Autolycus, which is several times referred to, though not by name[1]. But it is evident also that Euclid is here quoting some work on Spherical Geometry, by an unknown author. In the preface, for instance, he cites casually the fact that if on a sphere two circles bisect one another, they are both great circles, and in the proofs he very frequently assumes in his reader a knowledge of other such theorems[2]. A comparison of these with the later *Sphaerica* of Theodosius shows that both Euclid and his successor had recourse to the same original work, which perhaps was written by Eudoxus.

The *Optics*, as commonly printed[3], consists of 61 propositions preceded by a preface and a list of assumptions ($\theta\acute{\epsilon}\sigma\epsilon\iota\varsigma$). The book has often been suspected because these assumptions are absurdly wrong and some of the proofs are, in the present text, slovenly or defective[4]. There seems, however, no fair reason for denying its authenticity, which is attested by Theon in many passages of his commentary on the Almagest. Pappus, though he does not name the book, cites some propositions from it just before he passes to Euclid's *Phaenomena*[5]. The preface, which is obviously not by Euclid, is part of a report of a discourse on Optics. It begins, for instance, with the words "After proving the theorems concerning sight, *he* proceeded to advance some suggestions, arguing that light is carried in straight lines" etc. A scholiast has added at the beginning of a Paris MS.[6]

---

[1] E.g. Prop. 1 of Autolycus is cited in Euclid's 5th, Prop. 2 in Euclid's 4th and 6th, Prop. 10 in Euclid's 2nd. See Gregory's ed. pp. 564, 567—569. Heiberg, pp. 41, 42.

[2] A full collection in Heiberg, pp. 43—46. The instances are difficult to cite because Euclid does not actually state the theorems, but says, for instance, in the course of a proof, "since in a sphere the circles ABC, DEF touch one another and the great circle GHK passes through the poles of one circle and the point of contact of both, therefore GHK passes through the poles of DEF and is perpendicular to it." (Prop. II. p. 564.)

[3] In Gregory's edition with notes by Savile. Gregory suspects the book, Peyrard rejects it altogether. Heiberg (pp. 93—129) prints an improved text in 62 props. from a Vienna MS., which he thinks is genuine.

[4] Such suspicion is protested against by Kepler (*Epp.* ad I. Kepler CLII.) quoted by Heiberg, p. 90, from E. Wilde, *Optik der Griechen*, p. 9 n. On the Optics, see also Delambre, *loc. cit.*

[5] Pappus VI. p. 568. The propositions cited are Nos. 35, 36, 37 of Gregory's ed. See Gregory's preface and Heiberg, pp. 130, 131.

[6] Heiberg, p. 139.

the words "the preface is taken from the commentary of Theon,"
and this may well be true, for the preface is quoted by Neme-
sius[1] who lived as early as the year 400. It is merely a
number of notes on the Euclidean hypothesis that light proceeds from the eye and not from the object seen. The contrary
is shown to be absurd by such arguments as these, that, if light
proceeded from the object, then we should not, as we often
do, fail to observe a needle on the floor, and a circle seen edge-
ways would not appear to be a straight line. The assumptions
(θέσεις, *positiones*, 12 in number) are such as 1. Rays emitted
from the eye are carried in straight lines, distant by an interval
from one another: 2. The figure contained by such rays is
a cone, having its vertex in the eye, its base on the object
seen[2]: 5. Things seen under a greater angle seem greater:
8. Things seen by the higher rays seem higher, etc. The
propositions, which are proved from these assumptions with
the aid of the Elements and Data, are of the following kind.
I. No object is seen *in toto* at one time: VI. Parallel intervals
seen from a distance seem of unequal width: XVIII—XXI. To
measure a given altitude, depth or longitude (proved by similar
triangles in the manner attributed above, p. 141, to Thales): XL.
The wheels of chariots appear now circular, now elliptical (παρε-
σπασμένοι) etc. Prop. XXII. is 'If a circle be described in the
same plane as the eye, it will seem to be a straight line.'
The proof is as follows[3]. Suppose the eye
at Δ: the circle BZΓ in the same plane.
The rays ΔB, ΔZ, ΔΓ proceed from the
eye. Since (by prop. I.) no object is seen
in toto at once, the circumference BZ will
not be seen, but only its extreme points B
and Z, wherefore the circumference BZ will
appear to be a straight line. And similarly

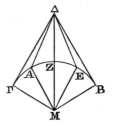

[1] Περὶ φύσεως ἀνθρώπου, ed. Matthaei
VII. p. 179.

[2] So also Arist. *Probl.* xv. 5.

[3] Another proof attributed to Pappus
in Gregory's ed. p. 617, but given as
Euclid's in Heiberg's text κγ', p. 102,

depends on the assumption that the
line MB, being seen under a greater
angle than MA, appears longer etc.
Aristotle *Problem.* xv. 5 gives a similar
explanation.

the circumference ΖΓ. Therefore the whole circumference ΒΓ will appear a straight line.

The *Catoptrica* is a book of 31 propositions on reflections in plane, convex and concave mirrors. It begins with assumptions of the same character as those in the *Optics*, to which are added four admitted *phaenomena*[1], the last of which is that a ring placed in a vase so as to be invisible from a certain position, may be made visible from the same position by filling the vase with water. The propositions start with proving that the angle of incidence is equal to the angle of reflection and go on to give reasons for such familiar facts as that in a convex mirror objects seem smaller and, in a concave, are seen upside down. But though Euclid certainly wrote a *Catoptrica*, which he mentions in the *Optics* (Prop. XX. of Heiberg's text, p. 101, l. 25), it is in the highest degree improbable that he wrote this one. The book is not cited by any ancient author. Heron's *Catoptrica* is cited for propositions which occur in Euclid and the explanation of the phenomenon, above mentioned, is expressly attributed to Archimedes, who suggested that the water acted as a mirror[2]. Probably Euclid's original work was superseded entirely by Archimedes and the extant *Catoptrica* is the work of a later compiler[3].

The *Sectio Canonis* is a work on musical intervals, which is probably Euclid's, who, according to Proclus (p. 69) and other commentators wrote an Elements of Music, but the *Introductio Harmonica* is mainly a collection of musical terms, not agreeing with the *Sectio Canonis*, and is generally rejected[4]. It remains only to mention the book περὶ διαιρέσεων, which is ascribed to Euclid by Proclus (pp. 69, 144). We have this in a Latin translation (*De Divisionibus*) made by John Dee, about 1563, from an incomplete Arabic copy attributed to Mohammed Bagdadinus. Woepcke subsequently found another and pro-

---

[1] The first three are false. So also are a great many propositions. The most curious slip is that Prop. 5 proves the contrary of Prop. 6. For a long list of errors and inconsistencies see Gregory's *Prefatio*.

[2] Olympiodorus *in Arist. Meteorol.* II. p. 94 (ed. Ideler). Euclid gives no explanation at all and does not allude further to the phenomenon.

[3] Heiberg, pp. 148—152.

[4] Heiberg, pp. 52—55.

bably complete Arabic text[1], containing some propositions
on the division of the circle, which, Proclus says, Euclid's book
contained but which are missing from Dee's translation. The
work is a collection of problems on the division of plane figures
into parts which have to one another a given ratio: e. g. Dee's
7th prop. is 'By a line drawn from an angle of a given trapezium, to divide the trapezium in a given ratio': Woepcke's
28th is 'To divide into two equal parts a given figure bounded
by an arc of a circle and by two straight lines containing
a given angle.' This, like the *Data*, may be regarded as a
collection of riders on the Elements[2].

124.   Beside these extant works, Euclid wrote others which
are lost.   One of these bore the title περὶ ψευδαρίων, *on*
Fallacies, but nothing is known of it save from a notice of
Proclus[3], who, in his usual wordy manner, explains that it consisted of exercises (apparently geometrical) in the detection
of fallacies.   The fact that Euclid wrote such a book renders it
more than ever probable that his *Elements* was composed solely
for educational purposes and that Euclid is responsible for the
whole style and arrangement of the latter work.   Beside the
Fallacies, we hear also of a treatise by Euclid on Τόποι πρὸς
ἐπιφανείᾳ or *Loci on a Surface* in two books.   The meaning
of this title has occasioned some controversy.   Prof. de Morgan
says frankly that he does not understand it and it is evident

---

[1] *Journ. Asiatique*, 1851, p. 233 sqq.
See Ofterdinger, *Beiträge zur Wieder-*
herstellung etc. über die Theilung der
Figuren, Ulm, 1853. Heiberg, pp. 13—
16, 36—38. Cantor, pp. 247, 248.

[2] There is appended to Gregory's
Euclid a Latin fragment of one page
only entitled *De levi et ponderoso*, of
the origin of which nothing is known.
It was printed in the Basle translation
of 1537, but the publisher Hervagius
says only that somebody brought it to
him during the progress of the work.
It consists of nine definitions and five
propositions.   The 4th definition is
"Bodies are equal *in power* (potentia)
which, in the same time and in the

same medium (air or water), move the
same distance." The 7th is "Bodies
are of the same kind which are equal
in magnitude and in power." The
book, if complete, would evidently furnish some interesting ideas on specific
gravity, but the language, especially
the use of *potentia* (δύναμις), is not
Euclid's or of Euclid's time, and is
indeed hardly in the Greek style.
Heiberg, pp. 9—11.

[3] p. 70 (ed. Friedlein). Heiberg, p.
38 *n.* suggests that there may be a
ref. to this book in the *Schol.* to *Theaet.*
191 B (VI. p. 248 of Hermann's ed.)
and in Alex. Aphr. in *Arist. Soph. El.*
fol. 25 b (Venet. 1520).

that Eutocius was in the same predicament, for he says, after
describing other *loci* well enough, that the τόποι πρὸς ἐπιφανείᾳ
derived their name 'from the peculiarity of them' (ἀπὸ τῆς περὶ
αὐτοὺς ἰδιοτῆτος) and so leaves them[1].   Prof. Chasles supposes
that the book contained propositions on "surfaces of the second
degree, of revolution, and sections therein made by a plane":
and he refers to the facts that Archimedes, at the end of Prop.
XII. of his *Conoids and Spheroids*, says that certain propositions
on sections of conoids φανεραί ἐντι (i.e. "are clear," not "are
well known" as Chasles takes it) and that the four lemmas
which Pappus gives on this book of Euclid[2] relate to conic
sections.   Heiberg, however, by a very elaborate analysis of all
the passages in which τόποι of various kinds are described[3],
comes to the conclusion that τόποι πρὸς ἐπιφανείᾳ means
simply *"loci which are surfaces,"* and that Euclid's treatise dealt
chiefly with the curved surfaces of the cylinder and the cone.
That such surfaces were regarded as *loci* before Euclid's time is
evident from Archytas' solution of the duplication problem
cited above p. 182[4].

Pappus[5] attributes to Euclid also a treatise on *Conic
Sections* (κωνικά) in four books, which formed the foundation of
the first four books of Apollonius' work on the same subject.
The former will more properly be considered when we come to
speak of the latter, but it may be mentioned here that the
names *ellipse, parabola* and *hyperbola* or the mode of producing
the conic sections which these names imply cannot have been
Euclid's, for not only are they expressly attributed to Apollonius,
but Euclid, in the preface to the *Phaenomena*[6], uses the old

---

[1] Prof. de Morgan in Smith's *Dic.*
Eutocius in *Apollon. Conic.* Halley's
ed. pp. 10—12.

[2] Pappus, VII. prop. 235 sqq. (Hultsch,
pp. 1004 sqq.).   Chasles, *Aperçu*, Note,
II. pp. 273, 274.   Montucla (I. p. 172)
says that τόποι πρὸς ἐπιφανείᾳ were
surfaces, and subsequently (p. 215)
that they were lines of double curvature described on curved surfaces, such
as a helix on a cylinder.

[3] pp. 79—83.

[4] Heiberg refers also to Pappus, pp.
258. 23, 260. 13, 262. 14.

[5] VII. p. 672 (Hultsch).

[6] Gregory's ed. p. 561.   Here Euclid
says that "any cone or cylinder, cut
by a plane which is not parallel to its
base, exhibits that section of an acute-
angled cone, which is like a shield"
(θυρεός).

expression 'section of an acute-angled cone' for the ellipse. The work of Euclid, therefore, must have been recast by Apollonius.

Lastly, a treatise on *Porisms* ($\pi o \rho \acute{\iota} \sigma \mu a \tau a$) in three books is attributed to Euclid by Pappus[1], and this has for more than two centuries provoked a lively controversy[2], partly because the definitions of 'Porisms' given by Pappus are very obscure and partly also because Pappus treats so largely of Euclid's book and gives so many lemmas to it that it has seemed possible, to many modern geometers, to restore the entire work. Of these the most recent, as well as the most successful, is the late Professor M. Chasles. The reconstruction of the book depends entirely upon a long passage of Pappus and a short one of Proclus, the effect of which is as follows. Proclus[3] says that $\pi \acute{o} \rho \iota \sigma \mu a$ is used, in geometry, in two senses, viz. a 'corollary,' for which it is the ordinary word, and also as the name of a proposition which is neither a theorem nor a problem, but partakes of the nature of both. Its aim is not, like a theorem, to describe a new characteristic nor, like a problem, to effect a construction or alter a given construction, but to find and bring to view ($\acute{v}\pi$' $\acute{o}\psi\iota\nu$ $\grave{a}\gamma a\gamma\epsilon\hat{\iota}\nu$) a thing which necessarily coexists with given numbers or a given construction, as, to find the centre of a given circle or to find the G.C.M. of two given numbers[4]. With this definition agrees also the ordinary use of the words $\pi o \rho \acute{\iota} \zeta \epsilon \sigma \theta a \iota$ (which means 'to find' but not 'to construct,' e.g. in Heron *to find* the length of a line) and $\pi \acute{o} \rho \iota \mu o \nu$ (which is synonymous with $\delta \epsilon \delta o \mu \acute{\epsilon} \nu o \nu$, and means 'discoverable')[5]. But the aim of the *porism* is not quite the same as that of a proposition in the *Data*. The latter is to the former as a theorem to a problem. A *datum* alleges, for

[1] VII. p. 648 (Hultsch).

[2] A very full bibliography is given by Heiberg, pp. 56, 57. It is necessary only to mention Fermat, 1655. Simson (posthumously published) 1776. Chasles, *Les trois livres de Porismes*, Paris, 1860. This also contains a bibliography pp. 8 and 9. See also Chasles, *Aperçu*, pp. 12—14, and Note

III. pp. 274 sqq.

[3] pp. 301—2 of Friedlein's ed.; cf. p. 212.

[4] The props. of the Elements III. 25, VI. 11, 12, 13, are 'porisms' in this sense. These ought to conclude with $\acute{o}\pi\epsilon\rho$ $\acute{\epsilon}\delta\epsilon\iota$ $\epsilon\dot{v}\rho\epsilon\hat{\iota}\nu$, *quod erat inveniendum*.

[5] See Heiberg, pp. 59, 60 and the note from Marinus *supra*, p. 209 n.

instance, that with a segment of a circle the angle in it is given, a corresponding *porism* is to find the ratio of the angle to a right angle. But though porisms occur in the Elements, they were used chiefly in higher geometry and Pappus says that Euclid's *Porismata* formed part of the collection Τόπος ἀναλυό- μενος, like the *Data*. He proceeds then[1] to discuss the nature of porisms, which he first defines, like Proclus, as intermediate between a problem and theorem, subsequently as "a proposition for the purpose of finding the thing proposed," afterwards again (but this, he asserts, is only a partial definition) as "that which is inferior by hypothesis to a local theorem" (τὸ λεῖπον ὑποθέσει τοπικοῦ θεωρήματος)[2] of which οἱ τόποι are the commonest examples. He then describes with some fulness two types of porisms contained in Euclid's book, but gives 28 more types with horrible brevity, e. g. in the first book, 'This line is given in position,' in the third book, 'The sum of these two straight lines has a given ratio to a straight line drawn from this point to a given point[3].' No figures are appended. The whole work contained, in three books, 171 propositions, to which Pappus sup- plies 38 lemmas. Upon these statements of Pappus, which Halley and Prof. de Morgan found unintelligible, Simson framed a defini- tion of a porism as "a proposition in which it is to be proved that one or several things is or are given which (like any one of an infinite number of things not given but having the same rela- tion to the things which are given) has or have a certain property, described in the proposition[4]." Chasles, who approves of this

[1] VII. p. 648. 18 sqq.

[2] The translation in the text is from Chasles. It seems, on authority, to be right. Heiberg explains it as "a local theorem with incomplete hypothesis." Whatever it may mean, it clearly is only intended to describe a special class of porisms, used by writers later than Euclid who, without attempting *to find* the thing proposed, merely declared that it was possible to do so (e.g. Archimedes, *De Spir.* propp. 5—9, cited by Heiberg, pp. 68, 69). Pappus then adds that οἱ τόποι belonged to this class of porisms but, owing to

their number, were collected in a sepa- rate work (κεχωρισμένον τῶν πορισμά- των ἤθροισται).

[3] See Nos. v. and xx. The whole list is given in Hultsch, pp. 654 sqq. Heiberg, pp. 73—77. The Greek of xx, is ὅτι λόγος συναμφοτέρου πρός τινα ἀπὸ τοῦδε ἕως δοθέντος. Halley, Simson and Heiberg interpret this dark saying as above: Chasles and Hultsch translate "the sum of these two rectangles has a given ratio to the segment lying between this point and a given point."

[4] *De Porismatibus*, p. 347, quoted by Chasles, *Le Livre de Porismes*, p. 27.

definition, then proceeds to show the similarity between porisms and the propositions called τόποι, for a τόπος " is a proposition in which it is declared that certain points subject to the same known law are on a line of which the nature is enunciated and of which it remains to find the magnitude and the position. Example : two points being given, as also a ratio, the locus of a point, the distances of which from the two given points are in the given ratio, is the circumference of a circle given both in magnitude and in position[1]." Hence, also, a connexion exists between the two meanings of 'porisma,' for every *porism* may be put as the *corollary* of a local theorem[2] and the close connexion between the porism and the *datum* is equally obvious[3]. Further, Chasles suggests a new definition of porism, which shall combine all the older definitions. Porisms, according to him, are incomplete theorems, " expressing certain relations between things variable according to a common law : relations indicated in the enunciation of the porism but requiring to be completed by the determination of the magnitude and the position of certain things which are the consequence of the hypothesis and which would be determined in the enunciation of a theorem properly so-called." In order to exhibit the similarity of porisms with the most usual propositions of modern geometry, Chasles gives the following example (among others) : " If in the diameter of a circle there be taken two points which divide it harmonically, the ratio of the distances between these two points and any point on the circumference will be constant." Substitute here " given " for " constant " and this proposition is a *porism*. Find the ratio and include it in the enunciation, and you have a complete *theorem*.

Upon the preliminary discourse of Chasles, from which these remarks are taken, Heiberg (pp. 56—79) has many criticisms, supported by much learning, to offer, but his observations are

Playfair (in *Trans. of R. S. of Edinburgh*, 1792), improving on Simson, suggested a def. of a porism as " a proposition affirming the possibility of finding such conditions as will render a certain problem indeterminate, or capable of innumerable solutions."

Chasles, pp. 31, 32, objects to this.

[1] Chasles, *Porismes*, pp. 33—36.
[2] *Ibid.* pp. 36—38.
[3] *Ibid.* pp. 42, 43. The *porisms* cited by Diophantus (*supra*, p. 121) are closely similar to *data*.

relevant mainly to the form of the *enunciation* of a porism and its relations, by virtue of its enunciation and hypothesis, to the τόπος and the local theorem[1]. The passage of Pappus, on which Chasles and Heiberg, and every other would-be restorer of Euclid's work must necessarily rely, is so obscure and is suspected of so many interpolations and mutilations[2], that I could not, save at inconvenient length, give the details of the controversy, which, after all, is of no practical importance. I have therefore preferred to accept Chasles's theories, which are founded on adequate learning and are followed by a restoration of Euclid's *Porisms* with which, at present, no serious fault has been found[3].

One of the types of porisms which Pappus describes at any length, is as follows: "If from two given points, two straight lines be drawn, which cut one another on a straight line given in position, and one of which intercepts on a straight line, given in position, a segment extending to a given point on it, the other will intercept on another straight line a segment which has a given ratio." This type was treated in one or more propositions early in the First Book, and this statement, together with the 38 lemmas of Pappus, gave Chasles his clue. The *Porisms* of the First Book, in his view, deal with propositions suggested by a hypothesis in which we suppose two straight lines to turn about two fixed points, to cut one another on a straight line given in position, and to make on two other fixed straight lines (or on one only) two segments which have to one another a certain constant relation. In the Second Book, the segments are, as a rule, formed on one line only. In the Third Book, the two fixed points are on the circumference of a circle and the two revolving straight lines cut one another on this circumference. "Almost, if not quite, all the relations of segments in the first two Books are

---

[1] E.g. according to Heiberg, a porism proper has nothing whatever to do with a corollary. A τόπος was, as Simson defined it, a proposition 'to find a locus,' and therefore τόποι were a kind of porisms. The propositions, which Chasles calls 'local problems' and distinguishes from 'loci' and 'local theorems,' are really identical with 'loci' and are porisms, etc.

[2] See Hultsch's edition. Heiberg accepts the whole of the text.

[3] Heiberg himself has very few criticisms to make, even on the enunciations, which, he admits, are generally of the true porismatic form. The one obvious error in Chasles' book is that his restored Porism xvii. (p. 119) is identical with the 8th Lemma of Pappus, which is only ancillary to a porism.

such as express that two variable points on two straight lines, or on one only, form two homographic divisions." It should be added that Chasles has had the good fortune to produce 201 *porisms*, or 30 more than Euclid himself composed[1]. The original *porisms* were used, as their place in the τόπος ἀναλυόμενος indicates, in the analysis, or in the synthesis, of a problem which was solved analytically. No doubt, a porism of the form 'it is possible to find' would be used in analysis, like the *Data;* a porism of the form ' to find' would be used in the synthesis.

125. The immediate successors of Euclid, as heads of the Alexandrian mathematical school, seem to have been **Conon** of Samos, who added " Berenice's hair " to the constellations[2], and **Dositheus** of Colonus. Perhaps also a certain **Zeuxippus** and **Nicoteles** of Cyrene were at Alexandria during this period. But nothing is known of these persons, save that Conon, Dositheus and Zeuxippus corresponded with Archimedes, who had a high opinion of their abilities (especially of Conon's[3]) and that Apollonius acknowledges some obligation to discoveries in conic sections by Conon and Nicoteles[4].

But **Archimedes**, the greatest mathematician of antiquity, lived not at Alexandria but at Syracuse. He is said by Tzetzes[5]

---

[1] A summary of the more interesting portion of Chasles' book is given in Taylor's *Ancient and Modern Conics*, pp. LII—LIV. Chasles himself says, p. 14, "Si ce livre de Porismes nous fût parvenu, il eût donné lieu depuis longtemps à la conception et au développement des théories élémentaires du *rapport anharmonique*, des *divisions homographiques* et de l'*involution*."

[2] Catullus LXVI. 7, 8, translating Callimachus. Delambre (I. p. 215) suggests that Callimachus invented the name of the constellation himself and attributed it to Conon. The Berenice in question was wife of Ptolemy III. (Euergetes). Ptolemy, the astronomer, cites some observations of Conon.

[3] See the prefaces to *Sph. et Cyl.* and *Arenarius*, ed. Torelli, pp. 63, 64, 319.

[4] *Conica*, Pref. to Bk. IV. Halley's ed. pp. 217, 218. A very important astronomer, Aristarchus of Samos, belongs to this interval. His extant work on the *Sizes and Distances of the Sun and Moon* is printed in the 3rd Vol. of Wallis's works. His proofs of course are geometrical (e.g. Prop. 2 is "If a greater sphere illuminate a less, more than half the latter is illuminated") but add nothing to geometry.

[5] *Chiliad.* II. 35, 105. Proclus, p. 68, cites Eratosthenes as witnessing that he was a contemporary of Archimedes. The chief authority on the life of Archimedes is Plutarch, *Vita Marcelli*, cc. 14—19. A biography, which was used by Eutocius, was written by one Heracleides who perhaps was the friend whom Archimedes mentions pp. 217, 318 (Torelli).

(an authority as late as the 12th century) 'to have died at the age of 75, and, as it is well attested that he was killed in the sack of Syracuse B. C. 212, he was probably born about 287 B. C. Diodorus[1] says that he visited Egypt and it is certain that he was a friend of Conon and Eratosthenes, who lived in Alexandria. His writings also show a most thorough acquaintance with all the work previously done in mathematics, and it may therefore be inferred that he was a disciple of the Alexandrian school. He returned, however, to Syracuse and lived there on intimate terms with King Hieron and his son Gelon, to whom possibly he was related by blood[2]. He made himself useful to his patrons by his extraordinary ingenuity of mechanical invention,—a gift by which he himself set little store[3]. He is said, by various contrivances, to have inflicted much loss on the Romans during the siege by Marcellus, but the city was ultimately taken and Archimedes perished in the indiscriminate slaughter. Marcellus wished to preserve his life but he was slain by accident[4]. The story is that he was contemplating a geometrical figure drawn on the ground when a Roman soldier entered. Archimedes bade him stand off and not spoil the diagram, but the soldier, insulted at this behaviour, fell upon him and killed him[5]. Marcellus raised in his honour a tomb bearing the figure of a sphere inscribed in a cylinder. Cicero had the honour of restoring this during his quaestorship in Sicily B. C. 75[6].

---

[1] Diod. v. 37.

[2] Plutarch, Marcell. 14.

[3] Ibid. 17, πᾶσαν ὅλως τέχνην χρείας ἐφαπτομένην ἀγεννῆ καὶ βάναυσον ἡγησάμενος, "thinking that every kind of art, which was connected with daily needs, was ignoble and vulgar."

[4] Cic. Verr. IV. 131, Livy xxv. 31, Plut. Marc. 19, Pliny, Hist. Nat. VII. 125.

[5] This tale is told in many slightly different forms. Plutarch loc. cit. Valerius Maximus VIII. 7, 7, Tzetzes II. 35. 135, Zonaras IX. 5.

[6] Cic. Tusc. Disp. v. 64, 65. The

authorities for Archimedes' life are collected and generally quoted in Torelli's Preface, pp. 11 and 12, and Heiberg's Quaestiones Archimedeae, Copenhagen, 1879, pp. 1—9. This little monograph deals chiefly with the text, but contains much very minute information on the arithmetic of Archimedes. Heiberg has since edited the text (Leipzig, 1880), but I have quoted always from Torelli, whose edition I happen to have. The errors and misprints which Heiberg points out in Torelli, are not such as to seriously affect his value for the present purpose.

126. The extant works of Archimedes seem to comprise almost all his more important contributions to mathematics. Internal evidence, derived from references in some books to proofs contained in others and from allusions in the prefatory letters which accompany many of the books, shows that the works are to be arranged in the following approximately chronological order[1]: viz.

(1) Book I. of '*Equiponderance of Planes or Centres of Plane Gravities*' (Περὶ ἐπιπέδων ἰσορροπιῶν ἢ κέντρα βαρῶν ἐπιπέδων), in 15 props. preceded by 8 (or 9) postulates[2].

(2) '*The Quadrature of the Parabola,*' in 24 props. (sent to Dositheus).

(3) Book II. of '*Equiponderance of Planes,*' etc., in 10 props.

(4) '*On the Sphere and the Cylinder,*' in two books, the first of 50 props., preceded by 5 postulates, the second of 10 props. (both sent to Dositheus).

(5) '*The Measurement of the Circle*' (κύκλου μέτρησις), in 3 props.

(6) '*On Spirals*' (περὶ ἑλίκων), in 28 props.

(7) '*On Conoids and Spheroids,*' in 40 props. (sent to Dositheus).

(8) '*The Sand-Counter*' (ψαμμιτής), an essay addressed to Gelon.

(9) '*On Floating Bodies*' (περὶ ὀχουμένων or περὶ τῶν ὕδατι ἐφισταμένων), in two books, the first of 9, the second of 10 props. (extant only in Latin)[3].

We have also, in a Latin translation from the Arabic, a collection of 15 *Lemmas,* which have certainly been tampered

---

[1] See Torelli's *Pref.* p. xiii. Heiberg, *Q. A.* pp. 10—13.

[2] Archimedes himself (*Quadr. Parab.* props. 6 and 10) refers to this book as τὰ μηχανικά. Proclus (p. 181) calls it αἱ ἀνισορροπίαι. Simplicius (*ad Arist. De Caelo,* IV. p. 508 *a.*) calls it Κεντρο-βαρικά.

[3] The Latin translation was made by Tartaglia (Venice, 1543 and 1565)

from a Greek codex which has not since been discovered. The title περὶ τῶν ὀχουμένων is cited by Strabo I. p. 54: τὰ ὀχούμενα in *Math. Vett.* p. 151, Pappus VIII. p. 1024. A fragment recently discovered has the other title, and Tzetzes evidently alludes to this book by the name ἐπιστασίδια (*Chil.* XII. 974). Torelli, *Pref.* xviii. Heiberg, *Quaest. Arch.* pp. 13, 22.

with (e.g. Archimedes is mentioned in the 4th and 14th) and
may not be authentic at all[1]. Those works, also, which are
extant in Greek, are evidently not now in precisely the same
form as when first written. Some of the titles for instance,
especially 'Quadrature of the *Parabola*,' are added by later
hands, and again, most of the books are written in inferior
Greek of the Attic dialect, whereas Archimedes wrote in Doric[2],
the dialect proper to Syracuse. Eutocius of Ascalon, a scholiast
of the 6th century, wrote commentaries still extant on the
books of the *Sphere and Cylinder, Measurement of the Circle*
and *Equiponderants*. These are valuable for the great number
of historical notices which they contain and of which very
frequent use has been made in these pages.

Beside the extant works, Archimedes is known to have
written several others and yet more are attributed to him. He
wrote a treatise on the half-regular polyhedra, i.e. the solids,
thirteen in number, which are bounded by regular but dissimilar
polygons of two or three kinds[3]. He himself refers (in the
Arenarius) to his arithmetical treatise called Ἀρχαί, 'First
Principles,' addressed to Zeuxippus. Pappus[4] quotes his work
Περὶ ζυγῶν, 'on Levers.' Theon quotes his *Catoptrica*[5]. Pappus[6]
quotes Carpus as an authority for the fact that Archimedes
wrote a mechanical treatise on the method of constructing a
globe or planetary (περὶ σφαιροποιίας). The Arabs ascribe to
him works on 'the heptagon in a circle,' on 'circles touching

---

[1] The translation in Borelli's edition
(Florence, 1661) is said to have been
made by Abraham Ecchellensis from
the Arabic of Tabit ibn Korra, with
notes by Almochtas Abulhasan. Torelli
reprints this (see his *Pref.* p. xix), but
there was another version by J. Gravius
(Foster's *Miscellan.* London, 1659).
Heiberg (p. 24) and Cantor (pp. 256,
257) are inclined to think that the book
contains some authentic propositions,
esp. the 4th and 14th, perhaps also
the 8th and 11th.

[2] Torelli's *Pref.* p. xv, Heiberg, *Q.
Archim.* ch. v., *De Dialecto Arch.* pp.

69 sqq.

[3] Pappus v. 19. Heron (*Deff.* 101)
says wrongly that Archimedes added
13 to the 5 Platonic regular solids.
Kepler resumed the study of such polyhedra in his *Harmonice Mundi.* Cantor,
p. 264.

[4] viii. 24, p. 1068.

[5] *Comm. in Ptol.* i. 3, p. 10 (Basle
ed.). Cf. Olympiodorus *in Arist.Meteor.*
ii. p. 94 (ed. Ideler). Apuleius, *Apol.*
16. Tzetzes, *Chil.* xii. 973. Heiberg,
Quaest. Arch. p. 33.

[6] viii. 3, p. 1026. Cf.Proclus, p. 41,
16.

one another,' on 'parallel lines,' on 'triangles,' on 'the properties of right-angled triangles,' on 'data[1].' Suidas says that Theodosius wrote a commentary on the 'Guide-book' or ἐφόδιον of Archimedes, perhaps a little treatise on geometrical methods. Beside these, it is possible that Archimedes wrote yet other books, for he on several occasions refers to propositions as already proved, which are not so in any extant work, or reduces a proposition to a problem which he does not solve (e.g. *Sph. et Cyl.* II. 5. p. 158), or uses a theorem which is not proved at all[2].

127. It is usual to divide the works of Archimedes into three groups, geometrical, arithmetical and mechanical, but these distinctions are not strictly maintained by Archimedes himself. Thus in *Quadrature of the Parabola,* propositions VI.—XIV. are founded on propositions proved in the preceding first book of *Equiponderance* (e.g. in props. VI. and VII. a triangle is suspended from one arm of a lever kept in equilibrium by another area suspended at the other end). So, also, the 3rd proposition of *Measurement of the Circle* is an attempt to find an arithmetical value for the ratio between the circumference

---

[1] Wenrich *De Auct. Graec. Versionibus,* pp. 194, 196, Heiberg *Q. A.* pp. 29, 30. Heiberg is inclined to reject these Arabic notices, save that on 'circles touching one another,' of which he thinks, some extracts may be preserved in the 15 Lemmas.

[2] E.g. in *De iis quæ in humido* II. 2, he uses, without a word of reference, a theorem that, in a segment of a parabolic conoid, the centre of gravity divides the axis into two parts such that the part on the side of the vertex is twice the other. The proposition *Sph. et Cyl.* II. 5 is to divide a sphere into two segments whose volumes are to one another in a given ratio. This is soluble only (to use algebraical symbols) if a line a can be so divided that $a - x : b :: c^2 : x^2$ i.e. if the cubic equation $x^3 - ax^2 + bc^2 = 0$, can be solved. Archimedes (who of course gives the pro-

portion only, not the equation) adds a *diorismus,* or determination of a condition under which this can be solved (for a positive root). If $c = 2(a - c)$, then $a - c$ must be greater than b. In other words, $x^3 - ax^2 + \frac{4}{9}a^2b = 0$, is soluble only if $b < \dfrac{a}{3}$. Archimedes promises a solution but does not give it. See Cantor, pp. 265, 270, 271. Archimedes is often said to have written a *Conics* (κωνικά), but it is now generally supposed that the *Conics* and the *Elements,* to both of which he often refers, are the works of Euclid; Cantor pp. 260, 261. Heiberg, *Q. A.* p. 31. Heracleides, however, the biographer of Archimedes, accused Apollonius of stealing from an unpublished work by his predecessor. (See Eutocius in Halley's *Apollonius,* p. 8.)

and its diameter, and the inquiry involves the extraction of $\sqrt{3}$. Nevertheless, the division first suggested is exact enough for most purposes, and may be adopted in the following brief summary of the contents of the various books. The geometrical are taken first.

The *Quadrature of the Parabola* begins with a letter to Dositheus announcing the chief contents of the book. It contains two solutions of the problem, the one mechanical, the other geometrical. Both involve the use of the method of exhaustion. Props. I.—III. are simple propositions in Conics without proofs: IV. V. are of the same kind, but are proved. Then props. VII.—XVII. contain the mechanical proof that "any segment which is contained by a straight line and the section of a right-angled cone is ⅘ (ἐπίτριτον) of a triangle which has the same base and the same altitude as the segment." Archimedes starts, as above mentioned, by suspending a triangle or trapezium and another area on opposite sides of a lever in equilibrium, the triangle or trapezium being suspended from two points, the area from one. The triangle or trapezium is then shewn to bear a certain ratio to the area[1]. Then if BΘΓ be a segment of a parabola, of which BΓ is the base and Θ the point on the curve most distant from the base[2], the *segment* BΘΓ is shewn by exhaustion to be one-third of the space of which the *triangle* BΘΓ is one-fourth. Props. XVIII.—XXIV.

---

[1] E.g. Prop. VI. ABΓ is a lever, of which B is the middle point. A right-angled triangle BΔΓ is suspended from B, Γ, the right angle being at B, the side BΓ being half the length of the lever. This is exactly balanced by an area Z, suspended from A. Then Z is one-third of the triangle. For in BΓ take E, so that EΓ = 2EB. Then the centre of gravity of the triangle (as previously proved in the 1st Book of *Equiponderance*) lies in the vertical line drawn from E, and the triangle may be suspended from E without disturbing the equilibrium. Suspend it from E and the triangle is to Z inversely as the arms of the lever, or as AB to BE, and AB = 3BE. A summary of the following propositions is given by Cantor, pp. 278—279.

[2] In Prop. XVII. Θ is called the vertex, κορυφή, of the curve. In Prop. XVIII. the first of the geometrical proof, it is shewn that if the base BΓ be bisected and BΘ be drawn parallel to the axis (called the 'diameter'), meeting the curve in Θ, then Θ is the point from which the greatest perpendicular can be drawn from the curve to BΓ, and is the κορυφή of the segment. The tangent at Θ is parallel to BΓ.

contain the geometrical proof. The triangle ΒΘΓ is half the parallelogram of the same altitude on ΒΓ, and is therefore more than half the segment. Inscribe triangles in the segments cut off by the lines ΒΘ, ΘΓ. Each of these is more than half the segment in which it is inscribed and is also one-eighth of the triangle ΒΘΓ: the two together are one-fourth of it. Take a series of magnitudes, $x, \frac{x}{4}, \frac{x}{16}, \frac{x}{64} \ldots\ldots \frac{x}{4^n}$, of which x is equal to the triangle ΒΘΓ. The sum of these is less than the segment. Their sum, again, *plus* $\frac{1}{3}$d of the least magnitude, is $\frac{4x}{3}$. Hence if the segment be exhausted by triangles in the manner above indicated, it is found by *reductio ad absurdum*, that the segment is $\frac{4}{3}$rds of the first triangle ΒΘΓ.

The treatise on the *Sphere and the Cylinder* is in two books. Book I. begins with another letter to Dositheus, announcing its principal contents[1]. Then follow some definitions (curiously called *ἀξιώματα*) and assumptions (*λαμβανόμενα*). Of the assumptions, the 1st is "a straight line is the shortest of all lines which have the same extremities." The book begins with 7 propositions, bearing on the theory of exhaustion, e.g. VI. is "a circle being given and also two unequal magnitudes, it is possible to describe about and within the circle two polygons, such that the circumscribed polygon shall have to the inscribed a less ratio than the greater given magnitude to the less." Props. VIII.—XVII. are on the surfaces of pyramids (described within and about cones), of cylinders and of cones (e.g. Prop. XVI. "The surface of an isosceles cone is to its base as the side of the cone to the radius of the base"). Props. XVIII.—XXI. are on the

[1] In this book Torelli numbers *fifty* propositions. Other editors, who do not count the first, number *forty-nine*. In Prop. III. Torelli omits a reference to Euclid by name which is given in all the MSS. Proclus (p. 68) says that Archimedes mentioned Euclid, and this is the only place in which such mention occurs. Heiberg (p. 157) thinks the words are genuine. They are merely "Take Δ equal to ΒΓ, by the Second of the First Book of Euclid's" (τῶν Εὐκλείδου). It is in the preface to this book that Archimedes states that the cubatures of the pyramid and cone (Euclid XII. 7, 10) were discovered by Eudoxus. The cubatures of the sphere and the cylinder are referred to that of the cone.

volumes of cones and of portions of cones. These propositions
are then used (XXII.—XXXIV.) in an exposition of the relations
of the surfaces and volumes of those solids, described within and
about a sphere, which are produced by the revolution of polygons
described in or about a great circle. Prop. XXXV. is selected
for mention in the prefatory letter. It is that "the surface of a
sphere is four times that of one of its great circles." Prop. XXXVI.
is "any sphere is four times a cone whose base is a great circle,
and whose altitude is a radius, of the sphere." This leads to
XXXVII. The volume and the surface of a sphere are ⅔rds of
the volume and surface, respectively, of a cylinder whose base is
a great circle, and whose altitude is the diameter, of the sphere
(the bases of the cylinder being included in its surface). This
discovery was the chief pride of its author. The figure of this
proposition is that which Marcellus, following an expressed wish
of Archimedes[1], inscribed on his tomb. Props. XXXVIII.—XLVII.
deal with segments of a sphere and the inscribed and circumscribed solids produced, as before, by the revolution of polygons
described within and about a great circle. Props. XLVIII.—XLIX.
prove that the surface of a segment of a sphere, whether less or
greater than a hemisphere, is equal to a circle whose radius
is the straight line drawn from the vertex of the segment
to the periphery of its basal circle. Prop. L. is on the volume
of a sector of a sphere, which is shewn to be equal to a cone
whose base is a circle equal to the surface of the segment, and
whose altitude is the radius of the sphere.

Book II. of the *Sphere and Cylinder* begins with another
prefatory letter to Dositheus, in which the chief glories of
Book I. are again recounted, and which says that the Second
Book contains some problems and theorems suggested by the
First. Prop. II. is a problem "To find a sphere equal to a given
cone or given cylinder." The analysis of this problem leads to
the discovery of two mean proportionals between two straight
lines. The synthesis, which is the analysis taken backwards,
of course, requires that two mean proportionals should be found.
Archimedes does not here shew how this is to be done, but it is

---

[1] Plutarch *Marcellus*, 17.

à propos of this passage that Eutocius introduces that historical account of the duplication problem which has been already so often cited[1]. Prop. III. is that " a segment of a sphere is equal to a cone whose base is that of the segment and whose altitude is to that of the segment as the radius of the sphere + the altitude of the remaining segment is to the altitude of the remaining segment." Some problems are founded on this, solved, as usual, first by analysis, then by synthesis. Prop. IX. is that " if a sphere be cut by a plane which does not pass through the centre, the greater segment is to the less in a ratio which is less than the duplicate but more than the sesquialter of the ratio which the surface of the greater bears to the surface of the less[2]". Lastly, Prop. X. is " of spherical segments with equal surfaces a hemisphere is the greatest[3]."

The book *De Spiralibus* begins with another letter to Dositheus, which, after deploring the death of Conon, who was studying the propositions[4], recounts the contents of the 2nd book of the *Sphere and Cylinder,* then points out the chief results of the treatise on *Spirals* and concludes with a note that Archimedes has used the ordinary lemma (Euclid X. or XII. i.) on which the method of exhaustion is founded. The definition of the spiral and the chief results of the book may be stated practically in the words of Archimedes himself. " If in a plane a straight line, fixed at one extremity, revolve evenly till it return to the position from which it started, and if along the revolving line a point moves evenly from the fixed extremity, this point will describe a spiral. I say that the

[1] The solutions which Eutocius records (Torelli, pp. 135—149) are those of Plato, Heron, Philon of Byzantium, Apollonius, Diocles, Pappus, Sporus, Menaechmus, Archytas, Eratosthenes, Nicomedes, in this order.

[2] It appears from the preface to *De Spiralibus* (p. 218) that Archimedes had wrongly stated this and the next proposition, in an earlier copy which he sent to Dositheus, for the express purpose of deceiving the boastful amateurs of geometry, " who say they have

found everything, but never produce a proof, and sometimes claim to have discovered the impossible."

[3] The treatise " Measurement of the Circle" is given in full in the next section. The quadratures of the spiral and ellipse depend upon a previous quadrature of the circle.

[4] Pappus says that Conon invented the spiral. Archimedes, however, only says that he had sent the enunciations of his propositions to Conon, who had been trying to prove them.

space which is included between the spiral and the straight
line after one complete revolution is one-third of a circle
described from the fixed extremity as centre, with radius
that part of the straight line over which the moving point
advances during one revolution (Prop. XXIV.). Again, if a
straight line touch the spiral at the last extremity of the
latter[1], and from the fixed point there be drawn a perpendicular
to the revolving line (after a complete revolution) produced to
meet the tangent, this perpendicular straight line is equal
to the circumference of a circle described from the fixed
point as centre with the revolving line at the end of a complete revolution as radius (Prop. XVIII.). Again, if the revolving line and the moving point thereon make several complete
revolutions, the space which is included by the second
revolution of the spiral is half that included by the third, a
third of that included by the fourth, a fourth of that in
the fifth and so on. But the space included by the first
revolution is one-sixth of that which is included by the second
(Prop. XXVII.). Again, if in the spiral of one revolution two
points be taken and straight lines be drawn from them to
the fixed point and two circles be drawn from the fixed point
as centre with these straight lines as radii, and the lesser of
these straight lines be produced (to meet the larger circle), the
half-crescents included between the circles, the spiral, and the
straight lines are to one another in a given ratio. (Prop. XXVIII.[2]).
The book begins with some lemmas on constructions (Props. I.—
IX.) and with two propositions, which are in effect the geometrical summation of the series $1 . 4 . 9 \ldots n^2$, (Prop. X.) and of the
series a, $2a$, $3a \ldots na$ (Prop. XI.). Then follow the definitions
and some propositions on tangents to the spiral and lines passing

---

[1] If AΘ be the revolving line, A
he fixed point, the last extremity (τὸ
ἔσχατον πέρας) of the spiral is Θ.

[2] The enunciation is extremely difficult to follow without a figure. Θ is
the fixed point, A, Γ are points on the
spiral. From centre Θ, describe circles
with radii ΘA, ΘΓ, and produce ΘA to

H. The space Ξ is to the space Π as
$ΘA + \frac{2}{3}HA$ is to $ΘA + \frac{1}{3}HA$.

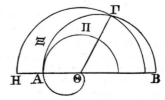

through the fixed point and cutting the curve (Props. XII.—XVII.).
The course of the remainder of the book is pretty well indicated
by the summary above given from the preface.   But a word
should be added on the way in which Archimedes arrives at the
area of the spiral.   The revolving line may be stopped anywhere.   The space included between the curve and the line
is divided into sectors having equal angles at the fixed point.
Each of these is shewn to be less than one, and greater than the
other, of two similar sectors of circles.   It follows, therefore,
that two plane figures (composed of similar sectors of circles)
can be described, one within, the other about, the spiral, such
that the difference between the two figures can be made as
small as we please, and exhaustion is thus effected[1].

The treatise on *Conoids and Spheroids* is also sent, as was
promised in the letter which accompanied the *De Spiralibus*, to
Dositheus.   A *conoid* is the solid produced by the revolution of a
parabola or a hyperbola about its axis.   *Spheroids* are produced
by the revolution of an ellipse, and are *long* (παραμάκεα) or
flat (ἐπιπλατέα) according as the ellipse revolves about its
major or its minor axis.   The first 3 propositions are certain
very complex arithmetical theorems[2].   Props. IV.—VII. deal
with conics, e. g. V. and VI. are on quadrature of the ellipse
by exhaustion; VII. shews that ellipses are to one another as the
products of their axes.   Props. VIII.—X. shew that an infinite
number of right cones and cylinders can be constructed so as
to contain a given ellipse.   Prop. XI. merely recapitulates some
well-known theorems on the ratios of cones and segments of
cones and cylinders to one another.   Props. XII.—XV. shew that
the plane sections of conoids and spheroids are conics; XVI.—XIX.
are on planes touching these solids, XX. is on the division of

---

[1] Compare the accompanying figure
to Prop. XXI. which deals with a spiral
of one revolution only.

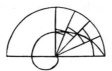

[2] They are of no intrinsic value. The
first is in effect that if

$$s = a + 2a + 3a \ldots\ldots + na,$$

then $2s > n^2a > 2 (s - na).$

The other two cannot be stated shortly,
even with symbols.   On these and the
other arithmetical propositions of Archimedes, see Heiberg, *Q.A.* Chap. IV.
pp. 44 sqq. esp. pp. 50, 51, 56, 57.

a spheroid into two equal parts. XXI.—XXII. are preparatory
to the cubature of the solids : if a conoid or a spheroid
be cut by two parallel planes, the segment so obtained
contains one cylinder and is contained in another, and the
difference between these two cylinders may be made as small
as we please by bringing the two planes of section closer and
closer together. Then follow the propositions selected for
mention in the preface : Props. XXIII.—XXIV. prove that every
parabolic " right-angled " conoid is to a cone on the same base
and of the same altitude as 3 : 2 ; XXV.—XXVI. shew that segments of a parabolic conoid (cut by planes in any direction) are to
one another as the squares of their axes. Props. XXVII.—XXVIII.
deal with the volume of hyperbolic (" obtuse-angled ") conoids;
and XXIX.—XXXIV. with the volume of sections of spheroids cut
by planes, whether passing through the centre or not.

Lastly, of the *Lemmas* which may be authentic, Nos. IV. and
XIV. are to find the area of two curvilinear figures, which
Archimedes calls respectively ἄρβηλος and σάλινον. The ἄρβηλος, which literally is the name of
a shoemaker's knife, is bounded by
three semicircles whose centres are in
a straight line. Its area is the circle
described about the perpendicular *DB*.

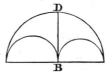

The σάλινον, which perhaps means a 'sieve', (cf. σάλαξ, κόσκινον)
is bounded by four semicircles, whose centres are in a straight
line, two having the same centre *A*. Its area is equal to a
circle described about *BC* as diameter[1].
No. XI. is that if in a circle two chords
cut one another at right angles, the
squares of the four segments of these
chords are together equal to the square
of the diameter. No. VIII. is as follows.

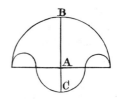

In a circle of which the centre is Δ draw any chord AB and
produce it to Γ, so that BΓ is equal to the radius. Join ΓΔ,

[1] Heiberg *Q. A.* p. 25, suggests that
these *Lemmas* IV. and XIV. are extracts
from the work of Archimedes on "circles touching one another." Pappus
IV. 14 (pp. 208—232, ed. Hultsch) treats
of the ἄρβηλος. See also Cantor pp.
256, 257.

cutting the circle in Z, and produce ΓΔ to meet the circle again
in E.   Then the arc AE will be three times the arc BZ.   The
figure which leads to the proof is appended.

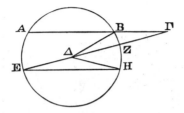

128.   The reader will see, from this brief summary, how
wide a range of subjects Archimedes studied and with what
astonishing ingenuity he treated them.   Nevertheless, quadrature
and cubature of curvilinear areas and solids bounded by curved
surfaces were his chief hobbies, and the process which he most
affects is exhaustion.   This he handles with consummate mastery,
and with it he obtains results for which we now look to the
infinitesimal calculus.   It is desirable, however, that an authentic
specimen of Archimedes' geometrical work should be given in full.
For this purpose, the little work on "Measurement of the Circle"
is especially well adapted, both because it is short in itself, and
does not appeal to any recondite propositions the proof of which
is too long to be admitted, and because it gives all the main
characteristics of Archimedes' style.   It will be seen, at once,
that Archimedes writes not with any educational purpose, like
Euclid, but for the *élite* of the mathematicians of his time.   He
does not confine himself to a stereotyped form of exposition, and
does not shrink from introducing, into a geometrical argument,
propositions of ἀριθμητική and operations of λογιστική.

The *Measurement of the Circle* is in three propositions only.
Prop. I. is "Every circle is equal to a right-angled triangle, such
that the sides containing the right angle one is equal to the
radius, the other to the circumference of the circle."   The
proof, literally translated, save for the introduction of symbols,
is as follows.

"Let the circle *ABCD* be related to the triangle *E* according to the hypothesis.   I say it is equal to the triangle *E*.

For, if possible, let the circle be greater and let the square AC be described in it, and let the circumferences be bisected, and

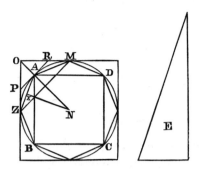

let the segments be finally less than the excess of the circle over the triangle.   Then the rectilineal figure is > the triangle. Take the centre N and the perpendicular NX.   Then NX is < the side of the triangle.   And the periphery of the rectilineal figure is < the other side, for it is < the circumference of the circle.   The rectilineal figure is therefore < the triangle, which is absurd.

But let the circle, if possible, be less than the triangle E. And let the square be circumscribed and let the circumferences be bisected, and let tangents be drawn through the points of bisection.   Then the angle OAR is a right angle : therefore OR is > MR, for $MR = RA$.   And the triangle ROP is > $\frac{1}{2}OZAM$. Let the segments similar to PZA be left less than the excess of the triangle E over the circle.   Then the circumscribed rectilineal figure is < E, which is absurd, for it is > E, since NA is equal to one side of the triangle and the perimeter is greater than the other.   The circle therefore is equal to the triangle E.

Prop. II. is "A circle has to the square on its diameter the ratio 11 : 14 very nearly."

The proof is as follows : "Take a circle, with diameter AB, and let the square CHD be circumscribed about it.   And let DE be double of the side CD, and EZ one seventh part of CD. Since then the triangle ACE has to ACD the ratio 21 : 7, and ACD has to AEZ the ratio 7 : 1, therefore the triangle ACZ is to

the triangle ACD as $22:1$. But the square CH is four times
the triangle ACD: therefore the triangle ACZ is to the square

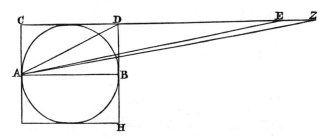

CH as $22:28$ or $11:14$. And the triangle ACZ is equal to the
circle, since AC is equal to the radius and CZ to the circumference (which will be shewn to be very nearly $3\frac{1}{7}$ of the
diameter). The circle therefore has to the square CH the
ratio $11:14$ very nearly [1].

Prop. III. is "The circumference of a circle exceeds 3 times
its diameter by a part which is less than $\frac{1}{7}$ but more than $\frac{10}{71}$ of
the diameter." The proof is:

"Let there be a circle with diameter AC and centre E
and tangent CLZ, and let the angle ZEC be a third of a right
angle. Then $EZ:ZC::306:153$ and
$EC:CZ>265:153$[2]. Draw EH, bisecting ZEC. Then $ZE:EC::ZH:HC$,
and *permutando* and *componendo*,
$ZE+EC:ZC::EC:CH$. Where-
fore $CE:CH>571:153$. Therefore
$EH^2:HC^2>349450:23409$ and
$EH:HC<591\frac{1}{8}:153$[3]. Again, bisect

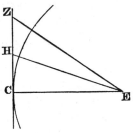

[1] The word ἔγγιστα "very nearly"
seems to have been added throughout by
Wallis. The proposition should possibly be placed third, but it must be
remembered that $\pi=3\frac{1}{7}$ was a very
common approximation in Archimedes'
time. Heron in his *Geometria* (ed.
Hultsch, pp. 115, 136) refers it first to
Euclid, then to Archimedes. The E-
gyptian value was $3\cdot1604$. Ptolemy
(ed. Halma VI. 7) uses $3\frac{17}{120}=3\cdot141666$.

[2] The omitted steps are $EZ=2ZC$
$\therefore EC^2=3ZC^2 \therefore \dfrac{EC}{ZC}=\sqrt{3}>\frac{265}{153}$. It is
not known how Archimedes obtained this
approximation. See *supra*, pp. 53—55.
But in fact $(\frac{265}{153})^2=\frac{70225}{23409}=3-\frac{2}{23409}$.

[3] N. B. $349450=(571^2+153^2)$
$=(326041+23409)$. This is greater
than $(591\frac{1}{8})^2=349,428\frac{49}{64}$. $(591\frac{1}{7})^2$ is
nearer.

the angle HEC by the line EP. On the same principle,
$EC : CP > 1162\frac{1}{8} : 153^{1}$. Therefore $PE : PC > 1172\frac{1}{8} : 153^{2}$.

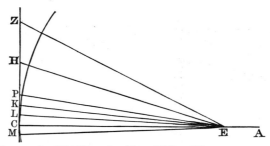

Bisect the angle PEC by the line EK. Then
$\quad EC : CK > 2334\frac{1}{4} : 153$. Therefore $EK : CK > 2339\frac{1}{4} : 153$.
$\quad$ Bisect the angle KEC by the line LE. Then
$$EC : LC > 4673\frac{1}{2} : 153.$$

The angle LEC is $\frac{1}{48}$th of a right angle. At E, make the
angle $CEM = LEC$ and produce zC to M. The angle LEM is
$\frac{1}{24}$th of a right angle. Therefore the line LM is the side of a
polygon of 96 sides (U_{96}) circumscribed about the circle.

Since it has been proved that $EC : CL > 4673\frac{1}{2} : 153$ and
$AC = 2EC$ and $LM = 2CL$, therefore $AC : LM > 4673\frac{1}{2} : 153$.
Therefore AC : periphery of $U_{96} > 4673\frac{1}{2}$: 14688. Of these
numbers, the latter is three times the first $+ 667\frac{1}{2}$, which is

$< \dfrac{4673\frac{1}{2}}{7}$. Wherefore the periphery of U_{96} is three times the

diameter $+$ a part less than $\frac{1}{7}$. Much more then is the circumference of the circle $< 3\frac{1}{7}$ of the diameter.

Secondly, Take a circle with diameter AC, and make the
angle BAC $\frac{1}{3}$rd of a right angle. Then $AB : BC < 1351 : 780$,
but $AC : CB :: 1560 : 780$.

Bisect BAC by HA. Then since $\angle BAH = \angle HCB$ and also
$= \angle HAC$, $\therefore \angle HCB = \angle HAC$. And the right angle AHC is
common. Therefore the third angle $HZC =$ the third $\angle ACH$.
Wherefore the triangles AHC, CHZ are equiangular and

[1] $EH : EC :: HP : PC$
$\therefore (EH + EC) : EC :: (HP + PC) : PC$
and $(EH + EC) : (HP + PC) :: EC : PC$.
But it was shewn above that
$\qquad CE : CH > 571 : 153$

and $EH : CH > 591\frac{1}{8} : 153$. Therefore
$\qquad EC : PC > (571 + 591\frac{1}{8}) : 153$.
[2] $PE^2 = PC^2 + CE^2 > 1373943\frac{1}{2} \; \frac{1}{64}$
$> (1172\frac{1}{8})^2$. $(1172\frac{1}{4})^2$ is nearer.

$AH : HC :: CH : HZ :: AC : CZ.$ But $AC : CZ :: CA + AB : BC.$

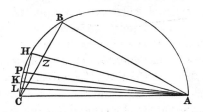

Therefore $CA + AB : BC :: AH : HC.$ Therefore
$$AH : HC < 2911 : 780 : \text{but } AC : CH < 3013\tfrac{11}{24} : 780.$$

Bisect the angle CAH by $AP.$ Then on the same principle
$AP: PC < 5924\tfrac{11}{24} : 780$ or $< 1823 : 240,$ which numbers are $\tfrac{4}{13}$ of
the preceding, respectively. Wherefore $AC : CP < 1838\tfrac{9}{11} : 240.$

Bisect the angle PAC by $KA.$ Then
$KA : KC < 3661\tfrac{9}{11} : 240,$ or (dividing by $\tfrac{11}{40}$) $< 1007 : 66.$ Therefore
$AC : CK < 1009\tfrac{1}{6} : 66.$

Lastly, bisect the angle KAC by $LA.$ Then
$$AL : LC < 2016\tfrac{1}{6} : 66 : \text{but } AC : CL < 2017\tfrac{1}{4} : 66.$$

Conversely $CL : AC > 66 : 2017\tfrac{1}{4},$ and the periphery of the
inscribed polygon : diameter $> 6336 : 2017\tfrac{1}{4}.$ Of these numbers,
the first is $> 3\tfrac{10}{71}$ of the second. Much more then is the circumference of the circle $> 3\tfrac{10}{71}$ of the diameter.

129. The arithmetical treatise of Archimedes (*Arenarius,*
ψαμμίτης) and also the cattle-problem have been summarised
above (pp. 57—61 and 99). It remains only to notice his works
on mechanics. For these he had fewer predecessors. Of the
simple machines two at least, the lever and wedge, were known
from a remote antiquity. Archytas is said to have invented the
screw (κοχλίας) and the pulley (τροχιλαία)[1]. Some kind of a
compound pulley seems to be described in Aristotle's *Mechanica
Problemata* (c. 18). The same work shews that, in the century
before Archimedes, the mathematical theory of the lever was
under consideration, and that it was known that the power and
the weight if applied perpendicularly to a straight lever, so as to
produce equilibrium, are to one another inversely as the arms of

[1] He invented also a child's rattle,
which Aristotle recommends (*Pol.* VIII.
5, 2) as a useful instrument to prevent
children "from breaking things about
the house."

the lever[1]. Some notion of the parallelogram of forces[2] and of the principle of virtual velocity also appears. Many intelligent questions in mechanics, moreover, are here asked, and Aristotle illustrates such explanations as he can give by geometrical figures[3]. The author of the fragment *De levi et ponderoso* (attributed to Euclid), if he lived before Archimedes, had some idea of specific gravity. Also somebody before Archimedes had invented the term "centre of gravity" (κέντρον βάρους) which Archimedes uses but does not define[4]. But there was not as yet any mathematical *proof* of any proposition in mechanics. This step is taken by Archimedes, who deals however only with statics. Book I. of the *Equiponderance of Planes* begins abruptly with some postulates[5], of which the second is "that equal weights suspended from unequal arms (longitudes, μάκεα) are not in equilibrium (μὴ ἰσορροπεῖν) but incline (*sic*) towards the weight which is suspended from the longer arm." A little further on, he assumes "that if equal and similar planes fit exactly upon one another, their centres of gravity also fit exactly upon one another" (ἐφαρμόζειν ἐπ'

---

[1] Aristotle says that cheating tradesmen would shift the centre of their balances towards the scale in which the weight lay (*Mech. Probl.* I. *fin.*). This practice, no doubt, led to the discovery of the law. Aristotle distinguishes the balance (ζυγόν) from the lever (μοχλός), and the σπάρτον (rope) by which the former is suspended from the ὑπομόχλιον (fulcrum) on which the latter is supported. He gives, however, the same explanation of both.

[2] See *Mech. Probl.* I. and XXIII., and Heller, *Geschichte der Physik*, pp. 63—66. Heller admits that "it would be foolish to attribute to Aristotle a clear knowledge" of the principle in question. All that Aristotle says is as follows. If a point *A* have two "motions" (φοραί) at the same time, the one along the straight line *AB* and the other along the straight line *AC*, and *AB*, *AC* represent in length the ratio of the mo-

tions, then the resulting motion is along the diagonal *AD* of the parallelogram *ABDC*. He shews this by supposing *A* to move along *AB*, while the whole line *AB* moves towards *CD*. There is a good note in Van Cappelle's Ed. (1812) pp. 150 sqq.

[3] Cantor, p. 219 and *supra*, pp. 105n. 189.

[4] Eutocius defines it at the beginning of his commentary (Torelli, p. 2). The κέντρον ῥοπῆς or βάρους of a plane figure is "the point from which it must be suspended, in order to remain parallel with the horizon." "The centre of gravity of two or more plane figures is the point from which the balance (ὁ ζυγός) must be suspended, in order to remain parallel with the horizon." Possibly Archimedes had given this definition in his lost treatise περὶ ζυγῶν.

[5] According to Eutocius, Geminus, who was a great purist in nomenclature, proposed to call these "axioms."

ἄλλαλα). "Of unequal but similar figures, the centres of gravity are similarly placed." "In similar figures points are similarly placed if the straight lines, making equal angles at such points, make also equal angles on the homologous sides." Lastly, "In any figure, of which the periphery is concave towards the same parts[1], the centre of gravity must fall within the figure." Props. I.—III. are of exactly the same kind as the initial postulates. Props IV.—V. shew how to find the centre of gravity of two or three equal magnitudes whose centres of gravity are in the same straight line. Props. VI. and VII. are "commensurable and incommensurable magnitudes hang in equilibrium from arms which are inversely as the magnitudes." Prop. VIII. is to find the centre of gravity of the remaining part of a magnitude, from which a portion not having the same centre of gravity as the whole, has been removed. Props. IX.—XV. shew how to find first the line in which the centre of gravity lies, and then the centre of gravity itself of a parallelogram, a triangle and a trapezium. Between Books I. and II. the *Quadrature of the Parabola* is interposed. Book II. begins (Prop. I.) by applying to parabolic segments the Props. VI.—VII. of the first book. Props II.—VII. deal with the centres of gravity of rectilineal figures inscribed in a parabolic segment, e.g. Prop. V. is "If a rectilineal figure be inscribed in a parabolic segment, the centre of gravity of the whole segment is nearer to the vertex than that of the inscribed figure." Prop. VIII. is "The centre of gravity of a parabolic segment divides the diameter so that the part towards the vertex is $\frac{3}{2}$ of the part towards the base." Prop. IX. is a very complicated proposition[2]

---

[1] This expression is not here explained. *De Sph. et Cyl.* Ax. 2 is "A line is concave (κοίλη) towards the same parts in which, if any two points be taken, the straight lines joining such points either all fall on the same side (ἐπὶ τὰ αὐτὰ πίπτουσι) of the line or some on the same side and some on the line itself (κατ' αὐτῆς) but none on the other side" (ἐπὶ τὰ ἕτερα).

[2] The enunciation is to this effect:

If four straight lines (a, b, c, d, of which a is the greatest) be in continued proportion, and $d : a - d :: e : \frac{2}{3}(a - c)$

and $\dfrac{2a + 4b + 6c + 3d}{5a + 10b + 10c + 5d} = \dfrac{f}{a - c}$, then

$e + f = \frac{2}{3}a$. This is worked out in a series of proportions obtained *permutando, componendo, dividendo.* The proof in modern symbols is given in Heiberg, *Q.A.* pp. 49, 50.

in arithmetic which is required for Prop. x. to find the centre of gravity of a truncated parabolic segment.

It is evident that, in the composition of this work, Archimedes' interest was not with mechanics but with mathematics. He does not care about weights or balances but about proofs. Some more practical propositions, perhaps, were contained in the lost book περὶ ζυγῶν, from which Pappus[1] seems to quote the problem "To move a given weight with a given power."

The two books on Hydrostatics, *De iis quae in humido vehuntur*, are similar in character to the *Equiponderance*, but in this department of mechanics Archimedes seems to have had no predecessors whatever. His attention seems to have been first called to the subject of specific gravity by the following circumstance. King Hiero, being anxious to discover whether a crown, which was ostensibly made of gold, might not perhaps be alloyed with silver, asked Archimedes to test it. The story relates that the philosopher was in the bath when the proper method of inquiry occurred to him, and that he immediately ran home naked, shouting Εὕρηκα, εὕρηκα, "I have found it." Our authorities, however, which agree thus far, now begin to diverge. One[2] says that Archimedes, having observed, on stepping into the bath, that bodies immersed in water displaced a quantity of water proportionate to their bulk and not to their weight, measured the *quantity* displaced by gold and silver masses of equal weight and thus obtained a ratio of bulk between the two metals. A later writer[3] on the other hand, states that Archimedes, by weighing two equal weights of gold and silver immersed in water, discovered not the quantity but the *weight* of the water displaced, and thus arrived at the specific gravity of the metals. Both methods may be authentic, but the latter leads more naturally to the treatise on Floating Bodies. Book I.[4] begins with

---

[1] viii. 19. p. 1060.

[2] Vitruvius, ix. 3.

[3] The author of a poem *De ponderibus et mensuris*, formerly attributed to Priscian but now supposed to be of about A.D. 500 (Hultsch, *Scriptt. Metrologici* p. 88 sqq.). The passage of Vitruvius and the lines of the poem

are printed in Torelli, p. 364.

[4] The definition of a fluid is given in *Positio* I. "Let it be assumed that the nature of a fluid is such that, all its parts lying evenly and continuous with one another, the part subject to less pressure is expelled by the part subject to greater pressure. But each

two propositions to the effect that the surface of every still fluid is spherical, the centre of the sphere being the centre of the earth. Prop. III. is that bodies of equal weight with an equal bulk of any fluid do not, if immersed in the fluid, rise above or sink below its surface. Props. IV.—VI. are on bodies lighter than a fluid. Prop. V. in particular contains the hydrostatic principle that "a body lighter than a fluid, when immersed therein, sinks so deep that the quantity of fluid displaced weighs as much as the whole body." Prop. VII. is on bodies heavier than a fluid and immersed therein. Props. VIII.—IX. are on segments of a sphere lighter than a fluid and immersed therein. These will float so that their axes are always vertical. Book II. begins with a proposition (I.), which gives a scientific definition of the specific gravity of bodies lighter than the fluid in which the unit of gravity is chosen. It is that "if a body, lighter than a fluid, floats therein, its weight is to that of an equal bulk of the fluid as the immersed part is to the whole." The remaining propositions II.—X. are on segments of parabolic conoids immersed in a fluid and the positions which they will assume under various conditions[1].

Although, in these works, it is evident that mathematical interest far exceeds the mechanical, and though Archimedes, as above mentioned, was of the opinion of Plato and Pythagoras that the employment of the intellect in the useful arts was degrading, yet it is certain that many of the most useful mechanical contrivances of antiquity were due to his ingenuity. Of these the most famous is the water-screw (κοχλίας), which

---

part is pressed perpendicularly by the fluid above it, if the fluid be falling (*descendens in aliquo*) or under any pressure." *Positio* II. occurs after Prop. VII. and is "Let it be assumed that a body which is borne upwards by a fluid, is so borne in the direction of the perpendicular line which passes through its centre of gravity."

[1] Two propositions of the *De Spiral.* (I. and II.) are of mechanical import- ance, though no mechanical use is there made of them. I. is "If in any line a point moves evenly and there be taken in the line two parts, these shall have to one another the ratio of the times in which the point traverses them respectively." II. is "If two points move evenly each in its own line and in each line there be taken two parts, of which the two first are traversed by the points in the same time and also the two second, the parts will be pro- portional."

is still used.   This apparently was invented by Archimedes
when in Egypt for the purpose of irrigating fields, but it was
used also for pumping water out of mines or from the hold of a
ship[1].   Further the problem " how to move a given weight with
a given power," above mentioned, was practically solved by
Archimedes[2] by the construction of a machine which is variously
described.   It is said by Athenæus (and Plutarch has a similar
tale), that Hiero was in a difficulty about the launching of a
certain very large ship.   Archimedes effected this very easily
by means of an apparatus of cogwheels, worked by an endless
screw ($\H{\epsilon}\lambda\iota\xi$)[3].   Plutarch, however, states that he used, for the
purpose, a compound pulley ($\pi o\lambda\acute{v}\sigma\pi a\sigma\tau o\varsigma$).   It is possible
that Athenæus has by some confusion attributed to Archimedes
the $\beta a\rho o\hat{v}\lambda\kappa o\varsigma$ which was invented by Heron[4], but many authorities concur in attributing to him a compound pulley of three
($\tau\rho\acute{\iota}\sigma\pi a\sigma\tau o\varsigma$) or more ($\pi o\lambda\acute{v}\sigma\pi a\sigma\tau o\varsigma$) wheels[5].   Perhaps this
machine was called by Archimedes himself a $\chi a\rho\iota\sigma\tau\acute{\iota}\omega\nu$, for
Tzetzes who, in one place (*Chil.* II. 130), records the proud boast
of the philosopher " Give me a place to stand on ($\delta\grave{o}\varsigma\ \pi o\hat{v}\ \sigma\tau\hat{\omega}$)
and I will move the whole earth with a $\chi a\rho\iota\sigma\tau\acute{\iota}\omega\nu$," elsewhere
(III. 61) repeats the same saying as referring to a $\tau\rho\acute{\iota}\sigma\pi a\sigma\tau o\varsigma$
(or $\pi o\lambda\acute{v}\sigma\pi a\sigma\tau o\varsigma$)[6].   It is well attested, again, that Archimedes
protracted the siege of Syracuse for a long time by his ingenuity
in constructing catapults which were equally serviceable for
long or short ranges, and others which could be applied to a
small loophole in a wall[7], but the tale that he set fire to the

---

[1] See the article 'Archimedean Screw'
with an illustration in *Encycl. Brit.*
The ancient authorities are Diodorus,
I. 34, v. 37, Vitruvius x. 6 (11), Philo
III. p. 330 (ed. Pfeiffer), Strabo XVII. p.
807, Athenæus v. 208 f.

[2] Plutarch, *Marcellus* 14, Athenæus
v. 207 *a, b.*

[3] Eustathius *ad Iliad* III. p. 114,
ed. Stallbaum.

[4] Pappus III. prop. 5. (Hultsch, p.
63) and VIII. props. 31 sqq.   So also
Tertullian (*De Anima*, 14) ascribes to
Archimedes the hydraulic organ which
everybody else attributes to Ctesibius,

the teacher of Heron.

[5] Beside Plutarch, Galen *in Hippocr.
De Artic.* IV. 27 (XVIII. p. 747, ed. Kühn),
Oribasius, *Coll. Med.* XLIX. 22 (IV. p.
407, ed. Bussemaker).   The latter writer
loc. cit. and Vitruvius x. 2, describe
the $\tau\rho\acute{\iota}\sigma\pi a\sigma\tau o\varsigma$.   Proclus (p. 63) only
gives the fact that Archimedes moved
a large ship.

[6] All the authorities are collected in
Heiberg *Q.A.*, pp. 36—38.

[7] Polybius VIII. 7, Livy XXIV. 34,
Plutarch, *Marcellus* 15.   More reff. in
Heiberg, *op. cit.* pp. 38, 39.

Roman ships, by means of burning-glasses or concave mirrors[1], though repeated by many late writers, is not found in any authority older than Lucian (*Hipp.* 2).

It is evident, again, both from the *Arenarius* itself and from many references in later authors, that Archimedes was much engaged in astronomical observations[2]. Hipparchus (*loc. cit.*) says " from these observations it is clear that the differences of the years are very small, but, as to the solstices, I almost think (οὐκ ἀπελπίζω) that both myself and Archimedes have erred, by a quarter of a day, both in the observation and in the calculation." It would seem from this, and Ammianus expressly states, that Archimedes was interested in the great question of the length of the year. Macrobius says that he discovered the distances of the planets. However this may be, it is certain that Archimedes not only wrote a treatise (mentioned above) on the constitution of a celestial globe (περὶ σφαιροποιίας) but himself actually made one and also a planetary, exhibiting the movements of the sun, moon and five planets. Both these were brought to Rome by Marcellus and were inspected by Cicero himself[3].

It is not difficult to understand how, in ancient times, Archimedes came to be considered as the prince of mathematicians, and that " an Archimedean problem " became a name for a difficulty insoluble to the ordinary intellect and an "Archi-

---

[1] The same story is told of Proclus by Zonaras (Montucla I. p. 334). Montucla, who has some rather amusing pages (I. 232—235) on this subject, shews the improbability of the tale about Archimedes. It appears that *le père* Kircher and also Buffon made some successful experiments with a great number of mirrors. Buffon, with 400 small mirrors, melted lead at a distance of 140 feet.

[2] Hipparchus in Ptol. *Almagest.* I. p. 153. Ammianus Marcell. XXVI. 1, 8, Macrobius, *Somn. Scip.* II. 3. Livy, *loc. cit.* calls Archimedes '*unicus spectator caeli siderumque.*'

[3] Cicero, *De Rep.* I. 21—22, Tusc.

I. 63, *Nat. D.* II. 88, Ovid, *Fasti*, VI. 277, etc. Most of the passages containing references to the mechanical contrivances of Archimedes are printed in Torelli's *Appendix*, pp. 363—370. Some further references are added by Heiberg, *Q.A.* cap. 3, pp. 35—44. The *loculus Archimedius*, mentioned by late Roman writers (Marius Victorin. *Art. gr.* 3, Atilius Fortun. *De Metr.* VI. p. 271), was a square of ivory cut into 14 pieces of various shapes. It was a common game to put these together again into the original square. There is no reason to suppose that Archimedes invented this toy.

medean proof" was the type of incontrovertible certainty[1]. The older men of the modern school, from Tartaglia to Leibnitz, while geometry and mechanics were still largely dependent for support on the discoveries and demonstrations of the Greeks, were as enthusiastic as the ancients about Archimedes. Even later writers, such as Gauss and De Morgan and Chasles[2], who were familiar with the highest modern methods, do not hesitate to rank him with Newton in the very forefront of the champions of science. But knowledge has lately advanced too fast for the fame of Archimedes to keep up with it, and, though his name is no doubt immortal, few readers now know upon what services his immortality depends. Possibly these few paragraphs will justify it at least to mathematicians who understand what difficulties the work of Archimedes involved.

130. The chief contemporary of Archimedes was the famous Eratosthenes. As he was eleven years younger than the mathematician of Syracuse, he was probably born B.C. 276 or 275. He was a son of Eglaus, a native of Cyrene, but lived almost all his life in Alexandria. He was a pupil of Callimachus, the poet, and after a visit to Athens, was invited to succeed his master as custodian of the Alexandrian library. He is said to have almost lost his sight by ophthalmia and on that account to have committed suicide, by voluntary starvation, about B.C. 194.

The multifarious activity of Eratosthenes may be guessed from the fact that, among other contributions to literature and science, he wrote works on *Good and Evil, Comedy, Geography, Chronology*, the *Measurement of the Earth* and the *Constellations*[3]. He was also a considerable poet. The students of the

---

[1] Cic. *ad Att.* XII. 4, XII. 28, *Pro Cluentio* 32, *Ac. Priora* 36.

[2] Chasles, *Aperçu*, p. 15, says of the discoveries of Archimedes that they are "for ever memorable for their novelty and the difficulty which they presented at that time, and because they are the germ of a great part of those which have since been made, chiefly in all branches of geometry which have for their object the measurement of the dimensions of lines and curved surfaces and which require the consideration of the infinite."

[3] See the article *Eratosthenes* in Smith's *Dic. of Gr. and Rom. Biogr.* for the authorities who mention these and other works, none of which are extant.

University used to call him *Pentathlus*, the champion in five sports[1]. It was Eratosthenes who first made a fairly accurate measurement of the obliquity of the ecliptic and an approximate measurement of a geographical degree[2]. It was certainly in his time also that the calendar[3], which we now call *Julian*, with an intercalary day every four years, was introduced. His arithmetical device for finding prime numbers has been described above (p. 87), but of the geometrical work of Eratosthenes only one fragment now remains, the letter which he addressed to Ptolemy Euergetes on the duplication-problem and which is preserved in the commentary of Eutocius on Archimedes, *Sph. et Cyl.* II. 5. This is mainly occupied with the description of a mechanical contrivance for effecting duplication, which Eratosthenes hence called a *mesolabium* or "mean-finder," and of which he was so proud that he dedicated a specimen of it in a temple to be a possession for ever to posterity. It consists of three oblong frames, with their diagonals, sliding in three grooves so that the second frame can slide under the first, the third under the second.

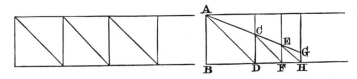

If *AB*, *GH* be the two lines between which it is required to find two mean proportionals, then slide the second frame under the first and the third under the second so that *AG* shall pass through the points *C*, *E*, at which the diameters of the

[1] They also called him *Beta*, as a little later they called a certain astronomer Apollonius *Epsilon*. I should think these were simply the numbers of certain lecture-rooms, but Ptolemy Hephaestio (in Photius, *Cod.* cxc.) says that Apollonius was called Epsilon because he studied the moon, of which the letter ε was a symbol. This Apollonius may be Apollonius of Perga, who certainly studied the stations and retrogradations of the planets (Ptol. *Almag.* XII. 1).

[2] On the astronomical and geodetical work of Eratosthenes see Delambre I. ch. VII. pp. 86—97.

[3] See the edict of Canopus, described by Lepsius in his *Zeitschrift* 1877. *Heft* I. Cantor, p. 283.

second and third frames, respectively, cease to be visible. Then *CD, EF* are the required two mean proportionals[1].

131. Contemporary with Eratosthenes and Archimedes, though younger than either, was Apollonius of Perga (in Pamphylia). He was born in the reign of Ptolemy III. (Euergetes 247—222 B.C.), and flourished under Ptolemy IV. (Philopator 222—205 B.C.). He came when quite young to Alexandria and studied under the successors of Euclid, though no special preceptor is named. He stayed for some time at Pergamum, where there was an university and library similar to the Alexandrian, and where he made the acquaintance of that Eudemus to whom the first three books of his *magnum opus,* the *Conic Sections,* are dedicated. The brilliance of this work gained for him the title of 'the great geometer,' but no more than these meagre facts[2] is known of his history.

Of the eight books which the treatise on Conic Sections originally contained, we possess only seven, and these again have come to us in two parts from two distinct sources. Sir Henry Savile had a Greek MS. of the first four books, but though the whole work seems to have remained for many centuries a text-book of the Greek schools[3], the last four books seem to have been ultimately abandoned as hopeless and the Greek text of them has wholly disappeared. The 8th was lost as early as the time of Tabit ibn Korra who (in the 9th century) translated the first seven books into Arabic. This translation remained the standard Arabic text of Apollonius[4]. The Persians,

---

[1] Pappus VII. Proem. pp. 636, 662 (Hultsch) mentions a work of Eratosthenes περὶ μεσοτήτων or τόποι πρὸς μεσότητας, which perhaps dealt with the duplication-problem or with conics. Montucla I. p. 280.

[2] These are obtained from the prefatory letter to Book I. of the *Conics,* and from Eutocius' Commentary thereon, Halley's ed. pp. 8 and 9.

[3] Geminus, Serenus, Pappus, Hypatia and Eutocius all wrote commentaries on Apollonius.

[4] There is some difficulty about the Arabic translations of Apollonius. Hankel (p. 234), quoting Casiri, says that a version was made, in the time of Al Mamun, of the first four Books: that this was edited by Muhammed, one of the Beni Moses (i.e. the three sons of Musa ibn Schakir), and that Tabit added a translation of the 5th, 6th, and 7th Books. But the writer of the Golian MS. (see Halley, p. 255) says that he has followed the version of Tabit, as emended by the Beni Moses. The Persians Abulphath and Abdulmelik, next mentioned, are not otherwise known.

Abulphath and Abdulmelik made an epitome of it, and the famous Nasir-Eddin edited the whole with a commentary about A. D. 1240. But, in Europe, only the first four books were known as late as the middle of the 17th century, when one Golius, a professor at Leyden, introduced an Arabic MS. written in 1248, containing the first four books in Nasir-Eddin's edition, but the last three from the translation of Tabit with emendations by the Beni Moses[1]. This MS. was bought by Dr Marsh, archbishop of Armagh, who lent it to Halley, the astronomer, who was then Savilian professor of geometry at Oxford. In 1710 Halley published the Greek text of four books and a Latin translation of the remaining three, together with the lemmas of Pappus to each book, the commentary of Eutocius and a conjectural restoration (by Halley himself) of the lost 8th book.

The contents of the eight books of *Conics* are stated in a very brief summary by Apollonius himself in the prefatory letter to Book I. The more interesting and material parts of this are as follows: "Apollonius to Eudemus, greeting. When I was in Pergamum with you, I noticed that you were eager to become acquainted with my *Conics;* so I send you now the first book with corrections and will forward the rest when I have leisure. I suppose you have not forgotten that I told you that I undertook these investigations at the request of Naucrates the geometer, when he came to Alexandria and stayed with me: and that, having arranged them in eight books, I let him have them at once, not correcting them very carefully (for he was on the point of sailing) but setting down everything that occurred to me, with the intention of returning to them later. Wherefore I now take the opportunity of publishing the needful emendations. But since it has happened that other people have obtained the first and second books of my collections before

---

[1] In 1656, almost simultaneously with the arrival of Golius' MS., another was found in the Medicean library at Florence. Galileo's pupil, Viviani, had then nearly completed his restoration of the *four* last books, all of which were supposed to be lost. It was found, on comparison, that he had omitted much that was in Apollonius but had improved on the real text in many respects. The restoration of lost works of Apollonius founded on the lemmas of Pappus and other authorities, was a favourite exercise of mathematicians from the 16th century onwards. See *infra*, pp. 261—263.

correction, do not wonder if you meet with copies which are different from this. Of the eight books, the first four are devoted to an elementary introduction. The 1st contains the mode of producing the three sections and the conjugate hyperbolas (ἀντικείμεναι, 'opposite') and their principal characteristics, more fully and generally worked out than in the writings of other authors. The 2nd Book treats of diameters and axes and asymptotes and other things of general and necessary use in *diorismi*. What I mean by diameters and axes you will learn from this book. The 3rd Book contains many curious theorems, most of which are pretty and new (καλὰ καὶ ξένα), useful for the synthesis of solid *loci* and for *diorismi*. In the invention of these, I observed that Euclid had not treated synthetically the *locus* ἐπὶ τρεῖς καὶ τέσσαρας γραμμάς ('the locus which is related to three or four lines')[1] but only a certain small portion of it, and that not happily, nor indeed was a complete treatise possible at all without my discoveries. The 4th Book shews in how many points the sections of a cone can coincide with one another or with the circumference of a circle[2] and some extra propositions (ἄλλα ἐκ περισσοῦ), none of which had been published by my predecessors. The rest (the last

---

[1] The τόπος ἐπὶ τρεῖς καὶ τέσσαρας γραμμάς would have been treated analytically in Euclid's lost *Conics*. Pappus VII. 36, p. 678 (Hultsch), defines this *locus* as follows: "If three straight lines be given in position and from a point straight lines be drawn to meet the given three at given angles, and the ratio of the rectangle under two of the lines so drawn to the square of the third be given, the point will lie on a *solid locus* given in position, i.e. on one of the three conics. If four straight lines be given in position and four straight lines be drawn as before, and the ratio of the rectangles under two pairs be given, similarly the point will lie on a conic." If five or six straight lines were drawn, whose products were in a given ratio, the *locus* of the point could not be described.

The conic as a *locus ad quattuor lineas* is used by Newton in the *Principia*. Chasles *Aperçu* p. 38 points out the importance of this aspect of conics.

[2] This sentence is only a paraphrase. The Greek has ποσαχῶς ("in how many ways") and κατὰ ποσὰ σημεῖα ("in how many points") in two distinct sentences, as if these were two different things. But the introduction to the 4th book has only κατὰ ποσὰ σημεῖα, and it is probable that these words were added as a gloss on ποσαχῶς by some commentator. The same introduction to the 4th Book says also that the subject here assigned to it had been treated already, but very badly, by Conon, whose work was severely criticised by Nicoteles of Cyrene, and that some props. of the 4th Book had been cursorily treated by this Nicoteles.

four books) is more advanced ($\pi\epsilon\rho\iota o\upsilon\sigma\iota\alpha\sigma\tau\iota\kappa\acute{\omega}\tau\epsilon\rho\alpha$). One is, for the most part, on *maxima* and *minima*: the next about equal and similar conics: the next about 'determinative' (dioristic) theorems; the last on some problems so 'determined' ($\delta\iota\omega\rho\iota\sigma$-$\mu\acute{\epsilon}\nu\alpha$)." The first three books were sent to Eudemus at intervals, the remainder (after Eudemus' death) to one Attalus. All (except the 3rd) are accompanied by little prefatory notes, which repeat in effect the remarks of the first letter. The preface to Book II. is interesting, as shewing the mode in which Greek books were "published" at this time. It runs "I have sent my son Apollonius to bring you the second book of my Conics. Read it carefully and communicate it to such others as are worthy of it. If Philonides the geometer, whom I introduced to you at Ephesus, comes into the neighbourhood of Pergamum, give it to him also."

It will be seen that Apollonius does not pretend that his first three books were entirely new, but only that they were improvements on his predecessors. The statement of Pappus, therefore, that Apollonius' first *four* books are founded on the *Conics* of Euclid is probably substantially true, and there may be some foundation for the accusation of Heracleides that Apollonius had stolen from the unpublished MSS. of Archimedes. But how far the study of conics had been carried before Apollonius cannot now be ascertained. Menaechmus, we know, first wrote on the subject and advanced far enough to apprehend the existence of asymptotes to the hyperbola. He was followed by Aristaeus the elder, whose work was used by Euclid at least in his treatment of the *locus ad tres et quattuor lineas*[1], which seem to have been partly discussed in his *Conics*[2]. But the *Conics* of Menaechmus, Aristaeus and Euclid were almost immediately driven out of the field by the superior book of Apollonius, and the only clue to their contents is to be found in those passages of Archimedes (especially in the *Quadrat. Paraboles* and *De Conoidibus*) in which propositions in conics

[1] Pappus VII. 34 (Hultsch p. 676).

[2] Eutocius (Halley p. 12) did not know where the passage of Euclid to which Apollonius refers in his preface, occurred. But it could hardly have been in the $\tau\acute{o}\pi o\iota$ $\pi\rho\grave{o}s$ $\dot{\epsilon}\pi\iota\phi\alpha\nu\epsilon\acute{\iota}\alpha$, because the *locus* in question was a conic. See note on preceding page.

are referred to as well known or assumed. A careful examination of these shews, in the first place positively, that almost all the propositions which Archimedes uses are to be found in the first three books of Apollonius[1], and, in the second place negatively, that no predecessor of Apollonius was acquainted with the names *parabola*, *ellipse* and *hyperbola*, and with the new treatment of conics which these names imply. It is evident, therefore, that almost the whole of Apollonius' work was original.

132. The completed work adheres closely to the lines indicated in the prefatory letter, but it is obviously difficult to give an intelligible or readable analysis of a huge treatise in which the propositions do not, as they generally do with Archimedes, lead gradually up to one crowning achievement. The theorems, of course, are in great measure identical with those of the modern text-books, but a summary of them, if stated in modern language, would lose historical suggestiveness, and, if stated in the language of Apollonius, would generally be tedious or incomprehensible. This paragraph, therefore, and the next are to be regarded only as containing some hints upon the matter and manner of Apollonius.

Book I. begins with a series of definitions. If a line be drawn from a fixed point to the circumference of a circle, which is not in the plane of the point, and the line revolve round the circumference of the circle, it describes a *cone*, of which the circle is the *base*, the fixed point the *vertex*. The *axis* is the line joining the vertex and the centre of the base. If the axis is at right angles to the base, the cone is *right*: if otherwise, *scalene*. "Of every curve in one plane, that straight line is a *diameter* which, being drawn from the curve, bisects all the straight lines drawn in the curve parallel to a certain straight line." The extremity of the diameter on the curve is the *vertex* of the curve: each of the parallels is drawn *ordinatim*

---

[1] See Heiberg in *Zeitschr. für Math. u. Phys. Hist. Lit. Abth.* xxv. pp. 41 sqq. and a summary of this in *Litterargesch. über Euclid*, pp. 86—88. He concludes that the props. I. 11, 17, 20, 21, 26, 33, 35, 36, 46, 49: II. 3, 12, 13, 27, 49: III. 17: VI. def. 7, props. 2 and 11, of Apollonius were known to his predecessors.

(τεταγμένως κατῆκται "is an *ordinate*") to the diameter. Of two curves in one plane, that straight line is a *transverse* (πλαγία) *diameter* which, cutting both curves, is a diameter of each; and that straight line is an *erect* (ὀρθία) diameter, which, lying between the curves, bisects all the lines intercepted between them which are parallel to a certain straight line. *Conjugate* (συζυγεῖς) diameters are straight lines of which each is a diameter and each bisects the straight lines parallel to the other. The *axis* of the curve (or of two curves) is the diameter which bisects the parallels at right angles, and *conjugate axes* are the conjugate diameters, each of which bisects the parallels to the other at right angles. The definitions of the centre of the ellipse and the conjugate hyperbolas and one or two more are added after Prop. XVI. Book VI. begins with the definitions of similar and dissimilar conics and segments of conics. But many of the most important definitions (e.g. of *parabola, ellipse* and *hyperbola, latus rectum* and *transversum, conjugate* hyperbolas and *asymptotes*) are contained in the propositions in which the things defined first appear. The seven extant books contain on an average about 50 propositions apiece.

The first and most striking of the novelties which are due to Apollonius himself is his mode of producing the three conic sections and the names and descriptions which he gives of them. It will be remembered that his predecessors had always cut the cone by a plane at right angles to one of its sides, and had therefore produced the parabola as the section of a "right-angled cone," the ellipse in an "acute-angled cone," the hyperbola in an "obtuse-angled cone." Apollonius produces all these sections in one and the same cone, whether right or scalene, cut by a plane which is parallel or not parallel to one of its sides. The old names, therefore, ceased to be appropriate, and new ones were required. It will be remembered, again, that a rectangle, applied to a straight line, was said παραβάλλεσθαι, if its base exactly coincided with the line, ὑπερβάλλειν, if it exceeded the line, ἐλλείπειν, if it fell short of it. From these technical terms, Apollonius derived his new names. Let *C* be any point on a conic of which *AB* is the axis, and from

C draw CD perpendicular to AB, cutting off (the *abscissa*) AD. From A draw AE at right angles to AB and equal in length to what we now call the *latus rectum* of the conic. Draw a rectangle equal to the square on CD and having AD for one of its sides. If this rectangle, applied to AE, has its other side exactly coinciding ($\pi\alpha\rho\alpha\beta\alpha\lambda\lambda\acute{o}\mu\epsilon\nu o\nu$) with AE, the conic is a *parabola*. If the side applied to AE is too short ($\grave{\epsilon}\lambda\lambda\epsilon\acute{\iota}\pi\epsilon\iota$), the conic is an *ellipse:* if it is too long ($\acute{v}\pi\epsilon\rho\beta\acute{\alpha}\lambda\lambda\epsilon\iota$), the conic is a *hyperbola*. (In the language of modern analytical conics, if p be the parameter, the Parabola is so-called because $y^2 = px$: the Hyperbola because $y^2 > px$: the Ellipse because $y^2 < px$.) It is in this way that Apollonius gets rid of the cone and exhibits the conic as a plane locus. But he does not define the conic with any reference whatever to a *focus* and *directrix*. The *focus* of an ellipse and hyperbola he discovers only incidentally (III. props. 45—52): he does not discover the *focus* of a parabola at all and has no notion of a *directrix* for any conic[1].

These remarks being premised, the critique of M. Chasles, which repeats some of them in another form, may be here substantially reproduced[2]. Almost the whole of the learned treatise of Apollonius, he says, "depends upon a single property of the conic sections, which is derived directly from the nature of the cone in which these curves are formed....Conceive an oblique cone on a circular base. A plane, passing through the axis, perpendicularly to the base, produces a triangular section, which is called *the triangle through the axis*. Apollonius supposes, in the formation of his conic sections, the cutting plane

---

[1] Pappus VII. 238 (p. 1013) first suggested the *focus* of a parabola and the *directrix*. The theory of *foci* was first worked out by Kepler; Newton first made any use of the *directrix*, which was adopted from him by Boscovich. Taylor, *Ancient and Mod. Conics*, LIV., LXV., LXXI.

[2] *Aperçu*, pp. 18—20. I select this passage because it rather happily combines some information on the nomenclature and elementary propositions of Apollonius, with some indications of the profounder part of his researches. A much fuller *summary* is given by Mr Taylor, *Ancient and Modern Conics*, pp. XLII.—L. Montucla (I. p. 247) is extremely brief. Cantor (pp. 290—296) is tolerably full, but gives no precise references. The fact is that Apollonius is tedious, as Prof. de Morgan found him (Art. "Apoll." in *Penny Cyclop.*).

to be perpendicular to the triangle through the axis. The points in which this plane meets the sides of the triangle are the *vertices* of the curve, and the straight line joining these points is a *diameter* of it. Apollonius calls this diameter *latus transversum* ($\pi\lambda\alpha\gamma\acute{\iota}\alpha$)[1].

" At one of the two vertices of the curve erect a perpendicular to the plane of the triangle through the axis, of a certain length, to be determined as herein-after mentioned : and from the extremity of this perpendicular draw a straight line to the other vertex of the curve. Now, from any point in the diameter of the curve draw at right angles an *ordinate :* the square of this ordinate, lying between the diameter and the curve, will be equal to the rectangle contained by the part of the ordinate comprised between the diameter and the straight line and the part of the diameter comprised between the first vertex and the foot of the ordinate. Such is the generic (*originaire*) and characteristic property which Apollonius recognises in his conic sections and which he uses for the purpose of inferring from it, by very adroit transformations and deductions, almost all the rest. It plays, as will be seen, in his hands, almost the same part as the equation of the second degree with two variables in the system of Analytical Geometry of Descartes.

" It will be observed that the diameter of the curve and the perpendicular raised at one of its extremities, suffice to construct the curve. These are the two elements which the ancients used to establish their theory of conics. The perpendicular in question was called by them *latus erectum* ($\grave{o}\rho\theta\acute{\iota}\alpha$) : the moderns first changed this name to that of *latus rectum,* which was long employed, and afterwards replaced it by *parameter,* which has remained. Apollonius and the geometers who wrote after him gave different geometrical expressions, found in the cone, for the length of this *latus rectum* for each section, but none has appeared to us so simple and elegant as that of Jacques Bernoulli. It is as follows : Take a plane parallel to the base of the cone and situate at the same distance from its vertex as the plane of the proposed conic : this plane will cut the cone in a

---

[1] A *parabola,* having only one vertex, has no *latus transversum.*

circle, the diameter of which will be the *latus rectum* of the conic[1]. From this it is easy to infer the mode of placing a given conic in a given cone.

"The most interesting properties of the conics are to be found in the treatise of Apollonius. We may cite those of the asymptotes, which form the chief part of Book II.: the constant ratio of the products of the segments made by a conic on two transversals parallel to two fixed axes and drawn through any point (props. 16—23 of Book III.): the principal properties of the *foci* of the ellipse and hyperbola (III. 45—52)[2]: the two pretty theorems on conjugate diameters (VII. 12 and 22: 30 and 31).

"We ought also to cite the following theorem, which has obtained so great importance in recent geometry as the basis of the theory of reciprocal polars, and which LaHire had, earlier, made the foundation of his theory of conics. 'If, through the point of concourse of two tangents to a conic section, a transversal be drawn which meets the curve in two points, and the chord which joins the points of contact of the two tangents in a third point, as the whole transversal to the part of it outside the curve, so are the segments of the chord to one another' (III. 37)[3].

"The first 23 propositions of Book IV. relate to the harmonic division of straight lines drawn in the plane of a conic. These are, for the most part, different cases of the theorem just enunciated. In the following propositions Apollonius considers the system of two conics and shews that these curves can cut one another only in four points. He examines what happens when they touch one another in one or in two points and treats

---

[1] *Novum theorema pro doctr. Sect. Conic.* in the Leipzig *Acta Eruditorum,* anno 1689, p. 586.

[2] The *foci* are called "points of application."

[3] Save for the use of the word "transversal" I give the enunciation practically as it stands in Apollonius. Chasles converts it into modern phraseology, concluding "ce troisième point et le point de concours des deux tangentes seront *conjugués harmoniques* par rapport aux deux premiers." So Mr Taylor, p. xLv. "Any chord through the intersection of two tangents to a conic is cut harmonically by their point of concourse and their chord of contact" (III. 37—40). Apollonius does not use the word "harmonic," but gives his proportions in full.

various other cases of the respective positions which they can present[1].

"Book V. is the most precious monument of the genius of Apollonius. Here, for the first time, appear questions of *maxima* and *minima*[2]. The book contains all that the analytical methods of to-day teach us on this subject, and we may recognise in it the germ of the beautiful theory of evolutes (*développées*)[3]. In fact, Apollonius proves that there is, on each side of the axis of a conic, a succession of points from which only one normal can be drawn to the opposite part of the curve: he gives the construction of these points and observes that their continuity separates two spaces which present this remarkable difference, viz.: from any point of the one two normals can be drawn to the curve and none can be drawn from any point of the other. Here then we have *centres of osculation* (curvature) and the *evolute* of a conic perfectly determined[4]. Apollonius makes use of an auxiliary hyperbola, of which he determines the elements, for the purpose of constructing the feet of the normals let fall, from a given point, on the proposed conic. All these investigations are conducted with admirable sagacity."

It should be added that Book VI. treats mainly of similar conics : Book VII. of conjugate diameters, Book VIII., as restored by Halley, consists of 33 problems (or *porisms*, as he might have called them) to find conjugate diameters which satisfy certain given conditions.

133. It will be obvious that, for the mere purpose of illustrating the style of Apollonius, one proposition will do almost as well as another. The proofs, which I shall give in this section, are those of two propositions of exceptional historical interest.

Prop. II. of Book I. exhibits the characteristic of the parabola above described. The enunciation (slightly abbreviated) is as

---

[1] Every proposition in Bk. IV. is proved by *reductio ad absurdum*.

[2] This is not quite true. Euclid VI. 27 (*supra* p. 84, *n.*), is the first known proposition in which a *maximum* is found. Compare also the determination given by Archimedes *De Sph. et Cyl.*

II. 5, given above, p. 225, *n.*

[3] Suggested first by Huyghens in 1673. Taylor, *Conics*, pp. 221, 222.

[4] The remarks of Chasles on Bk. V. are practically identical with Montucla's.

follows : " If a cone be cut by a plane through the axis and by another plane cutting its base along a straight line which is perpendicular to the triangle through the axis, and the diameter[1] of the section be parallel to one of the sides of the triangle through the axis: the square of the straight line which is drawn to the diameter from the section of the cone parallel to the common section of the cutting plane and the base of the cone will be equal to the area contained by the *abscissa* ($\dot{\eta}$ $\dot{a}\pi o\lambda a\mu\beta a$-$\nu o\mu\acute{e}\nu\eta$) of the diameter and a certain other line which has, to the straight line lying between the angle of the cone and the vertex of the segment, the same ratio which the square on the base of the axial triangle has to the rectangle under its sides. Let a section of this sort be called a *Parabola*."

The proof (somewhat abridged) is as follows :

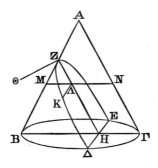

ABΓ is the axial triangle. Let the cone be cut also by a plane which cuts its base along ΔE, at right angles to BΓ. ΔZE is the conic, ZH its diameter, parallel to AΓ, one side of the axial triangle. From Z draw ZΘ at right angles to ZH, making ZΘ : ZA :: BΓ² : AB . AΓ. From any point K on the curve draw KΛ parallel to ΔE, meeting the diameter in Λ. Then KΛ² = ΘZ . ZΛ.

Through Λ draw MN, parallel to BΓ. Now KΛ is parallel to ΔE, therefore the plane through KΛ, MN is parallel to the plane through BΓ, ΔE, i.e. to the base of the cone. Therefore

---

[1] This diameter, which is in fact the axis, is called in the corollary to I. 46 the *principal* ($\dot{a}\rho\chi\iota\kappa\dot{\eta}$) diameter or the diameter $\dot{\epsilon}\kappa$ $\gamma\epsilon\nu\nu\dot{\eta}\sigma\epsilon\omega s$, "arising from the generation of the curve."

the plane through KΛ, MN is a circle, of which MN is a diameter. And KΛ is perpendicular to MN (as ΔE to BΓ). Therefore $KΛ^2 = MΛ . ΛN^1$.

And since $BΓ^2 : BA . AΓ :: ΘZ : ZA$, but $BΓ^2 : BA . AΓ$ is the ratio compounded of $BΓ : ΓA$ and $BΓ : ΓA$, therefore $ΘZ : ZA$ is the same compounded ratio. But

$$BΓ : ΓA :: MN : NΛ :: MΛ : ΛZ ;$$

and $BΓ : BA :: MN : MΛ :: MΛ : MZ ::$ the remainder $NΛ :$ the remainder ZA. Therefore $ΘZ : ZA$ (being compounded of the ratios $MΛ : ΛZ$ and $NΛ : ZA$) is $MΛ . NΛ : ΛZ . ZA$.

But $ΘZ : ZA :: ΘZ . ZΛ : ZA . ZΛ$. Therefore $MΛ . NΛ = ΘZ . ZΛ$. But $MΛ . NΛ = KΛ^2$, as already proved. Therefore $KΛ^2 = ΘZ . ZΛ$. Q. E. D.

The proof concludes with a direction that ΘZ may be called either the line related to the squares of the ordinates (παρ᾽ ἣν δύνανται) or *latus rectum* (ὀρθία).

The enunciation of I. 12 establishes a similar law for the hyperbola. "If a cone be cut by a plane through the axis and by another plane cutting its base along a straight line perpendicular to the base of the axial triangle, and the diameter of the section produced meet one side of the axial triangle produced on the other side of the vertex[2], the square of any ordinate (described as before) will be equal to an area {applied to a certain straight line, to which the portion of the diameter of the conic produced, which subtends the exterior angle of the triangle, has the same ratio as the square of the straight line which is drawn, parallel to the diameter, from the vertex of the cone to the base of the triangle has to the rectangle contained by the segments of that base made by it} having for its side the *abscissa* and excessive (ὑπέρβαλλον) by a figure similar and similar in position to that which is contained by the straight line subtending the external angle of the triangle and the straight line to which the area, equal to the square of the ordinates, is to be applied (ἡ εὐθεῖα παρ᾽ ἣν δύνανται αἱ καταγόμεναι). Let a section of this kind be called a *Hyperbola*." This rigmarole (abridged from the

---

[1] The nomenclature is remarkable. τὸ ἄρα ὑπὸ τῶν ΜΛΝ ἴσον ἐστὶ τῷ ἀπὸ τοῦ ΚΛ. Notice the use of ὑπὸ and ἀπό.

[2] Two cones, having a common vertex, may here be supposed.

original) will be easier to follow by reference to the figure[1],

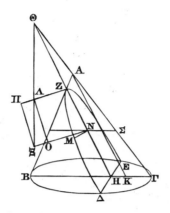

in which AΒΓ is the axial triangle: ΔZE the section: ZH the
diameter produced to meet ΓA in Θ.   MN is the ordinate.
Then MN² is equal to an area applied to ZΛ, which line is
perpendicular to ZH, and is such that ΘZ : ZΛ :: AK² : BK . ΚΓ,
the line AK being drawn parallel to the diameter ZH.   The
area in question, ZΞ, has the *abscissa* ZN for one side, and
is such that it "overlaps" (ὑπερβάλλει) the line ZΛ by the
figure ΛΞ which is similar and similar in position to the
rectangle ΘZ . ZΛ.   The line ZΛ is drawn at right angles to
ZH.   From N, NOΞ is drawn parallel to ZΛ, and the point Ξ is
that in which ΘΛ produced meets NOΞ.   The only addition
made to the figure for the purpose of the proof is that, through
N, PΣ is drawn parallel to BΓ.   The proof is of the same kind
as that for the parabola, but concludes with the additional
statement that ΘZ is to be called πλαγία, *latus transversum*.

The next proposition (I. 13) contains a similar theorem with
regard to the *ellipse*.   The *latus rectum* is determined precisely
as before.   The square of the ordinate is equal to an area
applied to the *latus rectum*, but *deficient* by a figure similar and
similar in position to the rectangle under the *latus transversum*
and the *latus rectum*.   The proposition contains also directions
for producing an elliptical section.

[1] In the figure ΛO and ΠΞ ought to be parallel to ZN.

Book III. prop. 45, first exhibits the *foci* of the ellipse and hyperbola. The enunciation is as follows: "If in a hyperbola or ellipse or circle or conjugate hyperbolas, from the extremities of the axis there be drawn straight lines at right angles, and a rectangle equal to a fourth part of *the figure*[1] be applied to the axis at either end, in the hyperbola or conjugate hyperbolas excessive by a square but in the ellipse deficient, and there be drawn a tangent to the curve meeting the straight lines drawn at right angles as aforesaid, the straight lines drawn from the points of concourse to *the points determined by the application aforesaid* (τὰ ἐκ τῆς παραβολῆς γενεθέντα σημεῖα) make right angles at those points."

The proof is as follows:

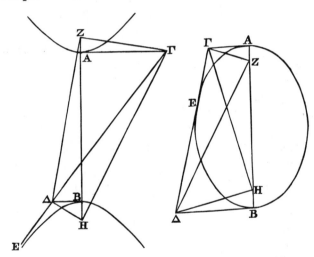

Let AB be the axis of any of the proposed sections, and draw AΓ, BΔ at right angles to this. ΓΕΔ is a tangent. And let a rectangle equal to a fourth part of the figure be applied at either end of AB, as above mentioned, viz. the rectangles

---

[1] The *figure* (τὸ εἶδος) is the rectangle contained by the *latus transversum* and the *latus rectum*. A rectangle, equal to one-fourth of this, is to be applied to the axis, so that in a hyperbola it overlaps by a square, and in an ellipse is deficient by a square. See Taylor, *Anc. and Mod. Conics*, pp. xliv. 81 *n.* 111 Schol. E.

AZ . ZB, AH . HB : and join ΓZ, ΓH, ΔZ, ΔH.   The angles
ΓZΔ and ΓHΔ are right angles.

It has been shewn (III. 42) that the rectangle AΓ . BΔ is
equal to the fourth part of "the figure" on AB.   Therefore the
rectangle AZ . ZB = the rectangle AΓ . BΔ.   Therefore

$$\text{ΓA} : \text{AZ} :: \text{ZB} : \text{BΔ}.$$

And the angles at A and B are right angles.   Therefore the
angle AΓZ = the angle BZΔ and angle AZΓ = angle ZΔB.
And the angles AΓZ, AZΓ are together equal to a right angle,
therefore the angles AZΓ, BZΔ are equal to a right angle :
therefore the remainder ΔZΓ is a right angle.   Similarly, ΓHΔ
may be proved to be a right angle.

The following propositions, XLVI.—LII., deal with some other
theorems suggested by the same construction.

The two proofs, here given, which are both comparatively
easy, will perhaps suffice to indicate to the reader the lack of
technical terms and symbols, and consequently the cumbrous
modes of proof, which characterise the higher Greek geometry.
It seems superfluous to add more specimens, which probably no
one would read.

134.   The century which produced Euclid, Archimedes and
Apollonius was, beyond question, the time at which Greek
mathematical genius attained its highest development.   For
many centuries afterwards geometry remained a favourite study,
but no substantive work fit to be compared with the *Sphere and
Cylinder* or the *Conics* was ever produced.   One great invention,
trigonometry, remains to be completed, but trigonometry with
the Greeks remained always the instrument of astronomy and
was not used[1] in any other branch of mathematics, pure or
applied.   The geometers who succeed to Apollonius are professors who signalised themselves by this or that pretty little
discovery or by some commentary on the classical treatises.

"The works of Archimedes and Apollonius," says M. Chasles[2],
"marked the most brilliant epoch of ancient geometry.   They
may be regarded, moreover, as the origin and foundation of two
questions which have occupied geometers at all periods.   The
greater part of their works are connected with these and are

[1] Except, perhaps, by Heron.   See below, pp. 283, 284.   [2] *Aperçu*, pp. 22, 23.

divided by them into two classes, so that they seem to share between them the domain of geometry.

"The first of these two great questions is the quadrature of curvilinear figures, which gave birth to the calculus of the infinite, conceived and brought to perfection successively by Kepler, Cavalieri, Fermat, Leibnitz and Newton.

"The second is the theory of conic sections, for which were invented first the geometrical analysis of the ancients, afterwards the methods of perspective and of transversals. This was the prelude to the theory of geometrical curves of all degrees, and to that considerable portion of geometry which considers, in the general properties of extension, only the forms and situations of figures, and uses only the intersection of lines or surfaces and the ratios of rectilineal distances.

"These two great divisions of geometry, which have each its peculiar character, may be designated by the names of *Geometry of Measurements* and *Geometry of Forms and Situations*, or Geometry of Archimedes and Geometry of Apollonius[1]."

135. It remains only to add a few words on a great number of other geometrical works which are attributed to Apollonius. Pappus[2] ascribes to him the following works (1) *On Contacts* (περὶ ἐπαφῶν), (2) *On Plane Loci* (ἐπίπεδοι τόποι), (3) *On Inclinations* (περὶ νεύσεων), (4) *On Section of an Area* (περὶ χωρίου ἀποτομῆς), (5) *On the Determinate Section* (περὶ διωρισμένης τομῆς), and gives a few *lemmas*, from which attempts have been made to reconstruct the lost originals[3]. Vieta restored the 1st in his *Apollonius Gallus*: Fermat in 1637 and Simson in 1746 attempted the 2nd: Ghetaldi the 3rd: Halley (in his edition of *De Sectione Rationis*) restored the 4th: Snellius, Ghetaldi, and Simson, again, worked at the 5th. All

---

[1] "These two divisions," he adds, "are those of all the mathematical sciences which have for their aim, to use Descartes' expression, the investigation of *order* and *measure*." Aristotle had already uttered the same thought in these terms: "With what are mathematicians concerned save with *order*

and *proportion?*" The quotations are from Descartes, *Règles pour la direction de l'Esprit*, 14ᵉ and 4ᵉ *Règle*, Aristotle, *Metaph.* xi. 3.

[2] vii. Nos. 298—311, pp. 990—1004 (Hultsch).

[3] See Montucla, i. pp. 251, 252 and notes F and G, pp. 285—288.

of these were certainly exercises in geometrical analysis, and an account of their supposed contents is given by Montucla, but does not seem worth citing. The passage, however, in which the same writer mentions Vieta's restoration of the work *On Contacts* is interesting as illustrating the manners and customs of mathematicians at a time when they were more dependent on Greek learning than they are now. Vieta (1540—1603) having a contention with one Adrianus Romanus, a clever geometer of the Low Countries, took occasion "to propose to him the principal problem, and the only difficult one in the book (*On Contacts*). It is this: Three circles being given, to find a fourth, which shall touch the three. Romanus solved this badly by adopting the expedient which presents itself at first sight and determining the centre of the desired circle by the intersection of two hyperbolas. The objection is that the problem is plane, and can consequently be solved by the aid of ordinary geometry. Vieta solved it in this way and very elegantly: his solution is the same as that in the *Arithmetica Universalis*[1] of Newton. Another is given in the 1st Book of the *Principia*[2], where this question is necessary for some determinations of physical astronomy. Here Newton, with remarkable skill, reduces the two solid *loci* of Romanus to the intersection of two straight lines. This problem, one of those to which algebraical analysis does not lend itself with facility, occupied Descartes also: and of two solutions which he found, he admits that one gave an expression so complicated that he would not undertake to construct it in a month. The other, though less crabbed, was sufficiently so to prevent Descartes from touching it. We may mention finally, on the subject of this problem, an anecdote which in a way illustrates it. The princess Elizabeth of Bohemia[3], who, as is well-known, honoured our philosopher (Descartes) with her correspondence, deigned to occupy herself with it and sent him a solution, but as this is derived from algebraic calculation, it is open to the same objection as that of Descartes."

But a work of Apollonius called *De Sectione Rationis* was translated from the Arabic and published by Halley in 1706. This deals with the cases of one problem, which is as follows.

---

[1] Prob. XLVII. [2] Lemma XVI. [3] Daughter of our James I.

Two straight lines of infinite length, MN, PQ, are in one plane, parallel to one another or intersecting in a point. On each any one point is taken (A and B respectively), and a point O is given outside them. It is required to draw from O a straight line meeting MN, PQ in the points C and D, so that the sequents AC, BD shall be in a given ratio. In the first book, 14 cases are treated, where the lines are parallel and where they intersect, but the points A, B upon them are the point of their intersection. The second book contains 63 cases. All are solved analytically with the aid of conics[1].

A work of Apollonius on *Unclassed Incommensurables* (ἄλο-γοι ἄτακτοι) is mentioned in an Arabic commentary on Euclid's 10th Book, which is translated from a Greek commentary, written perhaps by Vettius Valens, a Byzantine astronomer of the 2nd century. It is, however, impossible to discern from the commentary what these "unclassed incommensurables" were[2]. Hypsicles (see below) knew another work of Apollonius, and Proclus (p. 105) mentions a treatise on the screw.

136. Lastly, Eutocius, in his often-cited commentary to the *Sphere and Cylinder*, attributes to Apollonius, Heron and Philon of Byzantium, methods of duplication which are practically identical, and which Apollonius, as the oldest of these three mathematicians, must be taken to have invented[3]. This solution is in effect as follows. If AB, AC be the two straight lines between which it is required to find two mean proportionals, place them at right angles to one another, the right angle being at the common extremity A, and complete the parallelogram $ABDC$.

Join BC and bisect it in E. From the centre E describe a circle FG, cutting AB, AC produced in F and G, so that the

---

[1] On all these minor works of Apollonius, the *lemmas* upon them in Pappus' vɪɪth Book and the important anticipations of modern geometry which these contain, see Chasles, *Aperçu*, pp. 28—47.

[2] Cantor, pp. 299—301, quoting an essay of Woepcke's in *Mémoires présentés à l'Acad. des Sciences*, xɪv. 658—720,

Paris, 1856, and Chasles in *Comptes Rendus*, xxxvɪɪ. 553—568 (Oct. 17, 1853).

[3] Philon may be the oldest, for Vitruvius assigns him to Alexander's reign, but other authorities give him a much later date, about b.c. 150. Heron's solution is given first by Eutocius (Torelli, pp. 136—138). Philon constructs the figure a little more conveniently. But see p. 318.

points F, D, G are on the same straight line. "This may be effected if a ruler ($\kappa\alpha\nu\acute{o}\nu\iota o\nu$) cutting AF, AG be turned about D until EF, EG are equal." From E draw EH perpendicular to AC and bisecting it in H.

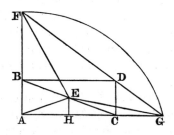

Then (by Euc. II. 6) $AG \cdot GC + HC^2 = HG^2$. Add EH^2 to each equal. Then $AG \cdot GC + EC^2 = EG^2$. In the same manner it may be shewn that $AF \cdot FB + EB^2 = EF^2 = EG^2$. And $EC^2 = EB^2$. Therefore $AG \cdot GC = AF \cdot FB$ and

$$AG : AF :: FB : GC.$$

But, by similar triangles, $AG : AF :: CG : CD :: BD : BF$. Therefore $BD : BF :: BF : CG :: CG : CD$.

CHAPTER VIII.

137. THE materials for a history of Greek geometry after Apollonius are both scanty in quantity and most unsatisfactory in quality. We know the names of many geometers who lived during the next three centuries, but very few indeed of their works have come down to us, and we are compelled to rely for the most part on such scraps of information as the later scholiasts, Pappus, Proclus, Eutocius and the like, have incidentally preserved. But this information, again, generally affords little clue to the date of the geometer in question. Thus, though we have abundant evidence that mathematics remained a chief constituent of the Greek liberal *curriculum*, we cannot tell with any accuracy what subjects were most in vogue or what mathematicians were most generally regarded at any particular time. It is certain, however, that during the whole period between Apollonius and Ptolemy only two mathematicians of real genius, Hipparchus and Heron, appeared, that both of these lived about the same time (120 B.C.), and that neither was interested in mathematics *per se*, for Hipparchus was above all things an astronomer, Heron above all things a surveyor and engineer. The result might have been different if some new methods had been introduced. The force of nature could go no further in the same direction than the ingenious applications of exhaustion by Archimedes and the portentous sentences in which Apollonius enunciates a proposition in conics. A briefer symbolism, an analytical geometry, an infinitesimal calculus were wanted, but against these there stood the tremendous authority of the Platonic and Euclidean tradition, and no discoveries were made in physics or astronomy which rendered them imperatively necessary. It remained only for mathematicians, as Cantor says, to descend from the height which they had reached and "in the descent to pause here and there and look around at details which had been passed by in the hasty ascent[1]." The elements of planimetry were exhausted, and the theory of conic sections. In stereometry something still remained to be done, and new curves, suggested by the spiral of Archimedes, could still be investigated. Finally, the arithmetical determination of geometrical ratios, in the style of the *Measurement of the Circle,* offered a considerable field of research, and to these subjects mathematicians now devoted themselves.

138. One of the first of the successors of Apollonius was perhaps **Nicomedes**, who invented the curve called *conchoid* or "mussel-like." At any rate the conchoid was known to Geminus about B.C. 70[2], and Eutocius[3] says that Nicomedes made sport of Eratosthenes' mesolabium, and boasted the superiority of his own invention. It is not likely that Eratosthenes had been long dead at this time.

The treatise on the conchoid which Nicomedes wrote is known to us only from Eutocius' commentary on duplication, from Pappus[4], and two or three casual remarks of Proclus. It

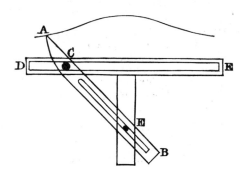

is a curve such that the straight line joining any point on the

[1] Cantor, p. 301, cf. p. 233.   [2] Proclus, p. 177.   [3] Torelli, p. 146.
[4] Book IV. (Hultsch) pp. 244—246.

curve with a given point is cut by a given straight line so that
the segment between the curve and the given straight line is of
a given length. Nicomedes invented a little machine for
describing it, of the form here depicted. It will be seen that
the arm AB can move only horizontally along DE, to which it
is confined by a button C sliding in a groove. The length AC
therefore is constant. The point E was called the pole (πόλος).

The method of duplication, with the aid of the conchoid,
may be thus described.

Let $a\lambda$ and $a\beta$ be the given straight lines between which it
is required to find two mean proportionals. Place these at right
angles to one another (as in the solution of Apollonius), and
complete the rectangle $a\beta\gamma\lambda$.

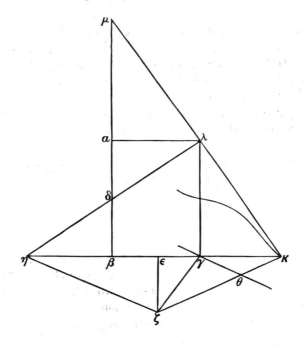

Produce $\gamma\beta$ to η, making $\beta\eta = \beta\gamma$, and join $\eta\lambda$, bisecting $a\beta$
in δ. Bisect $\beta\gamma$ in ϵ, and from ϵ draw $\epsilon\zeta$ at right angles to $\beta\gamma$,
so that $\gamma\zeta = \beta\delta$. Join $\eta\zeta$ and through γ draw $\gamma\theta$ parallel to $\eta\zeta$.
From ζ, as *pole*, with $\gamma\theta$ as fixed straight line and $\beta\delta$ as the

length of the constant segment describe a conchoid, cutting $\eta\beta$ produced in κ. Join $\kappa\lambda$ and produce it to meet βa produced in μ. Then $a\mu$ and $\gamma\kappa$ are the two mean proportionals required. By similar triangles $\dfrac{\mu a}{a\lambda} = \dfrac{\lambda\gamma}{\gamma\kappa}$ $\therefore \mu a = \dfrac{a\lambda \cdot \lambda\gamma}{\gamma\kappa}$. Now

$a\lambda \cdot \lambda\gamma = \eta\gamma \cdot \beta\delta$, (since $y \times 2x = 2y \times x$), $\therefore \mu a = \dfrac{\eta\gamma \cdot \theta\kappa}{\gamma\kappa}$. But

$\dfrac{\eta\gamma}{\gamma\kappa} = \dfrac{\zeta\theta}{\theta\kappa}$ $\therefore \mu a = \zeta\theta$, and $\zeta\kappa = \zeta\theta + \dfrac{a\beta}{2} = \mu\delta$.

By the use of Euc. II. 6 (precisely as in the solution of Apollonius) it may be shewn that $\zeta\kappa^2 = \beta\kappa \cdot \kappa\gamma + \gamma\zeta^2$ and $\mu\delta^2 = \beta\mu \cdot \mu a + a\delta^2$. $\therefore \beta\kappa \cdot \kappa\gamma + \gamma\zeta^2 = \beta\mu \cdot \mu a + a\delta^2$. But $\gamma\zeta^2 = a\delta^2$ $\therefore \beta\kappa \cdot \kappa\gamma = \beta\mu \cdot \mu a$. $\therefore \beta\mu : \beta\kappa = \kappa\gamma : \mu a$. But $\beta\mu : \beta\kappa = \gamma\lambda : \gamma\kappa = a\mu : a\lambda$. $\therefore \gamma\lambda : \gamma\kappa = \gamma\kappa : a\mu = a\mu : a\lambda$.

The conchoid was also used to solve the trisection of an angle in a way which closely resembles the 8th of the lemmas attributed to Archimedes (*supra*, p. 233). Proclus says that Nicomedes himself solved this problem, but Pappus claims the solution which he gives as his own[1].

Let $a\beta\gamma$ be the angle which it is required to trisect. From a draw $a\gamma$ perpendicular to $\beta\gamma$. Complete the parallelogram.

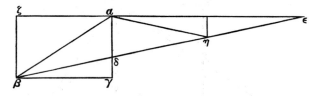

Now from β as *pole*, with $a\gamma$ as fixed straight line and $2a\beta$ as constant distance describe a conchoid which shall meet ζa produced in ϵ. The line $\beta\epsilon$ cuts $a\gamma$ in δ. Bisect $\delta\epsilon$ in η and join $a\eta$. It is then easy to see that $a\eta = \eta\epsilon = a\beta$ and the triangles $a\beta\eta$, $a\eta\epsilon$ are isosceles. Therefore the exterior angle $a\eta\beta = 2a\epsilon\eta = 2\eta\beta\gamma$, and the angle $a\eta\beta = a\beta\eta = 2\eta\beta\gamma$.

139. Probably at the same time as Nicomedes, say 180 B.C., lived **Diocles**, the inventor of the *cissoid* or "ivy-like" curve. His date can be approximately determined only by the two

---

[1] Proclus, p. 272; Pappus IV. 38, p. 274 (Hultsch).

facts that Geminus knew the cissoid by this name, and that *Diocles* lived after Archimedes, for he wrote a commentary on the unfinished problem (II. 5, *supra*, p. 225n) of the *Sphere and Cylinder*. The work in which this occurs was called περὶ πυρίων or πυρείων[1], whatever that may mean, and contained also a solution of the duplication problem which Eutocius cites with the rest[2]. This solution, which involves also the definition of the cissoid, may be described as follows. Let αβ and γδ be

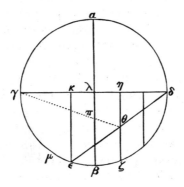

diameters of a circle at right angles to one another. On γδ, at equal distances on either side of the centre λ, take the points κ and η, and draw the ordinates κε, ηζ. Join εδ, cutting ηζ in θ. The point θ (as also all other points similarly determined) lies on the cissoid. Also γη : ηζ = ηζ : ηδ = ηδ : ηθ.

As ηζ is perpendicular to the diameter γδ, it is plain that γη : ηζ = ηζ : ηδ. For a similar reason[3], γκ : κε = κε : κδ. And by similar triangles κε : κδ = ηθ : ηδ. Therefore γκ : κε = ηθ : ηδ

[1] Eutocius in Torelli, p. 171. Πυρεῖον (which may be the right reading), Lat. *igniaria*, was an instrument for making fire, by turning a pointed perpendicular stick (τρύπανον) in a hole made in a flat board (ἐσχάρα). If this was worked by strings, like a drill (see the chapter on fire-drills in Tylor's *Early Hist. of Mankind*), then Diocles' book may have been a treatise on some geometrical theorems suggested by the machine.

[2] Torelli, p. 138. Cantor, pp. 306, 307. The solutions of Pappus and Sporus (an otherwise unknown geometer), which Eutocius gives next, are practically identical with this, though the constructions are not obtained with a cissoid.

[3] A shorter proof would run: 'And γη : ηζ = δκ : κε = δη : ηθ. Therefore γη : ηζ = ηζ : ηδ = ηδ : ηθ.'

and $\kappa\epsilon : \kappa\gamma = \eta\delta : \eta\theta$. But $\kappa\epsilon = \eta\zeta$, and $\kappa\gamma = \eta\delta$. Therefore $\eta\zeta : \eta\delta = \eta\delta : \eta\theta$. Thus $\eta\zeta$, $\eta\delta$ are two mean proportionals to $\gamma\eta$, $\eta\theta$.

Now, in any circle, with diameters $\alpha\beta$, $\gamma\delta$, at right angles to one another, draw the corresponding cissoid. On the diameter $\alpha\beta$, take a point π such that $\gamma\lambda : \lambda\pi = a : b$, where a and b are the two straight lines to which two mean proportionals are required. Join $\gamma\pi$ and produce it to meet the cissoid in θ. Then $\gamma\eta : \eta\theta = a : b$. It is now necessary only to alter the lines $\eta\zeta$, $\eta\delta$ (which are known to be mean proportionals to $\gamma\eta$, $\eta\theta$) in the ratio of $\gamma\eta : a$, and the solution is obtained.

140. In the same century, again, perhaps about the year 150 B.C. **Perseus**, a geometer who treated of the sections of the $\sigma\pi\epsilon\hat{\iota}\rho\alpha$[1], seems to have lived. His date can be guessed only from the facts that he is not included in the Eudemian summary, that no notice is taken of him by the classical geometers, that Heron describes the $\sigma\pi\epsilon\hat{\iota}\rho\alpha$ (110 B.C.), and that the work of Perseus was well known to Geminus[2]. The $\sigma\pi\epsilon\hat{\iota}\rho\alpha$ is somewhat imperfectly described by Heron[3] as the solid " produced by the revolution of a circle which has its centre on the circumference of another circle and which is perpendicular to the plane of that other circle. This is also called a $\kappa\rho\hat{\iota}\kappa o\varsigma$ (ring)." This solid varies in form according to the ratios between the radii of the two circles. It may resemble an anchor-ring or a modern tea-cake, with a dimple at the centre. Proclus describes three kinds of sections, which were obtained from it and which were the same as those described above (p. 185), à *propos* of the $\hat{\iota}\pi\pi o\pi\acute{\epsilon}\delta\eta$ of Eudoxus. Elsewhere (p. 356, 12) he seems to suggest that Perseus had treated the *spiral sections* as Apollonius had treated the conics. From this, perhaps, it may be inferred that whereas one or two sections of the $\sigma\pi\epsilon\hat{\iota}\rho\alpha$ were known before and were obtained from different forms of the solid, Perseus investigated all the sections and shewed that they

---

[1] Geminus in Proclus, pp. 111, 112.

[2] The dates of Perseus, Nicomedes, Diocles, Serenus and Hypsicles are all discussed by Bretschneider (*Anhang*, p. 175—end who is especially severe, in this connexion, on the errors of Montucla. Bretschneider is obviously right on all the dates except that of Serenus.

[3] Deff. 98, p. 27 of Hultsch's ed.

could be obtained from one $\sigma\pi\epsilon\hat{\iota}\rho a^{1}$. But the work of Perseus is wholly lost, and no extracts whatever from it are preserved by any later writer[2].

141. There is not so much reason for assigning **Zenodorus** to the 2nd century B.C. as there is for the other writers above mentioned. He is later than Archimedes, whom he names, and is older than Quintilian (A.D. 35—95) who names him. He is supposed to be an early successor of the former merely because his style recalls the classical period. He was the author of a geometrical treatise on *Figures of Equal Periphery*, fourteen propositions of which are preserved both by Pappus and Theon[3]. Both citations are almost verbally identical, but Theon does, and Pappus does not, name Zenodorus as the author. Theon's ascription is confirmed by Proclus, who says that Zenodorus called a quadrilateral with re-entrant angle a $\kappa o\iota\lambda o\gamma\acute{\omega}\nu\iota o\nu$, which word occurs in Theon's extract. Of these fourteen propositions five, Nos. 1, 2, 6, 7 and 14, are worth quoting. Prop. 1 is "of regular polygons with equal periphery, that is the greatest which has most angles." Prop. 2 is "The circle has a greater area than any polygon of equal periphery." Prop. 6 is "Two similar isosceles triangles on unequal bases are together greater than two dissimilar isosceles triangles which are upon the same bases and have together the same periphery as the two similar

---

[1] Bretschneider (pp. 179, 180) makes a great difficulty about this, owing to the fact that he mistranslated $\dot{\eta}\ \tauo\hat{\nu}$ $\ddot{\iota}\pi\pi o\nu\ \pi\acute{\epsilon}\delta\eta$ as "horse's hoof" (p. 177) instead of "horse-fetter." He conceived this apparently to be a curve of the form (o) instead of ∞ and could not understand how it was obtained from a $\sigma\pi\epsilon\hat{\iota}\rho a$ at all. His mistake is the more remarkable because Proclus afterwards twice (pp. 127, 128) refers to the curve as $\ddot{\iota}\pi\pi o$-$\pi\acute{\epsilon}\delta\eta$, which no decent scholar ought to render "horse-hoof."

[2] Chasles (pp. 8, 9) speaking of the spirals of Perseus and Geminus, says "il serait intéressant de voir leur théorie géométrique de ces spiriques, qui sont des courbes du quatrième degré, dont l'étude semble exiger aujourd'hui des équations de surfaces et un calcul analytique assez profond."

[3] Pappus v. pt. I. p. 301 sqq. (Hultsch). Theon. *Comm. Almag.* ed. Halma, p. 33 sqq. reprinted by Hultsch in *Pappus*, pp. 1190—1211, with a prefatory note on the date of Zenodorus. The fact that both Theon and Pappus cite the same props. seems to Hultsch (*Pappus*, Vol. III. p. xv.) to give colour to his theory that a large part of Theon's commentary was really taken from Pappus.

triangles." Prop. 7 is "Of polygons with equal periphery the regular is the greatest." Prop. 14 is "Of segments of circles, having equal arcs, the semicircle is the greatest." It is obvious that investigations of this kind were closely connected with and suggested by the work of Archimedes and Apollonius.

142. To the same century, again, **Hypsicles** is assigned. To him the 14th and 15th Books added to Euclid's *Elements* are attributed by many MSS., but recent critics are of opinion that these are by different authors[1], and that only the 14th is by Hypsicles. This is certainly not Euclid's, for it has a preface which cannot have been written by Euclid, and the *Elements* are expressly stated by Marinus, in his prolegomena to the *Data*, to consist of 13 books. The preface in question, which is addressed to one Protarchus, is as follows : "Basilides of Tyre, coming to Alexandria and making the acquaintance of my father through their common love of mathematics, stayed with him during the greater part of his visit. They were discussing at one time the writings of Apollonius on the comparison of the dodecahedron and the icosahedron inscribed in the same sphere[2], shewing what ratio these have to one another, and they came to the conclusion that Apollonius was wrong. They therefore emended the proof, as my father used to tell. But I afterwards came across another book of Apollonius[3] containing a sound proof on the subject, and was greatly incited to the investigation of the problem. The publication of Apollonius may be seen anywhere, for it has a large circulation, but I send you my lucubrations," etc. From this it is inferred, not very cogently, that Hypsicles' father died in the lifetime of Apollonius, or that, at any rate, Hypsicles cannot have lived long after the latter. But a more satisfactory determination of Hypsicles' date is obtained from the fact that his astronomical work, 'Αναφορικός, does not use the trigonometry which was certainly introduced by Hipparchus, and would have been absurdly antiquated if written after Hipparchus' time (B.C. 130)[4].

---

[1] See esp. Friedlein in *Bulletino Boncompagni* 1873, pp. 493—529.

[2] This is the only mention of such a treatise by Apollonius.

[3] This in XIV. prop. 2 is referred to as "a second edition."

[4] Bretschneider, p. 182, quoting Vossius and Delambre.

The 14th Book of the Elements, or the book of Hypsicles on 'the Regular Solids', consists of seven propositions, viz. 1. The perpendicular from the centre of a circle to a side of the inscribed regular pentagon is half the sum of the radius and the side of an inscribed decagon. 2. The same circle comprises the pentagon of a dodecahedron and the triangle of an icosahedron inscribed in the same sphere[1]. 3. If from the centre of a circle there be drawn a perpendicular to the side of the inscribed regular pentagon, thirty times the rectangle under the perpendicular and the side is equal to the superficies of the corresponding dodecahedron. 4. The surface of the dodecahedron is to that of the icosahedron as the side of the cube to the side of the icosahedron. 5. The side of the cube is to that of the icosahedron as $(x + y)^2 + x^2 : (x + y)^2 + y^2$, where x is the greater, and y the less, of the segments of a line cut in extreme and mean ratio. 6. The volume of the dodecahedron is to that of the icosahedron as the side of the cube to that of the icosahedron. Prop. 7 is really a lemma to 6 and is that two straight lines cut in extreme and mean ratio are to one another as their greater segments.

The ἀναφορικός, or treatise on 'Risings' (ἀναφοραί), contains only six propositions, of which the first three, dealing with arithmetical progressions, have been already cited. The only interesting proposition is the 4th, which is to the following effect[2]. Divide the zodiac into 360 *local* degrees and the time of its revolution into 360 *chronic* degrees. Then, given the ratio, for any place on the earth, of the longest day to the shortest, we can deduce the number of chronic degrees for each number of local degrees[3]. Here, for the first time in any Greek work, we find a circle divided in the Babylonian manner into 360 degrees. This division, perhaps, was used by Eratosthenes who is said to have calculated the length of a degree, but it is

---

[1] The proof of this is said to be given by Aristaeus in his work on "The Comparison of the Five figures" (πέντε σχημάτων σύγκρισις). This is not mentioned by Pappus, who (VII. *pref.*) alludes only to the στερεοὶ τόποι of Aristaeus.

[2] See Delambre *Astr. Anc.* I. pp. 246 sqq. The text was printed *Paris*, 1657, ed. J. Mentel.

[3] Delambre *loc. cit.* pp. 248, 9 shews that the proof is faulty.

18

not necessary to suppose that Eratosthenes actually performed this feat, though he undoubtedly shewed how it was to be done[1], and it is observable that Hypsicles introduces the division as if it were a novelty. He does not, however, take the next step, to trigonometry.

143. This was undoubtedly taken by **Hipparchus**, one of the greatest geniuses of antiquity, the observer and thinker upon whose work the whole system of Greek astronomy was founded. He was a native of Nicaea in Bithynia and made astronomical observations, certainly at Rhodes, possibly also at Alexandria, between 161 and 127 B.C.[2] But though the Almagest of Ptolemy is clearly derived almost entirely from writings of Hipparchus, none of the works of the earlier astronomer have survived, save a commentary in three books on the *Phenomena* of Aratus, a poor poet who copied Eudoxus. The criticisms of Hipparchus on his predecessors are founded chiefly on his own more accurate observations and have no mathematical interest. In the Second Book, however[3], he claims to have invented a method of solving spherical triangles for the purpose of finding the exact eastern point of the ecliptic. The treatise in which this was contained was called ἡ τῶν συνανατολῶν πραγματεία, but is lost. Theon, in his commentary on the Almagest, also states that Hipparchus calculated a "table of chords" (i.e. practically of sines) in

---

[1] Eratosthenes (Delambre, pp. 86—97) found that the distance between the tropics was $\frac{11}{83}$ of the circumference. This looks as if he had not, at that time, any division of the circle into degrees. Similarly, he found that the distance between Alexandria and Syene (which he believed to be on the same meridian) was $\frac{1}{50}$th of the circumference of the earth, from which it was easy to infer the length of a degree, though perhaps Eratosthenes did not do this.

[2] Delambre (I. p. 167, cf. p. 170) gives a very neat instance of the way in which Hipparchus' date can be ascertained. In the concluding chapter of Book III. of his commentary on Aratus, Hipparchus gives, in time, the distances between stars, obtained by observing their meridian passages. He begins with η *Canis*, in his time on the solstitial colure, longitude 90°. In 1750, this star was *long*. 116° 4′ 10″. The precession here is 93850″. This, at 50″ per annum, would make the date of the book about 130 B.C. Delambre doubts whether Hipparchus was ever at Alexandria, because Ptolemy does not distinguish observations made at Rhodes and Alexandria, which, he supposed, were on the same meridian.

[3] Delambre I. pp. 142, 3.

twelve books. It is evident therefore that Hipparchus was the founder of trigonometry, though we are obliged to look elsewhere for information as to the progress of the Greeks in this department of mathematics.

It is not intended, in these pages, to give a history of Greek astronomy or to describe any astronomical theories, which depend for their verification on observation and not on deduction. But *est modus in rebus* and I do not like to pass over Hipparchus with merely the customary eulogy. The following little summary, taken from Delambre, will shew what manner of man he was. It was he who determined (very nearly but not with absolute accuracy) the precession of the equinoxes, the inequality of the sun, and the place of its apogee, as well as its mean motion: the mean motion of the moon, its nodes and its apogee: the equation of the centre of the moon and the inclination of its orbit. He had discovered a second inequality of the moon (the evection), of which he could not, for want of proper observations, find the period and the law. He had commenced a more regular course of observations for the purpose of supplying his successors with the means of finding the theory of the planets. He had both a spherical and a plane trigonometry. He had traced a planisphere by stereographic projection: he knew how to calculate eclipses of the moon and to use them for the improvement of the tables: he had an approximate knowledge of parallaxes, more correct than Ptolemy's. He invented the method of describing the positions of places by reference to latitude and longitude. What he wanted was only better instruments. Yet in his determination of the equations of the centres of the sun and moon and of the inclination of the moon, he erred only by a few minutes. For 300 years after his time astronomy was stationary. Ptolemy followed him with little originality. Some 800 years later the Arabs added a few more discoveries and more accurate determinations and then the science is stationary again till Copernicus, Tycho and Kepler[1].

[1] Delambre i. pp. 184—186. See also his preliminary discourse pp. xxi —xxv, and De Morgan's article *Ptolemy* in Smith's *Dic. of Gr. and Rom. Biogr.*

144. The same century which gave birth to all these writers produced also the famous **Heron** of Alexandria[1]. He was the pupil of Ctesibius of Alexandria, who, though originally a barber, obtained great fame by his mechanical inventions, especially a water-clock, a hydraulic organ and a catapult, worked by compressed air. Ctesibius lived in the reign of Ptolemy Euergetes II. (or Physcon, 'pot-belly'), that is, between 170 and 117 B.C. His pupil Heron, therefore, may be taken to have flourished about 120—100 B.C.

A very considerable number of writings, now extant, and others not extant, but mentioned by ancient writers, are attributed to *a* Heron, but it happens that the extant writings are in an extraordinary state of corruption and confusion and also that a great many Herons are known to history. It is only within recent years that any attempt has been made to bring order into this chaos. First Theodore Henri Martin, in a monograph[2] which is a model of its kind, investigated all the facts concerning the life of the great Heron of Alexandria and ascertained what works were rightly attributed to him and which of them are extant and where. His biographical results have been stated, in effect, in the above few lines. But his essay deserves a closer analysis. He finds (pp. 10—18) *eighteen* undoubted Herons named in later Greek literature, mathematicians, doctors and monks. Of these, three only belong to the first class, viz. Heron of Alexandria our author, Heron the teacher of Proclus (who was possibly the same as one Heronas, who wrote a commentary on the *Arithmetic* of Nicomachus) and Heron of Constantinople, who lived in the 10th century[3]. Then, after commenting on the date of the first Heron (pp. 22—28), he passes (pp. 28—51) to the works which are rightly

---

[1] This writer is usually called by the Latinized name *Hero*. Perhaps I ought to use this (like *Plato*), but there is a special advantage in retaining the form *Heron*, because the more familiar Hero was a woman.

[2] *Recherches sur la vie et les ouvrages d'Héron d'Alexandrie, disciple de Ctesibius*, etc. in Vol. IV. of *Mémoires*

présentés etc. *à l'académie d'inscriptions etc.*, Paris, 1854.

[3] Later writers, as Vincent, cited below and Cantor, p. 315, deny that there was such a person as Heron of Constantinople and doubt whether Heron, the teacher of Proclus, was a mathematician at all.

assigned to him. These are (1) Μηχανικά or Μηχανικαὶ εἰσαγωγαί, from which extracts are given by Pappus (III. 5, p. 63 and VIII. 31—end). The book obviously treated of centres of gravity and of the theory of the five simple machines, the lever (μοχλός), wedge (σφῆν), screw (κοχλίας), pulley (πολύσπαστον), and wheel and axle or windlass (ἄξων ἐν περιτροχίῳ). The work perhaps exists in MS. at the Escurial or at Venice. (2) the Βαροῦλκος, in three books, which dealt with the problem of Archimedes to move a given weight with a given power, perhaps exhibited the practical uses of these machines. The first chapter of this is appended, perhaps by accident, to the treatise περὶ διόπτρας and some extracts from it are given by Pappus at the end of his Book VIII. It exists at Leyden in a Latin MS. translation made by Golius from the Arabic. It is perhaps in Greek at Rome. (3) The καταπελτικά, βελοποιητικά or βελοποιϊκά is printed in the *Mathematici Veteres*[1]. The solution of the duplication-problem here given is quoted in Pappus III. (4) χειροβαλίστρας κατασκευὴ καὶ συμμετρία, also in *Math. Vett.*, but obviously an appendix to (3). (5) καμαρικά also in *Math. Vett.* but obviously an appendix to (4). So is another fragment περὶ καμβεστρίων. Both exist in MS. at Vienna. (6) αὐτόματα and ζύγια, on certain toys. The former is in *Math. Vett.* The latter is lost. (7) Περὶ ὑδρίων ὡροσκοπείων. This is mentioned in the πνευματικά and also by Pappus and Proclus. It is lost now but existed in the 10th century. (8) πνευματικά, in the *Math. Vett.*[2]

---

[1] This is a collection of writers on engines of war, edited by Thevenot and De la Hire, *Paris*, 1693. It contains works of *Heron* and of *Athenaeus*, *Apollodorus* (? both *temp. Hadriani*), *Philon* (B.C. 330, acc. to Vitruvius VII. pref.), *Biton* (probably soon after Alexander the Great), *Sextus Julius Africanus* (κεστοί, about A.D. 220), and a treatise on siege-works, which Martin ascribes to Heron of Constantinople. Of the named authors (other than Heron) all deal almost entirely with catapults save Africanus, who has some other matter. The date of Athenaeus seems to me to be wrongly given (on the authority of Heron of Constantinople). He himself speaks of Ctesibius as a contemporary and dedicates his book to one Marcellus, who may be the conqueror of Syracuse. But see p. 318.

[2] An English translation of the Πνευματικά, with woodcuts, was published by J. G. Greenwood, *London*, 1851. The book contains an account of 78 ingenious machines, some mere toys as whistling birds, drinking figures

(9) on *Hydraulics* and the armillary *Astrolabe,* according to an Arabic compilation, now in the Bodleian (*Cod. Arab.* CMLIV.). The following also are probably Heron's, (10) κατοπ-τρικά, cited by Damianus who was not much later than Ptolemy. This is probably the same work as the κατοπτρικά printed at Venice, 1518, and then ascribed to Ptolemy. (11) Περὶ διόπτρας, on a kind of theodolite. This is ascribed to Heron by the MSS. and was certainly written at Alexandria. It has been edited by M. A. H. Vincent[1]. (12) Scholia on

etc., but some more useful as a fire-engine (27), a self-trimming lamp (33), a new kind of cupping-glass (56), a water-clock (63), two small organs (76 & 77). In most of these, the action depends on a vacuum into which water will flow. But no. 50 is a toy in which a metal sphere, filled with steam, is made to revolve by the action of the steam as it issues from two bent spouts fixed in the sphere. (Compare also no. 70). Heron does not claim all the discoveries as his own, and it is curious that Vitruvius (ix. 8 & x. 7) and Pliny (vii. 38), describing similar inventions, attribute them to Ctesibius and say nothing of Heron. The preface shews clearly that Heron did not understand the pressure of the air as causing the filling of the vacuum, but ascribed this result to nature's abhorrence.

[1] Text and translation in *Notices et Extraits des MSS. de la Biblioth. Impér.* Vol. xix. Pt. ii. *Paris*, 1858, p. 157 sqq. The book contains 33 props. of which the last is the first of the βαροῦλκος. The others are of the following kind (1) to find the difference of level between two points, (13) to cut a straight tunnel through a hill from one given point to another, (14) and (15) to sink a vertical shaft to meet a horizontal tunnel, (24) to measure a field without entering it. The dioptra was a straight plank, eight

or nine feet long, mounted on a stand but capable of turning through a semi-circle. It was adjusted by screws, turning cogwheels. There was an eyepiece at each end and a water-level at the side. With the *dioptra* two poles, bearing discs, were used, exactly as by modern surveyors. Two append-ed props. (34) and (35) describe a *hodometer,* an arrangement of cog-wheels attached to a carriage, so that eight revolutions of the wheel turn the first cogwheel once and the motion is then slackened down through a series of cogwheels of which the last moves a pointer on a measured disc. The proposition from the βαροῦλκος also describes a machine consisting of a series of cogwheels, started by a screw. The case supposed is that a power of five talents is to move a weight of 1000. In Pappus viii. 10 (Hultsch p. 1061) the power is four talents, the weight 160, and the wheels are of a less diameter. Vincent, who is later than Martin, thinks that there was no Heron of Constantinople at all, but that some writer produced a geodesy, founded on the *Dioptra,* which he called "*a Heron*", as we might say "*an Euclid*". He also remarks that Heron (p. 163, *n.*) is not a Greek name but in Egyptian "porte une signification qui revient à celle d'in-génieur".

Euclid, mentioned by Proclus. It exists probably in Arabic at Leyden. (13) Μετρικά mentioned by Eutocius, at the end of his commentary on the *Measurement of the Circle*, as an authority on the extraction of square roots. Parts of this work were (a) τὰ πρὸ τῆς ἀριθμητικῆς στοιχειώσεως (lost), (b) τὰ πρὸ τῆς γεωμετρικῆς στοιχειώσεως, which is also lost, but portions of which have been preserved in the ὅροι, (c) εἰσαγωγαὶ τῶν γεωμετρουμένων, parts of which are preserved in the γεωμετρούμενα, γεωδαισία, or γεωμετρία, περὶ μέτρων or στερεομετρικά, and γεηπονικὸν βιβλίον, (d) εἰσαγωγαὶ τῶν στερεομετρουμένων of which fragments are contained in a work of the same title and also in the last two books mentioned under (c). All these fragments are extant in MS. at Paris and most of them contain tabular statements, made at different dates but all later than our era, of weights and measures. These abridgements and compilations seem to have passed through more than one hand and were made at different dates. The γεηπονικὸν seems to be as late as the 10th century and to have been made at Constantinople.

All the works here mentioned which are of mathematical importance were collected and edited in 1864 by Dr F. Hultsch, the well-known authority on ancient metrology and mathematics. Hultsch's volume contains the ὅροι, or Definitions of geometrical names, with a table of measures appended, the γεωμετρία, which begins with similar definitions and measures, the γεωδαισία, the εἰσαγωγαὶ τῶν στερεομετρουμένων, *Stereometricorum collectio altera*, the μετρήσεις or περὶ μέτρων, the γεηπονικόν, which again has similar definitions and measures, and an extract from the *Dioptra* on the measurement of triangles[1]. But no two MSS. contain exactly the same collection, and the contents of these works shew fully the grounds of Martin's opinion upon them. The Heronic formula for the area of triangles is given in the *Dioptra* and the *Geodesy*: the *Geodesy* is practically the same as a large part of the *Geometry*: the two books on *Stereometry* contain much repetition of one

---

[1] He adds also *Didymi Alexandrini Mensurae marmorum et lignorum* and *Variae collectiones ex Euclide, Herone* etc.

another, and the *Measurements* reproduces all the preceding in a very confused manner.   On the other hand, in the *Geometry* the area of a pentagon is said to be the square of the side × $\frac{12}{7}$, and "*elsewhere*[1]" to be the same square × $\frac{5}{3}$ and there are other similar discrepancies which point, at the very least, to two editions of the original, if not to gross interpolations and unauthentic additions.   The probability is, as Martin suggests, that all these fragments formed part of one comprehensive work on all the knowledge necessary for land-surveying, from which subsequent compilers took, correctly and incorrectly, such matter as they required for their immediate purpose.   These extracts in passing from hand to hand, were annotated by many generations of surveyors and thus contradictory statements and extracts from such a late writer as Patricius and references to Roman measures[2] became incorporated in the text.

145.   The character of the contents of the Heronic collection may be indicated in a very few lines.   The ὅροι contains 128 definitions of all manner of geometrical terms, followed by a short table of measures.   The *Geometry* begins with a few definitions, followed by an account of the empirical origin of the science, then more definitions, then measures, and passes finally to the solution of problems to find the areas or some linear measurements of triangles, circles, parallelograms and polygons, of which the necessary linear measures or areas are given[3].   The *Geodesy*, a short extract, begins in the same way

---

[1] "Elsewhere" is ἐν ἄλλῳ βιβλίῳ τοῦ Ἥρωνος, not named, *Geom.* c. 102, p. 134 (Hultsch).   A similar alternative is given on the same page for the hexagon.   So on p. 115 the value $\pi = \frac{22}{7}$ is attributed to Euclid, on p. 136 to Archimedes, and this value is generally used, but in the *Measurements* $\pi = 3$ is alone used.   So, again, although Heron is cited by Eutocius as an authority for square-roots, in the extant works the roughest approximations are continually used.

[2] E.g. οὐγγία = *uncia :* so φοῦρνος = *furnus* is mentioned.

[3] E.g. chap. xv. § 87 is Περὶ κύκλων.

(1) "Let there be a circle with circumference 22, diameter 7 σχοινία.   To find its area (ἐμβαδόν).   Do as follows. $7 \times 22 = 154$ and $\frac{154}{4} = 38\frac{1}{2}$.   That is the area."   (2) An alternative method, ($\frac{7}{2} \times \frac{22}{2}$) is then added.   Then (3) "If you wish to find the area from the circumference only, do as follows. $(22)^2 \times 7 = 3388$ and $\frac{3388}{88} = 38\frac{1}{2}$."   Then (4) to find the area from the diameter only. (5) The same according to Euclid. (6) To find the circumference from the diameter etc.   All these examples are applied to circles of various given circumferences, diameters, or areas. Heron then treats similarly of semi-

but deals only with the areas of given triangles. The *Stereometry* I. has no definitions but plunges at once into problems to find the volume of given spheres, cubes, obelisks, pyramids and similar figures and next the contents of cups, theatres, dining-rooms, baths, etc. The *Stereometry* II. is chiefly concerned with the same matter as the last part of the preceding book, but in c. 31 (p. 180) suddenly the method of finding heights by measuring shadows is inserted. The *Measurements* and *Geëponicus* are a miscellaneous collection of problems similar to or identical with those in the preceding books.

The reader will see at once that Heron is chiefly engaged in arithmetical calculations which depend on geometrical formulae, which for the most part he does not, and has no occasion to, prove. Sometimes, however, he actually works out a geometrical theorem. Thus, in the βελοποιϊκά[1], he happens to suggest a method of increasing threefold the power of a catapult. This requires that a certain cylinder should be trebled and, as cylinders are to one another as the cubes of their diameters, we are face to face with a problem of *triplication* of the cube. Upon this, Heron inserts a solution of the *duplication*-problem, which is identical with that attributed above to Apollonius. In the last chapter of the *Geodesy* (p. 151), he gives a general formula for finding the area of a triangle. The sides being *a*, *b*, *c*, he says the area is

$$\sqrt{\frac{a+b+c}{2} \cdot \frac{a+b-c}{2} \cdot \frac{a-b+c}{2} \cdot \frac{b+c-a}{2}}.$$

But he works out the proof in the *Dioptra*[2]. It is as follows.

circles, and segments greater or less than a semicircle. On p. 133 occurs the problem, "Given in one number the diameter and the circumference and the area of a circle, to find each." This of course leads to a quadratic equation, of which the solution was given above p. 106.

[1] *Vett. Math.* p. 142. The same proof is given by Pappus (III. p. 63) and Eutocius (in Torelli, p. 136). The latter says it occurs in the μηχανικαί

εἰσαγωγαί as well as βελοποιϊκά.

[2] Reprinted by Hultsch (pp. 235—237) who thinks it is interpolated in the *Dioptra*. The formula, together with Heron's example of its application to a triangle whose sides are 13, 14, 15 (and therefore its area 84), was stolen bodily by Brahmegupta. See Cole-brooke pp. 295 sqq. and comments by Vincent *op. cit.* pp. 200—293, Chasles *Aperçu*, Note XII. pp. 429 sqq. Cantor pp. 550 sqq.

'Let $\alpha\beta\gamma$ be the given triangle. Inscribe in it the circle $\delta\epsilon\zeta$, having its centre η. Join $\eta\alpha$, $\eta\beta$, $\eta\gamma$, $\eta\delta$, $\eta\epsilon$, $\eta\zeta$. (Comp. Eucl

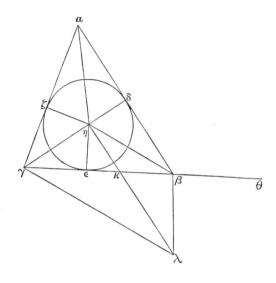

IV. 4). The rectangle $\beta\gamma \cdot \eta\epsilon$ is double of the triangle $\beta\eta\gamma$, and $\alpha\beta \cdot \eta\delta$ of $\alpha\eta\beta$, and $\alpha\gamma \cdot \eta\zeta$ of $\alpha\gamma\eta$. Therefore the rectangle under $\eta\epsilon$ and the perimeter of $\alpha\beta\gamma$ is double the area of $\alpha\beta\gamma$. Produce $\gamma\beta$ to θ. Make $\beta\theta = \alpha\delta$. Then $\theta\gamma$ is half the perimeter. Therefore the rectangle $\theta\gamma \cdot \epsilon\eta$ is equal to the area of the triangle $\alpha\beta\gamma$.

Draw $\eta\lambda$ at right angles to $\eta\gamma$, and $\beta\lambda$ to $\beta\gamma$ and join $\gamma\lambda$. Then, the angles $\gamma\eta\lambda$, $\gamma\beta\lambda$ being two right angles, the quadrilateral $\gamma\eta\beta\lambda$ is in a circle. Therefore the angles $\gamma\eta\beta$, $\gamma\lambda\beta$ are equal to two right angles and also equal to the angles $\gamma\eta\beta$, $\alpha\eta\delta$, which also = two right angles (since the angles at η were bisected by $\alpha\eta$, $\beta\eta$, $\gamma\eta$). Therefore the angle $\alpha\eta\delta$ = angle $\gamma\lambda\beta$, and the triangles $\alpha\eta\delta$, $\gamma\beta\lambda$ are similar. Therefore $\beta\gamma : \beta\lambda :: \alpha\delta : \delta\eta :: \theta\beta : \eta\epsilon$, and *permutando* ($\dot{\epsilon}\nu\alpha\lambda\lambda\dot{\alpha}\xi$) $\beta\gamma : \beta\theta :: \beta\lambda : \eta\epsilon :: \beta\kappa : \kappa\epsilon$, and *componendo* ($\sigma\upsilon\nu\theta\acute{\epsilon}\nu\tau\iota$) $\gamma\theta : \theta\beta :: \beta\epsilon : \epsilon\kappa$, and $\gamma\theta^2 : \gamma\theta \cdot \theta\beta :: \beta\epsilon \cdot \epsilon\gamma : \epsilon\gamma \cdot \epsilon\kappa$ (or $\eta\epsilon^2$). Therefore $\gamma\theta^2 \times \eta\epsilon^2 = \gamma\theta \cdot \theta\beta \times \beta\epsilon \cdot \epsilon\gamma$. But $\gamma\theta \cdot \eta\epsilon$, which is equal to the area of the triangle, is the square-root ($\pi\lambda\epsilon\upsilon\rho\acute{\alpha}$) of

$\gamma\theta^2 \times \eta\epsilon^2$.[1] Therefore the area of the triangle is the square-root of $\gamma\theta . \theta\beta . \times \beta\epsilon . \epsilon\gamma$. Each of these factors is given, for $\gamma\theta =$ half the periphery : $\theta\beta$ is half the periphery *minus* $\beta\gamma$, $\gamma\epsilon$ the same half *minus* $\alpha\beta$, $\epsilon\beta$ the same half *minus* $\alpha\gamma$. Therefore the area of the triangle is given." A triangle with sides 13, 14, 15 is selected as an illustration. Its area $\sqrt{\frac{4.2}{2} \times 6 \times 7 \times 8} = 84$.

But, though Heron's ability is sufficiently indicated by these proofs, as a general rule he confines himself merely to giving directions and formulae. From these also it is easy to perceive how readily he availed himself of the highest mathematics of his time[2]. Thus in the *Dioptra,* two chapters treat of the mode of drawing a plan of an irregular field and of restoring, from a plan, the boundaries of a field in which only a few landmarks remain. The method, in the former case is to draw a rectangle, three corners of which lie on three sides of the field. In the remaining spaces perpendicular co-ordinates are drawn to the sides of the rectangle and are measured off. The method is closely similar to the use of latitude and longitude introduced by Hipparchus. So, again, in three different places[3] Heron gives, for finding the area of a regular polygon from the square of its side, formulae which imply a knowledge of trigonometry. Suppose F_n to be the area of a regular polygon of which a_n is a side, and let c_n be the coefficient by which a_n^2 is to be multiplied in order to produce the equation $F_n = c_n a_n^2$, then it is easy to see that $c_n = \dfrac{n}{4} \cot \dfrac{180^0}{n}$. If we reckon the consecutive values of c to six decimal places, and give the

---

[1] This sentence is introduced earlier in the original. It will be seen that, though the expressions are geometrical, they are intended to indicate the algebraical rule that xy is $\sqrt{x^2y^2}$. No classical Greek geometer would have dared to multiply a square by a square. In his view this would have produced a figure of four dimensions, which would have been absurd. Pappus (p. 680 of Hultsch's *ed.*) expressly protests against the practice, which, he says, had come into use before his time.

[2] The following remarks are taken from Chaps. 18 and 19, pp. 313—343, of Cantor, who has made the ancient surveyors and Heron in particular a favourite study. Much more will be found in his pages than can be here given. See also his *Römische Agrimensoren,* Leipzig, 1875.

[3] *Geom.* 102, *Mens.* 51—53, *Geëpon.* 75—77. Hultsch pp. 134, 206, 218, 229. This repetition shews the authenticity of the formulae.

Heronic formula first in its original form and then in decimals, we find according to Heron,

$$c_3 = \tfrac{13}{30} = 0\;433,333 \quad \text{for the correct } 0\text{·}433,012.$$
$$c_4 = 1 = 1\text{·}000,000 \;\ldots\ldots\ldots\ldots\ldots\ldots\; 1\text{·}000,000.$$
$$c_5 = \tfrac{12}{7} = 1\text{·}714,285 \;(\text{or } \tfrac{5}{3} = 1\text{·}666,666)\; 1\text{·}720,477.$$
$$c_6 = \tfrac{13}{5} = 2\text{·}600,000 \;\ldots\ldots\ldots\ldots\ldots\; 2\text{·}598,176.$$
$$c_9 = \tfrac{51}{8} = 6\text{·}375,000 \;(\text{or } \tfrac{38}{6} = 6\text{·}333,333)\; 6\text{·}181,824.$$
$$c_{12} = \tfrac{45}{4} = 11\text{·}250,000 \;\ldots\ldots\ldots\ldots\; 11\text{·}196,152.$$

This table shews that his approximations are generally near enough. We need not be surprised that Heron could perform such calculations. We know that Hipparchus made a table of *chords*, that is to say, that the coefficients k_n were known, with the aid of which $a_n = k_n r$, where r is the radius. Then $c_n = \dfrac{n}{4}\sqrt{\dfrac{4}{k_n{}^2} - 1}$, and Heron was competent to extract such square roots. But Heron does not use the sexagesimal fractions, and it is clear, from this as from all other evidence, that sexagesimal fractions were always, as they were afterwards called, *astronomical* fractions; indeed, save by Heron, trigonometry was generally conceived to be a chapter of astronomy and was not used for the calculation of terrestrial triangles[1].

Some passages of Heron contain noticeable errors. Thus in *Geëponicus* (146—164, pp. 225—228) he gives a rule that the side of a polygon inscribed in a circle is equal to three diameters divided by the number of sides, which is true only of the hexagon, and in *Stereometrica* I. (35, p. 163) where he proposes to find the volume of a truncated pyramid on a triangular base, he gives dimensions for the upper and lower triangles which could not be found in similar triangles at all[2].

146. Enough perhaps has been said to shew that Heron was by no means a geometer of the Euclidean School. He is a practical man who will use any means to attain his end and is altogether untrammelled by the classical restrictions. He is also a mechanician who, unlike Archimedes, is clearly proud of

---

[1] Cantor pp. 335, 336 abridged. There is, in truth, no evidence to shew how Heron came by his formulae.

[2] Cantor pp. 337, 338 shews that the first error is probably not Heron's. The second is a mere slip.

his own ingenuity. He adds nothing, or almost nothing, to the
geometry of his time but he is learned in the ordinary book-
work. On the other hand, as was mentioned above (p. 106) he
is the first Greek writer who uses a geometrical nomenclature
and symbolism, without the geometrical limitations, for algebraical purposes, who adds lines to areas and multiplies squares by
squares and finds numerical roots for quadratic equations.
Hence, for a similar reason to that which led Prof. de Morgan
to suspect that Diophantus was not a Greek, it is now commonly
believed that Heron was an Egyptian. His name, if it is Greek
at all, is found only at a late era and belongs to persons of
Egyptian or Oriental birth. Further, the whole style of his
work recalls the book of Ahmes which has been described
earlier in these pages. His directions are introduced by the
same form of words, ποίει οὕτως, "Do as follows". Like Ahmes,
he gives few general rules, but a large collection of similar
examples. As Ahmes called the top-line of a figure *Merit*, so
Heron calls it κορυφή, *vertex*[1]. The isosceles parallel-trapezium
was a favorite figure of Ahmes: so it is of Heron[2]. Heron's
method of drawing a plan seems to have had its forerunner in
the method of Ahmes[3]. Ahmes gives tables of measures, so
does Heron. Lastly Heron treats equations in precisely the
style of Ahmes. "It will be remembered that the *hau*-problem
of Ahmes, no. 28, was literally '$\frac{2}{3}$ added, $\frac{1}{3}$ deducted, remainder
10', which was explained as meaning $(x + \frac{2}{3}x) - \frac{1}{3}(x + \frac{2}{3}x) = 10$.
Compare with this the problem of Heron. 'Given a segment
of a circle, with base 40 feet, height 10 feet : to find its circumference. Do as follows. Add base and height together. The
total is 50 feet. Take away a quarter. It is $12\frac{1}{2}$. Remainder
$37\frac{1}{2}$. Add a quarter. It is $9\frac{1}{4}\frac{1}{8}$. The total is $46\frac{1}{2}\frac{1}{4}\frac{1}{8}$. This
is the measure of the circumference. We added $\frac{1}{4}$ and sub-
tracted $\frac{1}{4}$, because the height is $\frac{1}{4}$ of the base[4].'" The style
here and the form of the fractions recall exactly the old

---

[1] *Geometria* 3 (p. 44). Other similari-
ties of nomenclature in Cantor, p. 331.

[2] Nine chapters of the *Geometry* are
devoted to it (pp. 103—108).

[3] The examples of Ahmes are mutilated. See above p. 127.

[4] Heron, pp. 199, 200, Cantor p.
332.

Egyptian. Such evidence as this goes a long way to confirm the suspicion not only that Heron was an Egyptian, but also that algebra was an Egyptian art and that the symbolism of Diophantus was of Egyptian origin. But it is obvious also that, if Heron was not a Greek, he relied almost entirely on Greek learning and did not resort to the stores of priestly tradition of which the contemporary Edfu inscriptions shew the miserable character. He is a man who writes in Greek upon Greek subjects, but who thinks in Egyptian[1].

[1] Let it be remembered that the *seqt*-calculation of Ahmes leads to trigonometry: his *hau*-calculation to algebra. Almost the first sign of both appears in Heron, whom there are other reasons for thinking to have been an Egyptian. An algebraic symbolism first appears in Diophantus, but the symbols are probably not Greek and probably are Egyptian. Both Heron and Diophantus were Alexandrians. This is all the evidence that trigonometry and algebra were of Egyptian origin, but does it not raise a shrewd suspicion? Proclus (p. 429) speaks of οἱ περὶ Ἥρωνα, as if Heron founded a school.

CHAPTER IX.

FROM GEMINUS TO PTOLEMY (B.C. 70—A.D. 150).

147. IF the materials for a history of Greek geometry in the second century B.C. are scanty, they become still more so for the next 250 years. Only a few works, and those not of a very valuable character, survive from this period.

About 70 B.C. lived **Geminus** of Rhodes[1] who seems to have been the freedman of a wealthy Roman and who wrote, beside the astronomical work εἰσαγωγὴ εἰς τὰ φαινόμενα, still extant[2], a book on the *Arrangement of Mathematics*, περὶ τῆς τῶν μαθη-μάτων τάξεως, which, without being expressly historical, con-

---

[1] Proclus always writes Γεμῖνος. Suidas has Γεμῖνος, ὄνομα κύριον. In the 6th chapter of his *Phaenomena* Geminus says "The Greeks suppose the feast of Isis to fall on the shortest day. So it did once, 120 years ago, but every four years the incidence is shifted a day and is now a month behind." If the feast of Isis here mentioned could be exactly identified, there would be no difficulty in finding the date of Geminus. But there are two dates in the Egyptian calendar (the 1st and 17th of Athyr) on both of which some sort of feast to Isis seems to have been held and calculations founded on both these give 77 B.C. and 137 B.C. as the dates

of Geminus. Cantor (pp. 344—6) gives excellent reasons for preferring the former, the chief of which is that Geminus edited an extract from ἐξή-γησις μετεωρολογικῶν of Posidonius, who can hardly be other than Cicero's teacher and Pompey's friend.

[2] It is printed in Halma's edition of Ptolemy's *Canon*, Paris, 1819. A very full abstract in Delambre *Astr. Anc.* I. c. XI. pp. 190—213. It is not like Euclid's *Phaenomena*, a geometrical treatise, illustrative of astronomical theory, but is an account of astronomical observations and of the theories by which they are explained.

tained abundant notices of the early history of Greek mathematics and from which Proclus and Eutocius[1] derived much of their most correct and valuable information on that subject. A book of this kind, written not long after the classical age by a competent geometer, would, if preserved, have cleared up a hundred difficulties which do not now admit of solution.

148. Probably near to the time of Geminus lived **Theodosius** (? of Tripolis), who is mentioned by Strabo and Vitruvius and must therefore be a pre-Christian writer, though Suidas attributes to him a commentary on one Theudas of Trajan's time[2]. He is the author of *Sphaerica*, a very complete treatise on the geometry of the sphere, in three books[3]. It was remarked above, however, on the subject of Euclid's *Phaenomena*, that both that and the treatise of Theodosius are evidently founded on some earlier work on Spherics, perhaps by Eudoxus. The work of Theodosius contains no trigonometry (a *spherical triangle* is not mentioned) and there is nothing particularly interesting either in his style or in his discoveries, if indeed he made any. The character of his propositions will be sufficiently indicated by the following enunciations. I. 13, "If in a sphere a great circle cut another circle at right angles, it bisects it and passes through its poles." (I. 14, 15 are the converse of this.) II. 22, "If in a sphere a great circle touch another (second) circle and cut a third which is parallel to the second and lies between it and the centre, and if the pole of the great circle lies between the two parallel circles, then any great circles which touch the third will be inclined to the (first) great circle, and that will be

---

[1] Eutocius in *Apoll. Conica*, p. 9, calls the book μαθημάτων θεωρία. The title τάξις is quoted by Pappus VIII. 3 (p. 1026 Hultsch). Proclus quotes the book sixteen times, especially on curves.

[2] Vitruvius (IX. 9) mentions a Theodosius who invented an universal sundial. Strabo (XII. 4, 9) mentions a mathematician Theodosius, but calls him a Bithynian, whereas Tripolis was on the Phoenician coast. Suidas (s.v.) expressly says that the author of the *Sphaerica* was a native of Tripolis but gives also another Theodosius of the same place, a poet. Probably Vitruvius refers to our Theodosius. Vitruvius and Strabo both lived under Augustus and earlier.

[3] This has been often printed. First in 1558 by Pena at Paris: in 1675 by Isaac Barrow, London: in 1852 at Berlin by Nizze with Latin trans. and an appendix of Arabic variant proofs. The figures are not given with the text.

at the greatest inclination (ὀρθότατος) which touches the third
at the point of bisection of its greater segment, and that will be
at least inclination (ταπεινότατος) which touches it at the
bisection of its lesser segment, etc.[1].

Strabo, also, (XII. 3) mentions **Dionysodorus**, a native of
Amisus in Pontus, who seems to be the mathematician who, like
Diocles, attempted to finish the problem (*Sph. et Cyl.* II. 5), 'to
cut a sphere so that its segments shall be in a given ratio', which
Archimedes had left incomplete. But Eutocius (Torelli p. 163,
169) complains of both that they did not fill up the gap in Archimedes' solution but produced entirely different proofs of their own.

149. Serenus of Antissa, in Lesbos, lived after Christ.
Bretschneider, indeed, who pointed out (pp. 183—184) that
Antissa was destroyed by the Romans B.C. 167[2], was inclined to
place Serenus about 200 B.C., but the name Serenus is Roman
and the town Antissa was restored in Strabo's time[3], so it is
probable that Serenus lived under the Roman *régime*[4]. He is
not mentioned, however, by any writer earlier than Marinus,
the pupil of Proclus (A.D. 500), and author of the preface to
Euclid's *Data*. His work, however, does not seem to be very
late and he may be placed here in default of better authority.
He is the author of two treatises, one on the *Section of the
Cylinder* in 35 propositions, the other on the *Section of the
Cone* in 63, both of which are printed as an appendix to Halley's
edition of Apollonius. The treatise on the Cone, which is
addressed to one Cyrus, deals entirely with the triangular
section. *E.g.* Props. 5 and 6 are "If a right cone be cut by
planes through the vertex and the axis be not less than the
radius of the base, then the triangle through the axis is the
greatest of the triangles so produced". Prop. 21, "To cut a

[1] Theodosius was also the author
of an extant astronomical treatise περὶ
ἡμερῶν καὶ νυκτῶν and another περὶ
οἰκήσεων, in the style of Euclid's
Phaenomena. The enunciations of
these were published by Dasypodius
(1572 *Strasburg*) along with the work
of Autolycus (Delambre I. pp. 234—
241). It is curious that both Geminus

and Theodosius seem to have been
ignorant even of the observations of
Hipparchus. There are a few lines on
both these mathematicians in Chasles
Aperçu, p. 25.

[2] Livy, XLV. 31.

[3] Strabo, XIII. 2.

[4] Cantor, p. 347, Blass in Fleckeisen's
Neue Jahrb. 1872, p. 34.

scalene cone through the vertex so that the section shall be an
isosceles triangle". Prop. 22, "Such isosceles triangle is the
greatest of the triangular sections of the scalene cone: the least
is that which is produced by a plane perpendicular to the base".
From this point onwards the book deals almost entirely with
maxima and *minima*. The treatise on the Cylinder, which is
addressed to the same friend Cyrus, deals with all the sections,
but chiefly the elliptical. Prop. 19 shews that the same ellipse
can be produced by sections of a cone and a cylinder. Props.
21 and 22 are "Given a cone (cylinder) and an ellipse in it, to
find the cylinder (cone) of which the same ellipse is a section".
Props. 22 and 23 are "Given a cone (cylinder), to find a
cylinder (cone), such that the section of both by the same plane
produces the same ellipse". Prop. 31 is "The straight lines
which are drawn from the same point to touch a cylinder have
their points of contact on the sides of a parallelogram". Prop.
33 is important as being the foundation of the modern theory
of harmonics. It is as follows:

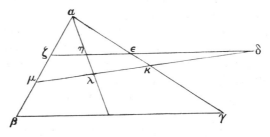

If from the point δ, outside the triangle αβγ, the straight
line δεηζ be drawn cutting the triangle in ε, ζ, and the point η
be taken so that δε : δζ = εη : ηζ, and λη be joined and produced
to meet the base, any other transversal δκλμ shall be so divided
by αη produced that δκ : δμ = κλ : λμ. With the aid of this it
is proved (Prop. 34) that all straight lines drawn from the same
point to touch a cone, have their points of contact on the sides
of a triangle. Then comes the last proposition (35 which is
similar in kind to Prop. 32). It is as follows. *ABC* is a triangle,
DE, FG are parallel to its base. From a point *H*, not in the
plane of the triangle draw *HD, HE, HF, HG* and produce them

to meet a plane $KLXMN$, which is at all points equidistant
from ABC. The plane $HDKE$ will cut
this second plane in KN, and the plane
$HFLG$ will cut it in LM, and KN, LM
are parallel to DE, FG. Also KL, DF
are parallel and MN, GE. Therefore KL,
NM produced will meet. Let them meet
in X. Then the triangles XKN, ABC are
similar. "Now if the point H be supposed
to be an illuminating point and the triangle ABC (whether *per se* or in a cone)
be opposite its rays, then the rays will

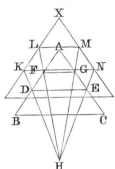

make the shadow KNX triangular and similar to ABC. Although
this consideration belongs to optics and on that account is
alien to our subject, yet it is clear that without the proofs here
given concerning the cone and the cylinder, I mean about the
ellipse and its tangents, it is impossible to solve a problem of
this kind: wherefore not carelessly but on purpose the subject
has here been introduced[1]."

A lemma of Serenus, on angles which stand on equal arcs
of a circle, is preserved in the *Astronomy* of Theon Smyrnæus[2],
but there is no evidence to shew how it came there. Theon
lived about 130 A.D. and may have himself used Serenus.

150. One date in the life of **Menelaus** is absolutely certain.
Ptolemy[3] records two astronomical observations made by him
in the first year of Trajan, A.D. 98. He was the author of a lost
work on the calculation of chords, but his *Sphaerica* in 3 books,
though not extant in Greek, is extant in Arabic and Hebrew
and has been often translated into Latin[4]. This is a treatise
on spherical triangles, describing their properties in much the
same way as Euclid, in Book I. of the Elements, treats plane

---

[1] On the question raised by this
proposition, whether the ancients were
acquainted with the method of perspective, see Taylor *Anc. & Mod.
Conics*, p. lv. See also Chasles *Aperçu*
pp. 47, 48, 74.

[2] Martin's *Ed.* p. 340.

[3] Almagest *ed.* Halma VII. 3. Vol. II.

pp. 25, 27.

[4] Halley made a translation which
was published by G. Costard, Oxford,
1758. Costard promises a preface,
but this is wanting from both copies
in the Brit. Mus. There is a full
summary, as usual, in Delambre I.
pp. 244—246.

triangles. But there is no attempt at solution of the triangles, and though in Book III. the first proposition is the foundation of the ancient method of solution, Menelaus makes no such use of it. His propositions are of the following kind. In every spherical triangle any two sides are greater than the third (I. 5): the sum of the three angles is greater than two right angles (I. 11): equal sides subtend equal angles and the greatest side the greatest angle (I. 8, 9): the arcs which bisect the angles meet in a point (III. 9): the arc which bisects any angle cuts the opposite side into two segments, such that the chords of twice the segments[1] are to one another as the chords of twice the other sides (III. 6). The chief proposition (III. 1) describes two properties of plane and spherical triangles, cut by a transversal. The property of plane triangles (stated in a *lemma*) is that if the three sides be cut by a straight line, the product of three segments which have no common extremity is equal to the product of the other three[2]. For spherical triangles, the rule is similar, but for "three segments" read "the chords of three segments doubled". "The proposition in plane geometry" says Chasles "of which we shall speak below in the article on Ptolemy (because it is in the Almagest that it has generally been noticed) has acquired a new and great importance in recent geometry, where the illustrious Carnot has introduced it, making it the base of his theory of transversals[3]." The theorem on spherical triangles was greatly admired by the Arabs, who called it "the rule of intersection": early mediaeval writers called it by its Arabic name *catha*, and it was known later by another name, *regula sex quantitatum*[4]. Pappus (IV. p. 270) says that Menelaus, and also two otherwise unknown geometers, **Demetrius** of Alexandria and **Philon** of Tyana, investigated curves on curved surfaces. One of these was called παράδοξος γραμμή, but Pappus does not describe it.

151. Practically all that we know of the trigonometry of

---

[1] Halley always translates "chord of twice the arc" by *sinus*, which of course properly is half the same chord.

[2] Menelaus does not say "product": he says that a_1 has to b_1 the ratio compounded of $b_2 : a_2$ and $b_3 : a_3$.

[3] *Aperçu*, pp. 25—27.

[4] This name is in Stifel's *Arithm. Integra*. Nuremberg, 1544. The names of the proposition are given in Costard's edition p. 82. A complete account of its history is in Chasles *Aperçu*, Note VI. pp. 291—293. Chasles thinks it was originally one of Euclid's porisms.

the Greeks, is derived from two chapters of the famous Μεγάλη
Σύνταξις[1] of **Claudius Ptolemæus**. This work contains many astronomical observations by Ptolemy himself, of which the earliest
was made in A.D. 125, the latest in A.D. 151. Beyond these facts
and also that Ptolemy certainly observed in Alexandria in
A.D. 139, we know nothing of his history. The Arabs indeed
have many details upon his personal appearance, etc., but these
statements betray the romancer by their minuteness[2]. The
common name μεγάλη Σύνταξις was altered by still more
fervent admirers into μεγίστη and this word was adopted by
the Arabs who got translations of the book earlier probably
than of any other Greek mathematical work. The Arabic article
was then added and the name corrupted into *Almidschisti*,
whence is derived its common mediaeval title *Almagest*[3].

Book I. chap. IX. of the Almagest, shews how to calculate a
table of chords[4]. The circle is divided into 360 degrees (τμήματα) each of which is halved: its diameter into 120 degrees
each of which is divided into 60 minutes, 3600 seconds (πρῶτα
ἑξηκοστά, δεύτερα ἑξηκοστά). Ptolemy does not pretend that
these divisions were new. The division of the circle was, among
Greeks, as old as Hypsicles and was of Babylonian origin : the
sexagesimal scale of the division of the diameter shews it also
to have been Babylonian, and, as such, it was no doubt known
at least to Hipparchus, though it is not now to be found before
Ptolemy[5]. But Ptolemy's method of calculating chords seems to

---

[1] Ptolemy's title is μαθηματική Σύνταξις.

[2] Boncompagni's *Gherardo Cremo-
nese*, pp. 16, 17 (cited by Cantor, p.
351) Weidler *Hist. Astr.* p. 177. The
Arabs say that Ptolemy was a fair
man, with a red mole on the right
side of his chin, etc.

[3] The whole of Delambre's second
volume is devoted to Ptolemy. There
is a splendid article on him by Prof.
de Morgan in Smith's *Dic. of Gr. and
Rom. Biogr.*, and a neat summary of
the Almagest in Wolf's *Gesch. der
Astronomie*, Munich, 1877. The great

edition is the Abbé Halma's, Paris,
1813—16.

[4] The chapter is introduced thus
early for the purpose of measuring the
arc of the solstitial colure which lies
between the poles of the equator and
the ecliptic. Our names "minutes"
and "seconds" are taken from the
Latin " partes minutae (primae)",
" partes minutae secundae".

[5] Ptolemy says merely "I shall use
the method of arithmetic with the
sexagesimal scale, because of the in-
convenience of fractions" (Halma, p.
26).

be his own. The measures of the sides of regular polygons, as chords of certain arcs, were known in terms of the diameter. Some of these Ptolemy first sets out. He next proves the proposition, now appended to Euclid VI. (D), that "the rectangle contained by the diagonals of a quadrilateral inscribed in a circle is equal to both the rectangles contained by its opposite sides[1]", and then proceeds to shew how from the chords of two arcs that of their sum and difference and how from the chord of any arc that of its half may be found. His proofs which are very pretty are as follows[2]:

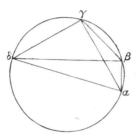

(1) Given the chords $\alpha\beta$, $\alpha\gamma$, it is required to find $\beta\gamma$. Draw the diameter $\alpha\delta$ and join $\gamma\delta$, $\beta\delta$. Then $\gamma\delta = \sqrt{120^2 - \alpha\gamma^2}$, $\beta\delta = \sqrt{120^2 - \alpha\beta^2}$, and $\alpha\gamma \cdot \beta\delta = \beta\gamma \cdot \alpha\delta + \alpha\beta \cdot \gamma\delta$. Therefore $\alpha\gamma \sqrt{120^2 - \alpha\beta^2} = 120\beta\gamma + \alpha\beta \sqrt{120^2 - \alpha\gamma^2}$, whence $\beta\gamma$ can be found.

(2) Given the chord $\beta\gamma$, it is required to find the chord $\gamma\delta$ of half the same arc. Draw the diameter $\alpha\gamma$ and join $\alpha\beta$, $\alpha\delta$, $\beta\delta$. In $\alpha\gamma$ take $\alpha\epsilon = \alpha\beta$. Join $\delta\epsilon$ and draw $\delta\zeta$ perpendicular to $\alpha\gamma$. The triangles $\alpha\beta\delta$,

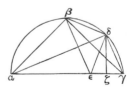

$\alpha\delta\epsilon$ are equal, and their sides $\beta\delta$, $\delta\epsilon$ are equal. But $\beta\delta = \delta\gamma$, therefore the triangles $\delta\epsilon\zeta$, $\delta\zeta\gamma$ are equal. Therefore

$$\zeta\gamma = \frac{\epsilon\gamma}{2} = \frac{\alpha\gamma - \alpha\epsilon}{2} = \frac{120 - o\beta}{2} = 60 - \tfrac{1}{2}\sqrt{120^2 - \beta\gamma^2}.$$

[1] Chasles Ap. p. 27 note I. says that Carnot in his $Géométrie de Position$ shewed that all rectilineal trigonometry could be deduced from this theorem.

[2] Halma, pp. 30—35. The proofs in the text are abridged after Cantor (pp. 352—354) with some corrections.

But the triangles $\gamma\zeta\delta$, $\gamma\alpha\delta$ are similar, therefore $\zeta\gamma : \gamma\delta :: \gamma\delta : \alpha\gamma$, whence $\gamma\delta^2 = \alpha\gamma \cdot \zeta\gamma = 120\,(60 - \tfrac{1}{2}\sqrt{120^2 - \beta\gamma^2})$. From this $\gamma\delta$ can be found.

(3) Given the chords $\alpha\beta$, $\beta\gamma$, it is required to find the chord $\alpha\gamma$. Draw the diameters $\alpha\delta$, $\beta\epsilon$ and join $\beta\delta$, $\delta\gamma$, $\gamma\epsilon$, $\delta\epsilon$. The triangles $\alpha\zeta\beta$, $\delta\zeta\epsilon$ are equal and $\alpha\beta = \epsilon\delta$. Then the diagonals $\beta\delta \cdot \gamma\epsilon = \beta\gamma \cdot \delta\epsilon + \gamma\delta \cdot \beta\epsilon$, or

$$\sqrt{120^2 - \alpha\beta^2} \times \sqrt{120^2 - \beta\gamma^2}$$
$$= \beta\gamma \cdot \alpha\beta + 120\,\sqrt{120^2 - \alpha\gamma^2},$$

whence $\alpha\gamma$ can be found.

Returning then to the known chords (or sides of polygons), Ptolemy finds from the chords of 72° and 60° the chord of 12°. From this the chord of 6°, 3°, $1\tfrac{1}{2}^\circ$, $\tfrac{3}{4}^\circ$. His intention, however, is to give a table of the chords of arcs, increasing successively by $\tfrac{1}{2}^\circ$. He requires therefore to find the chord of 1°. This he effects in the following manner.

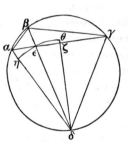

(4) $\alpha\beta$, $\beta\gamma$ are given arcs, of which $\beta\gamma$ is the greater. Draw their chords and also the chord $\alpha\gamma$. Bisect the angle at β by $\beta\delta$ cutting $\alpha\gamma$ in ϵ. Join $\alpha\delta$, $\delta\gamma$ and draw $\delta\zeta$ perpendicular to $\alpha\gamma$. From centre δ, with radius $\delta\epsilon$, describe a circle, cutting $\delta\alpha$, $\delta\zeta$ in η, θ respectively. Then (angle $\alpha\beta\gamma$ being bisected) $\alpha\beta : \beta\gamma :: \alpha\epsilon : \epsilon\gamma$, therefore $\alpha\epsilon < \epsilon\gamma$, i.e.

$\alpha\epsilon < \dfrac{\alpha\gamma}{2}$ and ϵ falls between α and ζ. Therefore $\delta\alpha > \delta\epsilon > \delta\zeta$, whence it is plain that η lies on $\delta\alpha$, θ on $\delta\zeta$ produced. Then sector $\delta\epsilon\eta <$ triangle $\delta\epsilon\alpha$, and sector $\delta\epsilon\theta >$ triangle $\delta\epsilon\zeta$. Therefore

$\dfrac{\text{tri. } \delta\epsilon\zeta}{\text{sec. } \delta\epsilon\eta} < \dfrac{\text{sec. } \delta\epsilon\theta}{\text{sec. } \delta\epsilon\eta}$ and $\dfrac{\text{tri. } \delta\epsilon\zeta}{\text{sec. } \delta\epsilon\eta} > \dfrac{\text{tri. } \delta\epsilon\zeta}{\text{tri. } \delta\epsilon\alpha}$. Therefore

$\dfrac{\text{tri. } \delta\epsilon\zeta}{\text{tri. } \delta\epsilon\alpha} < \dfrac{\text{sec. } \delta\epsilon\theta}{\text{sec. } \delta\epsilon\eta}$. But $\dfrac{\text{tri. } \delta\epsilon\zeta}{\text{tri. } \delta\epsilon\alpha} = \dfrac{\epsilon\zeta}{\epsilon\alpha}$ and $\dfrac{\text{sec. } \delta\epsilon\theta}{\text{sec. } \delta\epsilon\eta} = \dfrac{\text{arc. } \epsilon\theta}{\text{arc. } \epsilon\eta}$.

Therefore $\dfrac{\epsilon\zeta}{\epsilon\alpha} < \dfrac{\text{arc. } \epsilon\theta}{\text{arc. } \epsilon\eta}$. Add unity to each side and then double them. It follows that $\dfrac{\alpha\gamma}{\epsilon\varkappa} < \dfrac{2 \text{ arc } \eta\theta}{\text{arc } \epsilon\eta}$. Deduct unity

from each side and it follows that $\dfrac{\epsilon\gamma}{\epsilon\alpha} < \dfrac{\text{arc } \theta\epsilon + \text{arc } \theta\eta}{\text{arc } \epsilon\eta}$. But

$\dfrac{\epsilon\gamma}{\epsilon\alpha} = \dfrac{\beta\gamma}{\alpha\beta}$ and $\dfrac{\text{arc } \theta\epsilon + \text{arc } \theta\eta}{\text{arc } \epsilon\eta} = \dfrac{\text{angle } \beta\delta\gamma}{\text{angle } \beta\delta\alpha} = \dfrac{\text{arc } \beta\gamma}{\text{arc } \beta\alpha}$. That is to

say, the quotient of the greater chord by the less is smaller than
the quotient of the greater arc by the less. Now take the

chords of $1\frac{1}{2}^{0}$, 1^{0} and $\frac{3}{4}^{0}$, and we find that $\dfrac{\text{chord } 1^{0}}{\text{chord } \frac{3}{4}^{0}} < \dfrac{\text{arc } 1^{0}}{\text{arc } \frac{3}{4}^{0}}$ and

$\dfrac{\text{chord } 1\frac{1}{2}^{0}}{\text{chord } 1^{0}} < \dfrac{\text{arc } 1\frac{1}{2}^{0}}{\text{arc } 1^{0}}$. But $\dfrac{\text{arc } 1^{0}}{\text{arc } \frac{3}{4}^{0}} = \frac{4}{3}$ and $\dfrac{\text{arc } 1\frac{1}{2}^{0}}{\text{arc } 1^{0}} = \frac{3}{2}$. There-

fore $\frac{2}{3}$ chord $1\frac{1}{2}^{0} <$ chord $1^{0} < \frac{4}{3}$ chord $\frac{3}{4}^{0}$. From this is obtained
the approximation chord $1^{0} = 1 \cdot 2' \cdot 50''$. The chord $1\frac{1}{2}^{0}$ is
known and hence also the chord $\frac{1}{2}^{0}$, and the table of all chords,
rising by half a degree at a time, can be compiled. Ptolemy
goes only as far as 180^{0}, on the ground only that he never
requires arcs of greater magnitude. For arcs which lie between
any two given in the table, Ptolemy applies merely a proportion.
For instance, the arc 20^{0} has a chord $20 \cdot 50' \cdot 16''$, the arc $20\frac{1}{2}^{0}$,
has a chord $21 \cdot 21' \cdot 12''$. The addition of half a degree to the
arc corresponds to an addition of $30' 56''$ to the chord. This
increase, divided by 30, is $1' 1'' 52'''$ and this is taken to be the
increase in the chord for every increase of a minute in the arc
between 20^{0} and $20^{0} 30' {}^{1}$.

Chapter X., which follows, is on the obliquity of the ecliptic
as determined by observation. The next, XI., XII. contain
spherical geometry and trigonometry "enough for the determination of the connexion between the sun's right ascension,
declination and longitude and for the formation of a table of
declinations to each degree of longitude[2]." Chap. XI. contains
προλαμβανόμενα, "preliminaries to the spherical demonstra-

---

[1] These proportional increases are
stated in a third column by Ptolemy.
Ideler in Zachs' *Correspondenz*, Vol.
XXVI. July, 1812, pp. 3—38, finds that
Ptolemy's numbers are correct to 5
places of decimals.

[2] De Morgan. Ptolemy introduces the
subject by saying "It follows next to
shew the magnitudes of the arcs, comprised between the equator and the
ecliptic, of the great circles drawn
through the poles of the equator".

tions". These begin with the *lemma* of Menelaus, the *regula sex quantitatum*, borrowed without any acknowledgement. After proving this, he gives four proposi-
tions. If AB, BG be two arcs, each less than a semicircle ("a supposition which can be made of all arcs to be hereafter taken") and AG be joined and BD be drawn to the centre D, cutting AG in E,

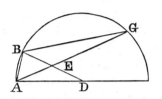

then the chord of $2AB$: chord of $2BG$ :: $AE : EG$. From this it follows that, given the arc AG and the ratios of the chords of $2AB$, $2BG$, the arcs AB, BG can be found. Produce GB to meet DA in F. Then chord $2GA$: chord $2AB$ :: $GF : BF$. From this it follows that, given arc GB only and the ratio between the chords of $2GA$, $2AB$, the arc AB can be found. These propositions being proved, Ptolemy then proves the *regula sex quantitatum* for a spherical triangle, and proceeds (Chap. XII.) to find the magnitudes of the arcs above-mentioned, and (Chap. XIII.) "the magnitudes of arcs of the equator which lie between circles which pass through its poles and through given points of the ecliptic". The method, in both cases, is founded on the rule of Menelaus.

$ABGD$ represents a great circle, passing through the poles of the equator AG. BED is the eclip-
tic: E is the vernal equinox; B the winter, and D the summer, solstice. Z is the pole of the equator. On the ecliptic take an arc HE, and through H describe the great circle ZHT. It is required to find the magnitudes of HT (Chap. XII.) and TE (Chap. XIII.). Ptolemy gives the solutions only for

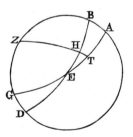

cases in which EH is 30° or 60°, and then adds tables. The lemma of Menelaus is, later on, applied in a great many ways to this same figure, for there are four triangles, EHT, ZHB, ZTA, EBA, which are cut by the following transversals respectively, ZBA, ETA, BHE, ZHT. One example (ch. XII.) will serve for an illustration. ETA is a transversal to the triangle ZHB. Therefore chord ZAZ : chord $2AB$ :: chord $2TZ \times$ chord $2HE$: chord $2TH \times$ chord $2EB$. HE is *ex hypothesi* 30°. Arc ZA is 90°, i.e. chord $2ZA = 120$. 2 arc AB is 47° 42′ 40″, its chord 48 . 31′ . 55″. 2 arc HE is 60°, its chord 60. 2 arc EB is 180°, its chord 120. "If from the ratio 120 : 48 . 31′ . 55″ we subtract ($\dot{\alpha}\phi\epsilon\lambda\hat{\omega}\mu\epsilon\nu$) that of 60 : 120, the remainder will be the ratio chord $2TZ$: chord $2TH$, i. e. 120 : 24 . 15′ 57″. But $2TZ = 180°$, its chord 120. Therefore chord $2TH = 24 . 15′ 57″$, its arc is 23° 9′ 59″ therefore the arc TH is half this, viz. 11′ 40″ very nearly[1]."

This paragraph contains in fact the whole of Greek trigonometry. The further progress of this department of geometry is due mainly to the Indians and after them to the Arabians. With the former, trigonometry seems, after its suggestion in Ptolemy, to have had quite a native development. The Indians never used "the chord of twice of the arc", as the Greeks always did, but half that chord. This they called *jyârdha* or *ardhajyâ*, but the name of the whole chord *jyâ* or *jivâ* was also used for shortness. The Arabs, taking the latter term, transliterated it to *dschîba*, which later was altered for the Arabic word *dschaib*, which is of nearly the same form. *Dschaib* means 'bosom' and was therefore translated 'sinus' by Plato of Tivoli in his Latin version ('De Motu Stellarum') of the astronomy of Albategnius[2].

---

[1] Delambre (in Halma) has some notes on the proof,

$$\frac{\text{chord } 2ZA}{\text{chord } 2AB} = \frac{\text{ch. } 2TZ}{\text{ch. } 2TH} . \frac{\text{ch. } 2HE}{\text{ch. } 2EB}$$

i.e. $\dfrac{\text{ch. } 180^0}{\text{ch. 2 obliq.}} = \dfrac{\text{ch. } 180^0}{\text{ch. 2 decl.}} . \dfrac{\text{ch.2 long.}}{\text{ch. } 180^0}$.

Then $\dfrac{\text{ch. 2 decl.}}{\text{ch. } 180^0} = \dfrac{\text{ch. 2 obliq.}}{\text{ch. } 180^0} . \dfrac{\text{ch.2 long.}}{\text{ch. 180}}$.

or $\dfrac{\text{ch. 2 decl.}}{120} = \dfrac{48 . 31′ \ 55″}{120} . \dfrac{60}{120}$.

"On voit par là que, dans le langage des anciens, retrancher une raison, c'étoit diviser par cette raison." Hankel (p. 285, 286 *n*.) has a very neat note on Ptolemy's procedure. He points out the four triangles and their transversals. All the arcs in the figure can be expressed in terms of HT (*a*),

BA (*a*), TE (*b*), HE (*h*), and their complements. The application of Menelaus' rule produces the following equations. The transversal ZBA gives $\cos h = \cos a . \cos b$: the transv. ETA gives $\sin a = \sin a \sin h$: the transv. BHE gives $\cos a . \sin b . \sin a = \cos a . \sin a$ (or $\tan a = \sin b . \tan a$) : the transv. ZHT gives $\sin b . \cos h = \cos b . \cos a . \sin h$ (or $\tan b = \cos a . \tan h$).

[2] Plato of Tivoli was certainly writing between A.D. 1116—1136. At the latter date he was at Barcelona. Albategnius (878—918) was Mohammed of Battân in Syria. On these persons see Cantor, pp. 560, 632, 778, where also the derivation of *sinus* is given. Also Hankel, pp. 217 sqq., 287 sqq.

In this way, *sine* came to be a technical term of modern trigono-metry. Further evidence of the distinct character of Indian trigonometry is to be seen in their division of the diameter. Ptolemy divided this into 120 parts with sexagesimal fractions and so did the Arabs. The Indians divided it in various ways. Dividing it into 120,000 parts they calculated the sides of regular polygons of 3, 4, 5, 6, 7, 8, 9 sides to be 103923, 84853, 70534, 60000, 52055 (for 52066), 45922, 41031 (for 41042) respectively. Ptolemy (*Almag.* VI. 7) has $\pi = 3 \cdot 8' \cdot 30''$ ($= 3 + \frac{8}{60} + \frac{30}{3600} = 3 \cdot 141$, 666....) The oldest Indian tradition makes $\pi = 3$ or, more exactly, $\sqrt{10}$. Aryabhatta has $\frac{31416}{10000}$. This value was obtained in the following way. If, in a circle with radius unity, S_n be the side of an inscribed regular polygon of n sides, S_{2n} that of a like polygon of $2n$ sides, then $S_{2n} = \sqrt{2 - \sqrt{4 - S_n{}^2}}$. From the side of the hexagon they calcu-lated the sides of polygons of 12, 24, 48, 96, 192, 384 sides. The periphery of the last (the diameter being taken $= 100$) is $\sqrt{98694}$. This square root or rather that of 986,940,000 is exactly Aryabhatta's value[1].

152. The applications of trigonometry in Book II. of the Almagest and the geometry of eccentric circles and epicycles in Book III. belong too distinctly, by language and purpose, to the history of astronomy to be described here. Besides the Alma-gest, Ptolemy wrote also many other works, most of which are extant. The *Geography* (edited by the Abbé Halma, Paris 1828) contains a description of the earth, defining the position of many thousand places by latitude and longitude. Book I., chap. 24 con-tains directions for drawing a map and various modes of projec-tion are here discussed. Ptolemy prefers the method by which

I have somewhere seen a statement that *sinus*, which in Latin means primarily 'a fold,' was applied to the 'folded' chord, i.e. half the chord.

[1] Hankel, pp. 215, 216. Pur-bach (1423—1461) and Regiomontanus (Müller of Königsberg, 1436—1476), both of whom made abstracts of the Almagest, but use sines, divided the radius into 600,000 parts. The latter afterwards substituted the value $r = 1,000,000$ (Montucla I. pp. 539—544). In Brigg and Gellibrand's *Tri-gonometria Britannica* (Goudae, 1633) cap. 2. sines are calculated to 15 places of decimals. Here also Ptolemy's propositions are given exactly.

the eye is supposed to be at the pole and points on the earth's surface are projected on the plane of the equator[1]. He wrote also a *Canon* or chronological list of kings of various countries (also ed. Halma), a treatise on *Sound* (Ἁρμονικά, ed. Wallis, Oxford, 1682) and another on *Optics*, extant only in a Latin translation from the Arabic. The 5th book of the *Optics* deals with refraction, in which, as an astronomer, Ptolemy was especially interested[2]. Cleomedes an earlier astronomer (A.D. 60) had already suggested that the reason why stars are still seen, though below the horizon, was due to the same cause as that which renders a ring, previously unseen, visible when the vessel, which contains it, is filled with water. But Ptolemy works up the subject carefully. He compares rays passing through air and water, air and glass, and water and glass. He finds as a general law that a ray, passing from a rarer to a denser medium, is refracted towards the perpendicular: if passing from a denser to a rarer medium, away from the perpendicular: and he invented a simple contrivance (a graduated circle with moveable spokes, the lower half of which is placed in water) for the purpose of ascertaining the amount of the refraction in water for various angles of incidence[3]. Some works in astrology and metaphysics probably not genuine, are also attributed to Ptolemy[4], but Proclus (pp. 362—368) has preserved some extracts from a work of his in pure geometry, from which it appears that he also discussed the propriety of Euclid's famous 12th Axiom, (sometimes printed as 11th), on parallel lines. He endeavoured to prove it as a theorem in the following way. If the straight

---

[1] Cantor p. 358 says that Aiguillon in 1613 gave to this method the name of "stereographic" projection. Modes of projection are also discussed by Ptolemy in his *Planisphere* and *Analemma* (trans. Commandinus 1558 and 1562). An *analemma* is a delineation on a plane of the circles of the heavenly sphere. See Hultsch's *Pappus*, III. *pref.* p. xi.

[2] Refraction is not mentioned in the Almagest. Delambre accounts for this by supposing that the *Optics* was written later.

[3] See Heller, *Gesch. der Physik.* pp. 136, 137, Delambre II. pp. 411—431. De Morgan doubts the authenticity of the *Optics*, chiefly on the ground that the geometry is bad.

[4] Simplicius (*in. Arist de Cælo*, Book I.) mentions a book on *dimensions* (περὶ διαστάσεων) and Pappus VIII, p. 1030 seems to mention a book on *Mechanics*. Both are lost.

line $\epsilon\zeta\eta\theta$ meet the straight lines
$\alpha\beta$, $\gamma\delta$ and make the two interior
angles equal to two right angles,
then $\alpha\beta$, $\gamma\delta$ are parallel. For if
not, let the interior angles $\beta\zeta\eta$,
$\zeta\eta\delta$ be two right-angles and let
the two straight lines, $\beta\zeta$, $\delta\eta$, meet
in κ. Then, the angles $\alpha\zeta\eta$, $\zeta\eta\gamma$,

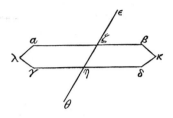

will also be equal to two right-angles and the straight lines $\alpha\zeta$,
$\gamma\eta$ will meet in λ and thus two straight lines will enclose a
space. Conversely, if the straight lines are parallel, the interior
angles are necessarily equal to two right-angles. For $\alpha\zeta$ and $\gamma\eta$
are not less parallel than $\zeta\beta$, $\eta\delta$ and therefore whatever the sum
of the angles $\beta\zeta\eta$, $\zeta\eta\delta$, whether greater or less than two right-
angles, such also must be the sum of the angle $\alpha\zeta\eta$, $\zeta\eta\delta$. But
the sum of the four cannot be more than four right-angles,
because they are two pairs of adjacent angles.

CHAPTER X.

Last Years.

153. The revival of Platonism and Pythagorean mysticism in Alexandria and the East, perhaps also the dispersion of the Jews and their introduction to Greek learning, led about Ptolemy's time to the revival of the theory of number and this in the hands of Nicomachus, Theon, Smyrnaeus and others became a favourite study[1]. No doubt geometry continued to be one of the most important parts of the Alexandrian course, but no important geometer appears for 150 years or so after Ptolemy. The sole occupant of this long gap is **Sextus Julius Africanus**, a Libyan by birth, who lived, however, most of his life in Palestine. He flourished about A.D. 200. Africanus has left a collection of papers similar to those of Heron, and entitled Κεστοί, i.e. 'Patchwork', 'miscellanies.' A portion of this dealing chiefly with catapults is printed in the *Mathematici Veteres*, but Chap. 31 contains some problems of strategy[2], to find the breadth of a river the opposite bank of which is occupied by the enemy, etc. Two solutions of this problem are given, both depending on similar triangles. The first is as follows. The point α being on the opposite bank, take a distance θβ, evidently greater than αθ, θ being on your own bank and let θβ be at right angles to the bank. With the dioptra determine

---

[1] Vide *supra*, p. 88 sqq.
[2] This is separately printed by Vin-cent along with Heron's *Dioptra* above mentioned.

$\beta\gamma$ at right angles to $\theta\beta$. From γ with the dioptra determine the angle $\beta\gamma\alpha$. Bisect $\beta\gamma$ in δ and from δ draw $\delta\epsilon$ parallel

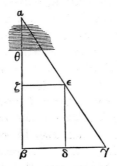

to $\theta\beta$, meeting $\alpha\gamma$ in ϵ and from ϵ draw $\epsilon\zeta$ to $\theta\beta$, parallel to $\beta\gamma$. Then $\zeta\beta$ is half $\alpha\beta$ and $\zeta\theta$ can be determined by measurement. The other method is as follows. The point α being on the opposite

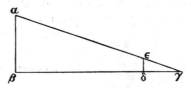

bank, determine $\alpha\beta$ crossing the river at right angles and measure $\beta\gamma$ parallel with the banks. At the point δ on $\beta\gamma$ lay a T-square $\gamma\delta\epsilon$, so that its extremity ϵ lies on the line $\gamma\alpha$, as determined by the dioptra. Then $\gamma\delta$: $\delta\epsilon$ :: $\gamma\beta$: $\beta\alpha$. The first three of these distances are known and thus $\beta\alpha$ is obtained. This procedure, the Roman *varatio*, was one of those which made the reputation of Heron and all the *gromatici* of antiquity. It could be applied of course, as Euclid applies it in the *Phœnomena*, to finding heights and depths as well as horizontal distances[1].

---

[1] Chap. 76 of the Κεστοί describes a curious system of telegraphy. Three posts were set up, each bearing 9 moveable arms. One represented units, the next tens, the third hundreds. The numbers thus exhibited were read in Greek alphabetic signs and in this manner a word was spelt out, e.g. 7 on the second post ($o=70$) and 4 on the third ($v=400$) would spell *ov*, 'No.' Cantor, pp. 372—374.

154. But the end of the third century produced one of the greatest of Greek mathematicians, **Pappus** of Alexandria. His date, indeed, though he certainly lived before Proclus and Eutocius who often mention him, is a matter of some doubt, because the two authorities for it contradict one another. A scholiast of the 10th century has written, in the margin of a MS. of Theon's manual tables (now at Leyden), opposite the name Diocletian, ἐπὶ τούτου ὁ Πάπος ἔγραφεν, which can hardly refer to anybody but our Pappus. Suidas, however, says that Pappus was a contemporary of Theon and as it is evident from the same tables that Theon lived about 372, in the time of Theodosius, the two accounts vary by nearly a hundred years. Now Suidas says that Pappus wrote a commentary on the *four* (instead of 13) books of the Almagest and it is in the highest degree unlikely that both Theon and Pappus, living in Alexandria at the same time, should both have written a commentary on the same work[1]. And Suidas, besides making a mistake about the Almagest itself, does not mention Pappus's great work, the συναγωγή, at all. From this it is inferred that Suidas knew hardly anything about Pappus and the other writer, who assigns him to Diocletian's time (A.D. 284-305) is deemed better worthy of credence[2].

Many writings are attributed to Pappus. Proclus (p. 429) speaks of Pappus' pupils (οἱ περὶ Πάππον), so he evidently was the head of a school. Eutocius (in Torelli, p. 208), and Suidas mention his commentary on the Almagest. The former also (Torelli, p. 90), mentions some notes on Euclid's Elements[3]. Suidas ascribes to him a description of the earth, a book on the rivers of Libya and another on the interpretation of dreams. Pappus himself (IV. 27, p. 246), speaks of his commentary on the *analemma* of Diodorus, a writer of whom nothing is known. Proclus (pp. 189, 190), perhaps quoting the notes on Euclid,

---

[1] Nevertheless Theon does not mention Pappus's commentary. Hultsch, as above stated, thinks he stole from it.

[2] See Hultsch's preface to the 3rd Vol. of his *Pappus* and Cantor, pp. 374—376.

[3] From these, no doubt, Axx. 4 and 5 were taken (Proclus, p. 197, 6).

says that Pappus pointed out that an angle may be equal to a right angle without being a right angle. In the annexed figure of two equal semicircles, for instance, the angle comprised between the two curves $a\delta$, $a\gamma$, is obviously equal to the right angle $\delta a\beta$.

But the one work by which Pappus is known is his συναγωγή, a collection of mathematical papers, originally in 8 books, of which the first and part of the second are missing. This is the work to which so frequent references have been made in these pages and which, of all extant Greek books, is the richest in information on the lost treatises of the foremost Greek geometers. The design of the collection is to give a brief account of the contents of most of the mathematical works which, in Pappus's day, enjoyed the highest repute and then to set out *lemmas* or auxiliary propositions to them. These lemmas, however, as is evident by a comparison of them with extant works, such as Euclid's *Phaenomena* or Apollonius' *Conics*, are selected in the freest possible manner, and have often no apparent bearing on the book which they are supposed to illustrate. On the other hand the same comparison shews that Pappus gives a very careful and correct summary of the works of which he treats, and for this reason it seemed possible to the mathematicians of the last century to reconstruct lost works on the authority of Pappus alone.

155. The contents of the συναγωγή may be here briefly indicated[1].

The fragment of Book II. deals entirely with the *tetrads* of Apollonius and has been described above (pp. 62—63).

Book III. contains four tracts, the first on the methods of duplication of Eratosthenes, Nicomedes, Heron and Pappus himself: the second on the theory of proportion, introduced by a problem to exhibit an arithmetical, a geometrical and a harmonic mean in the same figure[2]: the third is on Euclid I. 21, and shews that if the straight lines meeting within the triangle

[1] This summary is partly from Cantor, pp. 378—382.

[2] pp. 70, 72. Between a and c, b is an arithmetical, geometrical or harmonical mean, according as
$$a-b : b-c = a : a, \text{ or } a : b, \text{ or } a : c.$$

be drawn from two points *not* the extremities of the base, then "the sides of the included triangle may be greater than the sides of the triangle which includes it in any ratio which is less than that of two to one[1]": the fourth is on the five regular polyhedra inscribed in the sphere and uses the *Sphaerica* of Theodosius. Here Pappus proceeds by a method contrary to that of Euclid XIII. Euclid, who finds a ratio between a side of the polyhedron and the diameter of the sphere, constructs the polyhedron first and describes the sphere about it: Pappus constructs the sphere first and inscribes the polyhedron.

Book IV. begins with the theory of transversals to the circle, followed by the problem to describe a circle about three circles which touch one another. Then follow more problems on figures touching one another. Pappus next passes to the spiral of Archimedes, the conchoid of Nicomedes and the *quadratrix*, which last is very fully discussed. Various subjects are here incidentally treated, such as the rectification of the circle, the relations between the *quadratrix* and the spiral, the trisection of an angle, the division of a circle, into arcs which have to one another a given ratio, by means of the *quadratrix* and the spiral, the use of the *quadratrix* for the solution of the three problems (1) to describe in a circle a regular polygon of any number of sides, (2) to find for any given chord a circular arc which has a given ratio to the chord, and (3) to draw angles which shall be incommensurable with one another.

Book V. begins with an extract from the work of Zenodorus on plane figures of equal periphery, passes then to the treatise of Archimedes on the half-regular solids, then returns to Zenodorus on solids of equal surface and shews that, of the regular solids with equal surface, that is the greatest which has most angles.

Book VI. gives *lemmas* to the μικρὸς ἀστρονομούμενος (τόπος) or *Minor collection of Astronomy*[2]. This contained, according to the preface, the following works, viz., the *Sphaerica*

---

[1] Simson's note to Eucl. I. xxi.

[2] This collection, with some Arabic additions, constituted the "middle books" of the Arabs, i.e. the course of study intermediate between the Elements of Euclid and the Almagest. See Steinschneider in *Zeitschr. Math. Phys.* for 1865, x. pp. 456—498.

of Theodosius, the *Data, Optica, Catoptrica* and *Phaenomena* of Euclid, the περὶ διοικήσεων (*De Habitationibus*) and *De Noctibus et Diebus* of Theodosius, the *Moving Sphere* of Autolycus, the *De Magnitudinibus etc.* of Aristarchus, the ἀναφορικός of Hypsicles, the *Sphaerica* of Menelaus. The books, perhaps, were not studied in this order, for it is difficult to see why Autolycus should be taken after Euclid, but on the other hand the τόπος ἀναλυόμενος was studied in the order of its books[1] and there seems no reason otherwise for dividing the works of Theodosius. Pappus omits the *Catoptrica*, the ἀναφορικός and the Spherics of Menelaus, but as he promises (p. 602 *lin.* 1) some lemmas to a commentary by Menelaus on Euclid's *Phaenomena*, which are not now included in the book, it may be that some mutilation has taken place.

Book VII. deals, in like manner, with the τόπος ἀναλυόμενος or *Collection of Analysis*. This contained Euclid's *Data*, Apollonius' *Sectio Rationis, Sectio Spatii, Sectio Determinata, De Tactionibus*, Euclid's *Porisms*, Apollonius' *De Inclinationibus, Plane Loci*, and *Conics*, the *Solid Loci* of Aristaeus, the τόποι πρὸς ἐπιφανείᾳ of Euclid and lastly Eratosthenes' περὶ μεσοτήτων. The contents of these, down to the *Conics*, are described in a long preface and then follow *lemmas* to all the books except the *Data* and those of Aristaeus and Eratosthenes. The *Porisms* of Euclid are taken between the *Plane Loci* and the *Conics* of Apollonius, but otherwise the above order is preserved.

Book VIII. begins by announcing that it will deal with some mechanical questions "more tersely and clearly and in a better manner" than they had been handled by the ancients. To these belong the theory of the centre of gravity and of the inclined plane, and the problem, by means of cogwheels whose diameters are in a given ratio, to move a given weight with a given power. Here, again, arises the duplication-problem, or rather the problem to construct a cube which has a given ratio to another cube. This is solved by a mechanical device. Pappus then discusses the method of finding the diameter of a cylinder which is broken so that an exact measurement can-

[1] Cf. p. 636. 18. "Of the above mentioned books of analysis the order (τάξις) is as follows."

not be taken on either base. Suddenly he passes thence to problems (or porisms) to find given points on a sphere, e.g. the point which is nearest to a given plane or the points in which a given straight line will cut the sphere. Then he shews how to inscribe seven similar regular hexagons in a circle, one having the same centre as the circle, the other six standing each on one side of the first. This problem serves for the construction of cogwheels and extracts from the βαροῦλκος and the *Mechanics* of Heron, added perhaps by a later hand, conclude the collection.

156. To the development of Greek geometry the *Collection* of Pappus can hardly be deemed really important. It is evidence, indeed, that the geometrical school of Alexandria was still flourishing after 600 years and it shews what subjects were studied there. But among his contemporaries Pappus is like the peak of Teneriffe in the Atlantic. He looks back, from a distance of 500 years, to find his peer in Apollonius. In the long interval, only two or three writers, Zenodorus and Serenus and Menelaus, had produced in pure geometry a little work of the best order, and there are none such to follow. The *Collection* of Pappus is not cited by any of his successors[1], and none of them attempted to make the slightest use of the proofs and *aperçus* in which the book abounds. It becomes interesting only in the history of mathematics during the 17th and 18th centuries, when there were again geometers capable of using it and others who independently struck out and pursued lines of investigation which were more or less clearly anticipated by Pappus. To give here an elaborate account of Pappus would be to create a false impression. His work is only the last convulsive effort of Greek geometry which was now nearly dead and was never effectually revived. It is not so with Ptolemy or Diophantus. The trigonometry of the former is the foundation of a new study which was handed on to other nations indeed but which has thenceforth a continuous history of progress. Diophantus also represents the outbreak of a movement which probably was not Greek in its origin, and

---

[1] Hultsch's Preface to Vol. III. p. 3. Eutocius however, (in Torelli p. 139) referring to the μηχανικαὶ εἰσαγωγαί of Pappus, cites the proposition VIII. 11 of the *Collectio*. (This is also in Bk. III. pp. 64—69 of Hultsch).

which the Greek genius long resisted, but which was especially adapted to the tastes of the people who, after the extinction of Greek schools, received their heritage and kept their memory green. But no Indian or Arab ever studied Pappus or cared in the least for his style or his matter. When geometry came once more up to his level, the invention of analytical methods gave it a sudden push which sent it far beyond him and he was out of date at the very moment when he seemed to be taking a new lease of life.

A few lines only will be sufficient to call attention to some passages of Pappus in which modern geometers still take an antiquarian interest[1]. These occur mostly in Book VII. Here (p. 682) occurs the theorem, afterwards re-discovered or stolen by Guldin (1577-1643), that the volume of a solid of revolution is equal to the product of the area of the revolving figure and the length of the path of its centre of gravity. Here also (p. 1013) Pappus first found the focus of a parabola and suggested the use of the directrix. Here in the lemmas to the *Sectio Determinata* the theory of points in involution is propounded : and among those to the *De Tactionibus* the problem is solved, to draw through three points lying in the same straight line, three straight lines which shall form a triangle inscribed in a given circle[2]. Here also (p. 678) occurs the problem "given several straight lines, to find the locus of a point such that the perpendiculars, or more generally straight lines at given angles, drawn from the point to the given lines shall satisfy the condition that the product of certain of them shall be in a given ratio to the

---

[1] Some of these have been mentioned before à *propos* of the books to which the lemmas of Pappus refer. A summary of a kind more satisfying to the modern geometer will be found in Chasles *Aperçu* pp. 28—44. Cantor pp. 382—386 cites generally the same propositions as Chasles, but adds some remarks on hints of algebraical symbolism in Pappus. Taylor (*Anc. and Mod. Conics*. pp. lii—liv) gives little more than the lemmas to Euclid's porisms from Book VII.

[2] On this problem (no. 117) Chasles has the following remarks. "The props. 105, 107, 108 are particular cases of it. One of the points is there supposed to be at infinity. The problem, generalised by placing the points anywhere, has become celebrated by its difficulty, by the fame of the geometers who solved it and especially by the solution, as general and simple as possible, given by a boy of 16, Ottaiano of Naples." *Aperçu*, pp. 44, 328.

product of the rest"[1]. Descartes and Newton brought this into celebrity as the "problem of Pappus." But though the seventh Book, which contains the lemmas to the τόπος ἀναλυόμενος is by far the most important, there is matter in the other books of a very surprising character. The 4th Book, which deals with curves, contains a great number of brilliant propositions, especially on the *quadratrix* and the Archimedean spiral. Pappus supplements the latter by producing (p. 261 *sqq.*), a spiral on a sphere, in which a great circle revolves uniformly about a diameter, while a point on the circle moves uniformly along its circumference. He then finds the area of the surface so determined, "a *complanation* which claims the more lively admiration, if we remember that, though the whole spherical superficies was known since Archimedes' time, to measure portions of it, such as spherical triangles, was then and for long afterwards an unsolved problem[2]". The 8th Book (p. 1034 *sqq.*) contains a proposition to the effect that the centre of gravity of a triangle is that of another triangle of which the vertices lie on the sides of the first and divide them all in the same ratio[3]. All these, and many more of equal difficulty, seem to be new and of Pappus' own invention. It ought not, however, to be forgotten that in at least three cases, which have been noticed above in their proper places, Pappus seems to have assumed credit to which he is not entitled. In Book III. he gives as his own a solution of the trisection-problem with a conchoid, which can hardly be other than the solution which Proclus ascribes to Nicomedes: in Book IV. he gives 14 propositions of Zenodorus without so much as naming that author: and in Book VIII. he solves the problem 'to move a given weight with a given power' in a manner which differs only accidentally from Heron's[4]. It is probable that many

---

[1] It is in this problem that Pappus objects to having more than 4 straight lines, on the ground that a geometry of more than three dimensions was absurd.

[2] Cantor p. 384.

[3] Pappus supposes points, starting simultaneously from the three vertices, to move along the sides with velocities proportionate to the length of the sides.

[4] In Heron the weight is 1000 talents, the power 5, and he solves the problem by a series of cogwheels, the diameters of each pair being in the ratio 5 : 1. Pappus takes the weight 160, power 4 and the diameters 2 : 1. See Pappus VIII. prop. 10 (p. 1061 sqq.) and Vincent's Heron cited *supra*, p. 278 *n*.

works of ancient geometers were, in Pappus' time, becoming rare. Pappus himself, for instance, does not seem to have seen Euclid's *Conics* and Eutocius and Proclus (much later) had certainly not seen many books which they knew by name[1]. It was therefore possible to appropriate many proofs without much chance of detection and it may be that Pappus used this opportunity.

157. It was suggested at the beginning of this chapter, that possibly the Jews had something to do with the revival of the arithmetical investigations which culminate about this time in the Algebra of Diophantus. It is possible also that the decay of Greek geometry was due to the gradual advance of peoples who have never, at any time, cared much for this branch of mathematics, though they have a surprising natural talent for the other. At any rate, nearly all the leading writers of the Neo-Platonic and Neo-Pythagorean schools were not Greeks. Philo was a Jew: Nicomachus was an Arabian: Ammonius the founder of Neo-Platonism was an Egyptian: so was Plotinus: Porphyrius came from Tyre: the name of Anatolius, wherever he was born, means literally 'Oriental': Iamblichus was a native of Chalcis, in Cœlesyria. These are the philosophers who, in the first four centuries of our era, commanded the largest influence and not one of them was a geometer. Nevertheless, the world is wide and the geometrical school at Alexandria was still largely attended, though it produced no brilliant professors after Pappus. Perhaps **Patricius**, the author of two rules now inserted in Heron's works (*Geom.* 104* and *Stereom.* I. 22) belonged to this time, but there are two persons of this name, one a Lydian of about A.D. 374, the other somewhat later, a Lycian and the father of Proclus. **Theon** of Alexandria was certainly making astronomical observations in A.D. 365 and 372, and he as certainly held classes (συνουσίαι) for which he prepared his edition of Euclid. We have seen also that the preface to Euclid's *Optics* consists of notes from Theon's lectures. He also wrote a commentary on the Almagest, (*ed.* Halma. 1821) most of which is extant and which is perhaps in

---

[1] Heiberg, *Litterargesch. Euklid.* p. 89.

great part founded on the similar work of Pappus[1]. This also
contains many little historical notices which have been extracted
above in their proper places, and the commentary to Book I. of
Ptolemy is especially valuable for its specimens of Greek
arithmetic. Theon's daughter **Hypatia** (*ob.* A.D. 415), seems to
have been a better mathematician than her father. The story
of her life and her tragical death are familiar to English readers
through Kingsley's novel. None of her works are extant, but
Suidas (*sub voce*) says she wrote "ὑπόμνημα εἰς Διοφάντην τὸν
ἀστρονομικὸν κανόνα εἰς τὰ κωνικὰ Ἀπολλωνίου ὑπόμνημα".
This may mean three works, viz.: notes to Diophantus, the
astronomical canon and notes to Apollonius' conics, or (altering
Διοφάντην to Διοφάντου) may refer to two only, notes to the
astronomical canon of Diophantus and notes to the conics.
Hypatia was the last of the Alexandrian professors who attained
any fame. The Neo-Platonic school in Athens, under Syrianus,
now began to attract more attention, and in the interests of
Platonism the historical study of geometry was for a time
revived. **Proclus** the *successor* (διάδοχος) of Syrianus at the
Athenian school (A.D. 410—485), studied in Alexandria and
there acquired that general acquaintance with Greek geometry
which enabled him to write his commentary on Euclid's
Elements. His notes on the first Book are still extant[2], and
contain a very large proportion of all the most valuable information we possess on the history of Greek geometry. But Proclus
himself is a wordy and obscure writer and his best things are
taken from Geminus and Eudemus. Proclus' pupil **Marinus** of
Neapolis (i.e. Flavia Neapolis, the ancient Sychem in Palestine)
wrote the life of his master and is the author of the preface to
Euclid's *Data.* He also was at the head of the Athenian
school. Isidorus succeeded him and was the teacher of
Damascius of Damascus, who appended the 15th Book to

---

[1] The MSS. have a fragment of
Pappus's commentary at the beginning
of Theon's to Book V. and in Theon's
to Book I. occurs a tractate on calculation with sexagesimal fractions
which is, in some MSS, attributed to

Pappus or Diophantus.

[2] Some of the extant *scholia* to the
other books are thought to be by
Proclus. See *Knoche's* essay, cited
above p. 74 *n.*

Euclid's elements[1], and also of **Eutocius** of Ascalon, the commentator on Archimedes and Apollonius. Along with Damascius,
Simplicius, the author of the commentary to Aristotle's *De
Coelo*, taught in the Athenian school, but the Emperor
Justinian, who was by way of being a Christian, did not approve
of the heathen learning and, after many annoying decrees,
finally in 529 closed the school altogether. Meanwhile in
Alexandria the study of mathematics was still in some sort
maintained, but it may be conjectured that there was no great
zeal for geometry since the only mathematical works of which
we hear anything are three commentaries on the *Arithmetic* of
Nicomachus, by Hermas, Asclepius of Tralles and Johannes
Philoponus. The end was rapidly approaching. Mahomet fled
from Mecca in September 622 and died in 632, and his successors
prepared to enlarge the realm of Islam with the sword.
In 640 Alexandria fell and then "with one stride comes
the dark ".

---

158. A summary of the history of Greek mathematics,
which has been given in these pages, can be rendered effective
only by being so condensed that conjecture is indistinguishable
from fact.

At first the higher mathematics were cultivated only in the
service of philosophy and it was part of every philosophical creed
to despise the aims and arts of the vulgar. The same prejudice
remained after mathematics had come to be studied for their
own sake, and thus the attention of competent mathematicians
was always diverted from the ordinary methods of calculation
and Greek arithmetic remained to the last hampered by a vile
symbolism and consequently cumbrous procedure.

Geometry was introduced to the Greeks by Thales from
Egypt, but the same knowledge was, somewhat later, imported

---

[1] This supposition is founded on the
fact that the author of Bk. xv. mentions
(prop. 7) his great teacher Isidorus.
Cantor (p. 426) points out that there was
another Isidorus of Miletus, in this century, who along with Anthemius of Tralles built the *San Sofia* church at Constantinople. Book xv. appended to the
Elements contains only 7 props, chiefly
problems to inscribe one regular solid
in another.

elsewhere by Pythagoras and led in his hands to far more important results. He also, by insisting that every proposition on the relations of lines, or continuous magnitudes, has its analogue in the relations of numbers, or discrete magnitudes, and *vice versâ* started the investigation of the theory of numbers and gave to this inquiry its deductive style and the geometrical nomenclature which it always retained. From his time both these studies advance almost *pari passu*, but the history of the theory of numbers is far more obscure than that of geometry.

In the fifth century B.C. the head-quarters of mathematics shift from Italy to Athens. Here Hippocrates opened the geometry of the circle, which Pythagoras had neglected for that of rectilineal figures, and he also recast the problem of duplication of the cube into one of plane geometry. Plato revived stereometry and raised analysis to the position of a recognized geometrical method. The Athenian successors of Plato began the study of conics and other curves.

Then, about B.C. 300, the head-quarters are removed to Alexandria and in the following century Greek mathematics reach their highest development. Stereometry, the geometry of conics and theory of *loci* were now practically complete, so far as the Greeks were able to finish them. Succeeding centuries do no more than treat of isolated cases which the great geometers had overlooked.

But during this time practical astronomy had been making rapid strides in the hands of Eudoxus, Aristarchus, Eratosthenes and others down to Hipparchus. Now the needs of the practical astronomer are in many respects similar to those of the surveyor, the engineer and the architect. Each of these is chiefly concerned, not to find the general rules which govern all similar cases, but to find under what general rules a particular case, presented to them, falls. But the question whether an angle is acute, or a triangle isosceles, can be determined only by measurement, and hence about 130 B.C., in the time of Heron and Hipparchus, we find the results of geometry applied to measured figures, for the purpose of finding some other measurement as yet unknown. Trigonometry and an elementary algebraical method are thus introduced. For such calculations the Egyptians and Semites, who had now secured the grand results of Greek deductive science, had an especial aptitude, and the study of the theory of numbers, which was revived by Neo-Platonists and Neo-Pythagoreans, mostly of Semitic and Egyptian origin, changes its character accordingly. With Nicomachus, in effect, propositions no longer run "All numbers, having the same characteristic, have such or such another characteristic", but, "The following numbers have the same characteristics". The equations of Diophantus, in which for the first time algebraical symbols appear, and which are intended to find numbers which satisfy given conditions, are the inevitable consequence.

The learning of the Greeks passed over in the 9th century to the Arabs and with them came round into the West of Europe. But no material advance was made by the Arabs in geometry and it was their arithmetic, trigonometry and algebra which chiefly interested the mediaeval Universities. In the 16th century Greek geometry again became known in the original and was studied with intense zeal for about 100 years, until Descartes and Leibnitz and Newton, the best of its scholars, superseded it.

AN ALPHABETICAL LIST OF

THE CHIEF GREEK MATHEMATICIANS

WITH THEIR APPROXIMATE DATES.

(N.B. All dates are B.C., unless A.D. is expressly prefixed.)

	Flor. cir.		*Flor. cir.*
Anaxagoras,	460	Joh. Philoponus,	A.D. 650
Anaximander,	560	Menaechmus,	340
Anaximenes,	530	Menelaus,	A.D. 100
Antiphon,	430	Metrodorus,	A.D. 320
Apollonius,	230	Nicomachus,	A.D. 100
Archimedes,	250	Nicomedes,	180
Archytas,	400	Nicoteles,	250
Aristaeus,	320	Œnopides,	460
Aristotle,	340	Pappus,	A.D. 300
Asclepius Trall.,	A.D. 600	Perseus,	150
Autolycus,	340	Philippus,	320
Bryson,	430	Philolaus,	430
Conon,	250	Philon Byz.,	150
Democritus,	410	Plato,	380
Dinostratus,	320	Proclus,	A.D. 450
Diocles,	180	Ptolemy,	A.D. 150
Diophantus,	A.D. 360	Pythagoras,	530
Eratosthenes,	250	Serenus,	A.D. 50
Euclid,	290	Sextus J. Afric.,	A.D. 200
Eudemus,	320	Simplicius,	A.D. 550
Eudoxus,	360	Thales,	600
Eutocius,	A.D. 550	Theaetetus,	380
Geminus,	70	Theodorus,	420
Heron,	120	Theodosius,	60
Hipparchus,	130	Theon, Alex.,	A.D. 380
Hippias,	430	Theon, Smyrn.,	A.D. 100
Hippocrates,	430	Theudius,	320
Hypatia,	A.D. 410	Thymaridas,	A.D. 250
Hypsicles,	180	Zeno,	450
Iamblichus,	A.D. 340	Zenodorus,	A.D. 150

ADDENDA.

P. 24. THE relevant passages of Nicolaus Smyrnaeus and Bede are printed, with an interesting plate, by M. Froehner in an article on certain Roman *tesserae* in *Annuaire de la Soc. Numismatique*, Paris, 1884. Mr A. S. Murray gave me the article in pamphlet form, newly paged.

P. 44. Prof. Robertson Smith informs me that *gematria* is certainly from γεωμετρία, by a common Semitic transliteration.

P. 108 *n.* 3. In the *Journal of Philology*, XIII. No. 25. pp. 107—113, Mr T. L. Heath, after proving by new evidence that the algebraic ς of Diophantus is not the final *sigma*, shows that ς⁰ occurs in cursive MSS. as an abbreviation of ἀριθμός, used in its ordinary sense, for which also ἀρ. is sometimes found. Hence he suggests that Diophantus' ς is merely a contraction of ἀρ. This theory is pretty but I do not think it is true, for three reasons. (1) The contraction must be supposed to be as old as the time of Diophantus, for he describes the symbol as τὸ ς instead of τὰ or τὼ ἀρ. Yet Diophantus can hardly (as Mr Heath admits) have used cursive characters. (2) The abbreviation ς⁰ for ἀριθμός in its ordinary sense is very rare indeed. It is not found in the MSS. of Nicomachus or Pappus, where it might most readily be expected. It may therefore be due only to a scribe who had some reminiscence of Diophantus. (3) If ς is for ἀρ., then, by analogy, the full symbol should be ςⁱ (like δʸ, κ͞) and not ς⁰.

Pp. 110, 111. *n.* In the *Göttingen Nachrichten*, 1882. pp. 409—413, Prof. P. De Lagarde suggests that the *x* of modern algebra is simply the regular Spanish representative of the Arabic letter, which is the initial of *shai*, the Arabic name of the unknown quantity. This may be (but I believe is not) true of Luca Pacioli, Tartaglia and other early Italian algebraists. The accounts which I have seen of their works are inconsistent and inconclusive. But their more important successors had no prepossession whatever in favour of *x*. Wallis (in his *Algebra*,

1685. p. 127.), says "Whereas it was usual with Harriot (as before with Vieta and Oughtred) to put consonants B, C, D, &c., for known quantities and vowels, A, E, I &c., for unknown, Descartes chooseth to express his unknown quantities by the latter letters of the alphabet (as z, y, x) and the known by the former letters of it as a, b, c, &c." Thus Descartes probably set the fashion, but he may have resumed an old tradition.

P. 129.   There seems to be a reference to a Hebrew *harpedonaptes* in Micah ii. 5.

Pp. 181—185.   Dr Allman, in *Hermathena* No. x., has another paper on *Greek Geometry from Thales to Euclid*. This deals very elaborately with Archytas and Eudoxus.

P. 189.   A statement that the parallelogram of *forces* was known to Aristotle was struck out of p. 189 as incorrect, but by accident, no substitute was inserted. The omission is rectified on p. 238.

P. 204.   There is a very remarkable article by Dr Klamroth "über den Arabischen Euclid" in *Zeitschr. Deutsch. Morgenländ. Gesellsch.* 1881, pp. 270—326. This gives a most careful account of the Arabic texts of Euclid. It would appear that Euclid's Elements was the first Greek book translated into Arabic.

P. 208.   In the *American Journal of Math.* ii. pp. 46—48, Mr G. B. Halsted has a 'Note on the First English Euclid' from which it appears, among other things, that Billingsley became Sir Henry Billingsley, and was Lord Mayor of London in 1591.

Pp. 263 and 277 *n.*   I have wrongly followed Thevenot and Fabricius in the note on p. 277. The Philon mentioned by Vitruvius was an Athenian architect and is clearly not the engineer Philon, part of whose work is in the *Veteres Mathematici.* The latter Philon seems to be identical with Philon of Byzantium, who is mentioned on p. 263. If so, then Philon of Byzantium had certainly heard Ctesibius lecture and must be assigned to a date about 150 B.C.

Philon's construction should have been given on p. 263. He describes a circle about the rectangle $ABDC$. A ruler, cutting AB produced in F, AC produced in G, and the circle in H, D, is turned about the point D until FH equals DG. The line $FHDG$ is called "Philo's line" in modern geometry, but its author did not know its singular property.

INDEX.

(References are to *pages* and notes thereon.)